ENCYCLOPEDIA OF MATHEMATICS AND ITS APPLICATIONS
Volume 24

The Banach-Tarski Paradox

ENCYCLOPEDIA OF MATHEMATICS
and Its Applications

GIAN-CARLO ROTA, Editor
Massachusetts Institute of Technology

Editorial Board

For other books in this series see page 252

The Banach-Tarski Paradox

Stan Wagon

Smith College
Northampton, Massachusetts

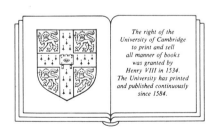

*The right of the
University of Cambridge
to print and sell
all manner of books
was granted by
Henry VIII in 1534.
The University has printed
and published continuously
since 1584.*

CAMBRIDGE UNIVERSITY PRESS

Cambridge

London   New York   New Rochelle

Melbourne   Sydney

Published by the Press Syndicate of the University of Cambridge
The Pitt Building, Trumpington Street, Cambridge CB2 1RP
32 East 57th Street, New York, NY 10022, USA
10 Stamford Road, Oakleigh, Melbourne 3166, Australia

First published 1985

Printed in Great Britain

Library of Congress Cataloging in Publication Data
Wagon, Stan

The Banach-Tarski paradox.

Bibliography: p.

Includes index.

1. Banach-Tarski paradox.   2. Measure theory.
3. Decomposition (Mathematics) I. Title.
QA248.W22   1984   511.3′22   84–9033
ISBN 0 521 30244 7

Delians: "How can we be rid of the plague?"
Delphic Oracle: "Construct a cubic altar of double the size
of the existing one."
Banach and Tarski: "Can we use the Axiom of Choice?"

Contents

Foreword

This book is motivated by the following theorem of Hausdorff, Banach, and Tarski: Given any two bounded sets A and B in three-dimensional space $\mathbf{R}^3$, each having nonempty interior, one can partition A into finitely many disjoint parts and rearrange them by rigid motions to form B. This, I believe, is the most surprising result of theoretical mathematics. It shows the imaginary character of the unrestricted idea of a set in $\mathbf{R}^3$. It precludes the existence of finitely additive, congruence-invariant measures over all bounded subsets of $\mathbf{R}^3$ and it shows the necessity of more restricted constructions such as Lebesgue's measure.

In the 1950s, the years of my mathematical education in Poland, this result was often discussed. J. F. Adams, R. M. Robinson, and W. Sierpiński wrote about it; my Ph.D. thesis was motivated by it. (All this is referenced in this monograph.) Thus it is a great pleasure to introduce you to this book, where this striking theorem and many related results in geometry and measure theory, and the underlying tools of group theory, are presented with care and enthusiasm. The reader will also find some applications of the most recent advances of group theory to measure theory—work of Gromov, Margulis, Rosenblatt, Sullivan, Tits, and others.

But to me the interest of mathematics lies no more in its theorems and theories than in the challenge of its surprising problems. And, on the pages of this book, you will find many old and new open problems. So let me conclude this foreword by turning your attention to one of them, due to my teacher E. Marczewski (before 1939 he published under the name Szpilrajn): Does there exist a finite sequence $A_1, \ldots, A_n$ of pairwise disjoint open subsets of the unit cube and isometries $\sigma_1, \ldots, \sigma_n$ of $\mathbf{R}^3$ such that the unit cube is a proper subset of the topological closure of the union $\sigma_1(A_1) \cup \cdots \cup \sigma_n(A_n)$? This remarkable problem is discussed in Chapters 3, 8, 9, and 11 of this book.

I wish you the most pleasant reading and many fruitful thoughts.

Jan Mycielski

Preface

While many properties of infinite sets and their subsets were considered to be paradoxical when they were discovered, the development of paradoxical decompositions really began with the formalization of measure theory at the beginning of the twentieth century. The now classic example (due to Vitali in 1905) of a non-Lebesgue measurable set was the first instance of the use of a paradoxical decomposition to show the nonexistence of a certain type of measure. Ten years later, Hausdorff constructed a truly surprising paradox on the surface of the sphere (again, to show the nonexistence of a measure), and this inspired some important work in the 1920s. Namely, there was Banach's construction of invariant measures on the line and in the plane (which required the discovery of the main ideas of the Hahn-Banach Theorem) and the famous Banach-Tarski Paradox on duplicating, or enlarging, spheres and balls. This latter result, which at first seems patently impossible, is often stated as: It is possible to cut up a pea into finitely many pieces that can be rearranged to form a ball the size of the sun!

Their construction has turned out to be much more than a curiosity. Ideas arising from the Banach-Tarski Paradox have become the foundation of a theory of finitely additive measures, a theory that involves much interplay between analysis (measure theory and linear functionals), algebra (combinatorial group theory), geometry (isometry groups), and topology (locally compact topological groups). Moreover, the Banach-Tarski Paradox itself has been useful in recent work on the uniqueness of Lebesgue measure: It shows that certain measures necessarily vanish on the sets of Lebesgue measure zero.

The purpose of this volume is twofold. The first aim is to present proofs that are as simple as possible of the two main classical results—the Banach-Tarski Paradox in $\mathbf{R}^3$ (and $\mathbf{R}^n$, $n > 3$) and Banach's theorem that no such paradox exists in $\mathbf{R}^1$ or $\mathbf{R}^2$. The first three chapters (and part of Chapter 5) are devoted to the Paradox, and are accessible to anyone familiar with the rudiments of linear algebra, group theory, and countable sets. (Some specific facts about the Euclidean isometry groups are included in Appendix A). Chapter 10, which contains Banach's theorem in $\mathbf{R}^1$ and $\mathbf{R}^2$, can be read independently of Chapters 4–9, but requires a little more background in measure theory (Lebesgue measure) and general topology (Tychonoff compactness theorem). Although isolated proofs use some special techniques, for example, transfinite induction or analytic functions, most of the book is accessible to a first- or second-year graduate student.

The book's other purpose is to serve as a background source for those interested in current work that has some connection with paradoxical decompositions. The period since 1980 has been especially active, and several problems in this area have been solved, some by using the deepest techniques of modern mathematics. This volume contains a unified and modern treatment of the fundamental results about amenable groups, finitely additive measures, and free groups of isometries, and so should prove useful to someone in any field who is interested in these modern results and their historical context.

The group theory connections arise from an analysis of the difference between the isometry groups of $\mathbf{R}^2$ and $\mathbf{R}^3$, a difference that explains the presence of the Banach-Tarski Paradox in $\mathbf{R}^3$ and its absence in the plane. This analysis led to the study of the class of groups that are not paradoxical, that is, groups that cannot be duplicated by left translation of finitely many pairwise disjoint subsets. This class, denoted by AG for amenable groups, contains all solvable and finite groups, but excludes free non-Abelian groups. A famous problem is whether AG equals NF, the class of groups not having a free non-Abelian subgroup. This has been solved only recently (Ol'shanskii [178]), using ideas connected with growth conditions in groups (Cohen [35]) and the solution of Burnside's Problem (Adian [2]). However, the classes AG and NF do coincide when restricted to linear groups (a deep result of Tits [237]) or to connected, locally compact topological groups (Balcerzyk and Mycielski [5]). Growth conditions in groups, first studied in depth by Milnor and Wolf [147, 148, 258], also elucidate a weaker sort of paradox, the Sierpiński-Mazurkiewicz Paradox, which exists in $\mathbf{R}^2$ but not $\mathbf{R}^1$. The class AG has also led to the study of (topological) amenability in topological groups (where only Borel sets are considered). Amenability and the related notion of an invariant mean have proven to be useful tools in the study of topological groups (Greenleaf [80]). Chapter 10 contains an introduction to the theory of amenable groups, and Chapter 12 discusses the relevance of growth conditions for the theory of amenability and paradoxical decompositions.

In analysis, recent work solving the Ruziewicz Problem has its roots in Banach's results about $\mathbf{R}^1$ and $\mathbf{R}^2$. Banach showed that Lebesgue measure is not the only finitely additive, isometry-invariant measure on the bounded, measurable subsets of the plane (or line) that normalizes the unit square (or interval). The analogous problem for $\mathbf{R}^3$ and beyond was unsolved for over fifty years. But using Kazhdan's Property T and techniques of functional analysis, Margulis [137, 138], Sullivan [228], Rosenblatt [195], and Drinfeld recently settled this question in the expected way: No "exotic" measures exist except in the cases Banach considered. The construction of exotic measures in $\mathbf{R}^1$ and $\mathbf{R}^2$ and a discussion of the higher-dimensional situation is presented in Chapter 11.

Actions of free groups are central to the whole theory, and a general problem is to determine which naturally occurring groups have certain sorts of free subgroups. Chapter 5 presents the classical results on the isometry groups of spheres, Euclidean spaces, and hyperbolic spaces. Here too, some problems were solved only recently. Until the recent paper of Deligne and Sullivan [57], it was not known that SO_6 (or SO_{4n+2}) contained a free non-Abelian subgroup, no element of which (except the identity) has $+1$ as an eigenvalue. And a similar problem about locally commutative free subgroups was solved in all SO_n except SO_5 in 1956 (Dekker [49]), with the remaining case solved by A. Borel [20] in a paper generalizing the work of Deligne and Sullivan. Chapter 6 presents a technique for improving this type of result to get uncountable free subgroups, and discusses the geometrical consequences of these larger free groups of isometries.

There are many interesting problems in this area having a geometrical flavor. The Banach-Tarski Paradox makes use of the abstract properties of the equidecomposability relation, a concept that goes back to Greek methods of computing areas. The most famous unsolved problem here is Tarski's Circle-Squaring Problem, which is discussed in Chapter 7. A three-dimensional version of this problem is related to Hilbert's Third Problem on whether a regular tetrahedron is equidecomposable (using subpolyhedra) to a cube. Geometrical ideas are also brought to the fore by questions of whether a paradoxical decomposition of an entire space ($\mathbf{R}^n$ or H^n, hyperbolic n-space) using isometries is possible using pieces that are geometrically nice (e.g., measurable). Although not possible in Euclidean spaces, a quite constructive paradoxical decomposition of H^n ($n \geqslant 2$) exists (see Figures 5.2 and 5.3). Moreover, deeper considerations (such as Property T again) come into play if one returns to the Euclidean case but allows area-preserving affine transformations (see 11.17–18).

The book is divided into two parts: The first deals with the construction of paradoxical decompositions (which imply that certain sorts of finitely additive measures do not exist) and the second deals with the construction of measures (which show why certain paradoxical decompositions do not exist). Chapter 9 ties the two parts together by presenting a theorem of Tarski that asserts that the existence of a paradoxical decomposition is equivalent to the nonexistence of an invariant, finitely additive measure. The final chapter, Chapter 13, discusses some technical and philosophical points relevant to the foundational discussion engendered by the use of the Axiom of Choice in the Banach-Tarski Paradox and related results.

The author would like to express his gratitude to Jan Mycielski and Joseph Rosenblatt for their support and advice during the preparation of this volume. Moreover, a book such as this, touching on many mathematical disciplines, would not have been possible without the willingness of many people to share their expertise. The help of the following mathematicians is gratefully acknowledged: W. Barker, A. Borel, B. Grünbaum, W. Hanf, J. Hutchinson, V. Klee, R. Laver, R. Macrae, A. Ramsay, R. Riley, R. Solovay, D. Sullivan and A. Taylor.

                                                                Stan Wagon

PART I

Paradoxical Decompositions, or the Nonexistence of Finitely Additive Measures

Chapter 1

Introduction

It has been known since antiquity that the notion of infinity leads very quickly to seemingly paradoxical constructions, many of which seem to change the size of objects by operations that appear to preserve size. In a famous example, Galileo observed that the set of positive integers can be put into a one-one correspondence with the set of square integers, even though the set of nonsquares, and hence the set of all integers, seems more numerous than the squares. He deduced from this that "the attributes 'equal,' 'greater' and 'less' are not applicable to infinite ... quantities," anticipating developments in the twentieth century, when paradoxes of this sort were used to prove the nonexistence of certain measures.

An important feature of Galileo's observation is its resemblance to a duplicating machine; his construction shows how, starting with the positive integers, one can produce two sets, each of which has the same size as the set of positive integers. The idea of duplication inherent in this example will be the main object of study in this book. The reason that this concept is so fascinating is that, soon after paradoxes such as Galileo's were being clarified by Cantor's theory of cardinality, it was discovered that even more bizarre duplications could be produced using rigid motions, which *are* distance-preserving (and hence also area-preserving) transformations. I refer to the Banach-Tarski Paradox on duplicating spheres or balls, which is often stated in the fanciful form: a pea may be taken apart into finitely many pieces that may be rearranged using rotations and translations to form a ball the size of the sun. The fact that the Axiom of Choice is used in the construction makes it quite distant from physical reality, though there are interesting examples that do not need the Axiom of Choice (see Theorems 1.7 and 5.9).

Two distinct themes arise when considering the refinements and ramifications of the Banach-Tarski Paradox. First is the use of ingenious geometric and algebraic methods to construct paradoxes in situations where they seem impossible, and thereby getting proofs of the nonexistence of certain measures. Second, and this comprises Part II of this book, is the construction of measures and their use in showing that some paradoxical decompositions are not possible.

We begin with a formal definition of the idea of duplicating a set using certain transformations. The general theory is much simplified if the transformations used are all bijections of a single set, and the easiest way to do this is to work in the context of group actions. Recall that a group G is said to act on a set X if to each $g \in G$ there corresponds a function (necessarily a bijection) from X to X, also denoted by g, such that for any $g, h \in G$ and $x \in X$, $g(h(x)) = (gh)(x)$ and $1(x) = x$, where 1 denotes the identity of G.

Definition 1.1.   *Let G be a group acting on a set X and suppose $E \subseteq X$. E is G-paradoxical (or, paradoxical with respect to G) if for some positive integers m, n there are pairwise disjoint subsets $A_1, \ldots, A_n, B_1, \ldots, B_m$ of E and $g_1, \ldots, g_n, h_1, \ldots, h_m \in G$ such that $E = \bigcup g_i(A_i)$ and $E = \bigcup h_j(B_j)$.*

Loosely speaking, the set E has two disjoint subsets $(\bigcup A_i, \bigcup B_j)$ each of which can be taken apart and rearranged via G to cover all of E. If E is G-paradoxical, then the sets witnessing that may be chosen so that $\{g_i(A_i)\}$, $\{h_j(B_j)\}$, and $\{A_i\} \cup \{B_j\}$ are each partitions of E. For the first two, one need only replace A_i, B_j by smaller sets to ensure pairwise disjointness of $\{g_i(A_i)\}$ and $\{h_j(B_j)\}$, but the proof that, in addition, $\{A_i\} \cup \{B_j\}$ may be taken to be all of E is more intricate, and will be given in Chapter 3 (Corollary 3.6).

EXAMPLES OF PARADOXICAL ACTIONS

The Banach-Tarski Paradox

Any ball in $\mathbf{R}^3$ is paradoxical with respect to the group of isometries of $\mathbf{R}^3$.

This result, a paradigm of the whole theory, will be proved in Chapter 3. More generally, we shall consider the possibility of paradoxes when X is a metric space and G is a subgroup of the group of isometries of X (an *isometry* is a bijection from X to X that preserves distance). In the case that G is the group of all isometries of X, we shall suppress G, using simply, E is *paradoxical*. We shall be concerned mostly with the case that X is one of the Euclidean spaces $\mathbf{R}^n$.

Free Non-Abelian Groups

Any group acts naturally on itself by left translation. The question of which groups are paradoxical with respect to this action turns out to be quite fascinating and is discussed at the end of this chapter. In this context the central example is the free group on two generators. Recall that the free group F with generating set M is the group of all finite words using letters from $\{\sigma, \sigma^{-1} : \sigma \in M\}$, where two words are equivalent if one can be transformed to the other by the removal or addition of finite pairs of adjacent letters of the form $\sigma\sigma^{-1}$ or $\sigma^{-1}\sigma$. A word with no such adjacent pairs is called a reduced word and to avoid the use of equivalence classes, F may be taken to consist of all reduced words, with the group operation being concatenation; the concatenation of two words is equivalent to a unique reduced word. (From now on, all words will be assumed to be reduced.) The identity of F, which is denoted by 1, is the empty word. Any two free generating sets for a free group have the same size, which is called the rank of the free group. Free groups of the same rank are isomorphic; any group that is isomorphic to a free group will also be called a free group. See [135] for further details about free groups and their properties.

Theorem 1.2.   *A free group F of rank 2 is F-paradoxical, where F acts on itself by left multiplication.*

Proof.   Suppose σ, τ are free generators of F. If ρ is one of $\sigma^{\pm 1}, \tau^{\pm 1}$, let $W(\rho)$ be the set of elements of F whose representation as a word in σ, $\sigma^{-1}, \tau, \tau^{-1}$ begins, on the left, with ρ. Then $F = \{1\} \cup W(\sigma) \cup W(\sigma^{-1}) \cup W(\tau) \cup W(\tau^{-1})$, and these subsets are pairwise disjoint. Furthermore, $W(\sigma) \cup \sigma W(\sigma^{-1}) = F$ and $W(\tau) \cup \tau W(\tau^{-1}) = F$. For if $h \in F \setminus W(\sigma)$, then $\sigma^{-1}h \in W(\sigma^{-1})$ and $h = \sigma(\sigma^{-1}h) \in \sigma W(\sigma^{-1})$. Note that this proof uses only four pieces.   □

The preceding proof can be improved so that the four sets in the paradoxical decomposition cover all of F, rather than just $F \setminus \{1\}$. The reader might enjoy trying to find such a neat four-piece paradoxical decomposition of a rank 2 free group (or, see Figure 4.1). When we say that a group is paradoxical, we shall be referring to the action of left translation; this should cause no confusion with the usage mentioned in Example 1.

Free Semigroups

We shall on occasion be interested in the action of a semigroup S (a set with an associative binary operation and an identity) on a set X. Because of the lack of inverses in a semigroup, the function on X induced by some $\sigma \in S$ may not be a bijection; thus it is inappropriate to apply Definition 1.1 to such actions. Nonetheless, there are similarities between free semigroups and free groups, as the following proposition shows. A free semigroup with free generating set T is simply the set of all words using elements of T as letters, with concatenation being the semigroup operation. The rank of a free semigroup is the number of elements in T.

Proposition 1.3.   *A free semigroup S, with free generators τ and ρ, contains two disjoint sets A, B such that $\tau S = A$ and $\rho S = B$. Hence, any group with a free subsemigroup* of rank 2 contains a nonempty paradoxical set.*

Proof. Simply let A be the set of words in τ, ρ that begin, on the left, with τ, and let B be the set of words beginning with ρ. If S is embedded in a group, then S itself is a paradoxical subset of the group since $S = \tau^{-1}A = \rho^{-1}B$. □

Arbitrary Bijections

The following result, showing that any infinite set is paradoxical with respect to arbitrary bijections, is the modern version of Galileo's observation about the integers.

Theorem 1.4.   *The following are equivalent, where $|X|$ denotes the cardinality of the set X. (The implications with (c) as hypothesis use the Axiom of Choice.)*

(a) $|X| = 2|X|$;
(b) *X is paradoxical with respect to the group of all permutations of X, that is, all bijections from X to X;*
(c) *X is infinite (or empty).*

Proof. Since bijections preserve cardinality, the transformations used to witness (b) yield (a). To prove (a) $\Rightarrow$ (b), suppose $f : X \to A$ and $g : X \to B$ are bijections, where $A \cup B = X$ and $A \cap B = \emptyset$, as guaranteed by (a). Let f_1, f_2 be bijections of X to itself that, respectively, extend f on A, B; such bijections exist by an easy application of the Schröder-Bernstein Theorem (a proof of which is a consequence of the proof of Theorem 3.5).

---

* By this is meant a subset of the group that contains the identity and is closed under the group operation (i.e., a subsemigroup), and that is isomorphic to the free semigroup of rank two.

Then $X = f_1^{-1}(f(A)) \cup f_2^{-1}(f(B))$. Similarly, X can be realized as a union of permutations restricted to $g(A)$ and $g(B)$. Since $f(A)$, $f(B)$, $g(A)$, and $g(B)$ are pairwise disjoint, this shows that X is paradoxical.

That (a) implies (c) is clear. The converse is a consequence of the Axiom of Choice. First it is proved for cardinals, by transfinite induction, and then the Axiom of Choice (in the form, every set may be mapped bijectively onto a cardinal) is invoked (see [118, chap. 8] or [113, p. 32]). Alternatively, one can give a more direct proof using Zorn's Lemma (see [68, p. 162]). □

Note that if $X = \mathbf{N}$ (or, indeed, any infinite cardinal), then the Axiom of Choice is not needed, since $\mathbf{N}$ splits into the even and odd numbers, and the rest of the proof may be carried out as above. For arbitrary X, however, it is known that Choice is required. More precisely, let ZF denote the axioms of Zermelo-Fraenkel set theory, and let ZFC denote ZF with the addition of the Axiom of Choice. Then Theorem 1.4 is a theorem of ZFC, but is not a consequence of ZF alone (see [98, prob. 11.10 and 5.18]). This is not to say, however, that the assertion that all infinite sets are paradoxical with respect to bijections (or, equivalently, assertion (a) of Theorem 1.4) implies the Axiom of Choice. Indeed, it does not, as was proved in 1973 by G. Sageev [200]. This result solved a problem posed in 1924 by Tarski [231], who proved that the Axiom of Choice *is* equivalent to the assertion that $|X| = |X|^2$ for all infinite sets X.

GEOMETRICAL PARADOXES

The first example of a geometrical paradox, that is, one using congruences (isometries), arose in connection with the existence of a non-Lebesgue measurable set. The well-known construction of such a set fits into our context if Definition 1.1 is modified to allow countably many pieces. Thus, E is *countably G-paradoxical* means that

$$E = \bigcup_{i=1}^{\infty} g_i A_i = \bigcup_{i=1}^{\infty} h_i B_i$$

where $\{A_1, A_2 \ldots, B_1, B_2 \ldots\}$ is a countable collection of pairwise disjoint subsets of E, and $g_i, h_i \in G$. Recall that S^1 denotes the unit circle and SO_2 denotes the group of rotations of the circle.

Theorem 1.5 (AC).* S^1 *is countably* SO_2-*paradoxical. If* G *denotes the group of translations modulo* 1 *acting on* $[0, 1)$, *then* $[0, 1)$ *is countably G-paradoxical.*

---

* In the sequel, theorems whose proof uses the Axiom of Choice will be followed by (AC).

Proof. Let M be a choice set for the equivalence classes of the relation on S^1 given by calling two points equivalent if one is obtainable from the other by a rotation about the origin through a (positive or negative) rational multiple of 2π radians. Since the rationals are countable, these rotations may be enumerated as $\{\rho_i : i = 1, 2, \ldots\}$; let $M_i = \rho_i(M)$. Then $\{M_i\}$ partitions S^1 and, since any two of the M_i are congruent by rotation, the even-indexed of these sets may be (individually) rotated to yield all the M_i, that is, to cover the whole circle. The same is true of $\{M_i : i \text{ odd}\}$. This construction is easily transferred to $[0, 1)$ using the bijection taking $(\cos\theta, \sin\theta)$ to $\theta/2\pi$, which induces an isomorphism of SO_2 with G. $\square$

Corollary 1.6 (AC)

(a) *There is no countably additive, rotation-invariant measure of total measure 1, defined for all subsets of S^1.*
(b) *There is a subset of $[0, 1]$ that is not Lebesgue measurable.*
(c) *There is no countably additive, translation-invariant measure* defined on all subsets of $\mathbf{R}^n$ and normalizing $[0, 1]^n$.*

Proof

(a) Suppose μ is such a measure. Then, if $A = \bigcup\{M_i : i \text{ even}\}$ and $B = \bigcup\{M_i : i \text{ odd}\}$ (M_i as in proof of 1.5), $1 = \mu(S^1) = \mu(A) + \mu(B) \geq \mu(S^1) + \mu(S^1) = 2$, a contradiction.

(b) This follows from (c); in fact, $\{\alpha \in [0, 1) : (\cos\alpha, \sin\alpha) \in M\}$ is not Lebesgue measurable.

(c) For $\mathbf{R}^1$ such a measure cannot exist since its restriction to subsets of $[0, 1]$ would be invariant under translations modulo 1, contradicting Theorem 1.5. Such a measure in $\mathbf{R}^n$ would induce one on the subsets of $\mathbf{R}$, by the correspondence $A \leftrightarrow A \times [0, 1]^{n-1}$. $\square$

The connection between the Axiom of Choice and the existence of nonmeasurable sets is complex, involving the theory of large cardinals and forcing—two branches of contemporary set theory. We shall consider these connections in more detail in Chapter 13. For now we note only that (without assuming Choice) the following two assertions are *not* equivalent.

· All sets of reals are Lebesgue measurable, and
· There is a countably additive, translation-invariant extension of Lebesgue measure to all sets of reals,

It is known that the second assertion does not imply the first.

---

* Measures are allowed to have values in $[0, \infty]$.

It comes as a bit of a surprise that even with the restriction to finitely many pieces, paradoxes can be constructed using isometries. The following construction, the first of its kind, does not require any form of the Axiom of Choice, which adds some weight to the comment of Eves [70, p. 278] that the result is "contrary to the dictates of common sense." Recall that when no group is explicitly mentioned, it is understood that the isometry group is being used.

Theorem 1.7 (Sierpiński-Mazurkiewicz Paradox). *There is a non-empty, paradoxical subset of the plane* **R**^{2}.

This theorem is a consequence of the next two results. The reason such a subset of the plane exists is that the planar isometry group, G_2, has a free non-Abelian subsemigroup that acts in a particularly nice way.

Theorem 1.8. *G_2 contains two isometries, τ, ρ, that freely generate a free subsemigroup S of G_2. Moreover, if $w_1, w_2 \in S$ are words beginning, on the left, with τ, ρ, respectively, then $w_1(0,0) \neq w_2(0,0)$.*

Proof. To simplify the proof, identify **R**^{2} with the complex plane. Choose θ so that $u = e^{i\theta}$ is a transcendental complex number; there are only countably many algebraic numbers on the unit circle, so such a θ exists (in fact $\theta = 1$ works, but then we must appeal to the difficult Hermite-Lindemann Theorem to see that e^i is transcendental). Let τ be the translation $\tau(z) = z + 1$ and ρ the rotation $\rho(z) = uz$. We need only prove that τ and ρ satisfy the second assertion, since their freeness follows from that. For if $w_1 = w_2$ where w_1 and w_2 are distinct (semigroup) words and one of them is (the identity or) an initial segment of the other, then left cancellation yields $w' = 1$; but then $w'\rho(0) = \rho(0)$ and $w'\tau(0) = \tau(0)$, one of which contradicts the second assertion. And if neither is an initial segment of the other, then left cancellation yields w_1' and w_2' which are equal in G_2 but have different leftmost terms.

So, suppose $w_1 = \tau^{j_1}\rho^{j_2}\cdots\tau^{j_m}$ and $w_2 = \rho^{k_1}\tau^{k_2}\cdots\tau^{k_\ell}$ where $m, \ell \geqslant 1$ and each exponent is a positive integer; since $\rho(0) = 0$, it is all right to assume that w_1 and w_2 both end in a power of τ, unless w_2 is simply ρ^{k_1}. Then

$$w_1(0) = j_1 + j_3 u^{j_2} + j_5 u^{j_2 + j_4} + \cdots + j_m u^{j_2 + j_4 + \cdots + j_{m-1}}$$

and

$$w_2(0) = k_2 u^{k_1} + k_4 u^{k_1 + k_3} + \cdots + k_\ell u^{k_1 + k_3 + \cdots + k_{\ell-1}} (= 0 \text{ if } w_2 = \rho^{k_1}).$$

If $w_1(0) = w_2(0)$ these two expressions may be subtracted to yield a nonconstant polynomial with integer coefficients that vanishes for the value $e^{i\theta}$, and this contradicts the choice of θ. $\square$

Perhaps the single most important technique in constructing a paradoxical decomposition is the transfer of a replicating partition from a group or semigroup to a set on which it acts. (This technique was first used, independently, by Hausdorff and by Sierpiński and Mazurkiewicz.) The next proposition shows how the simple decomposition of a free semigroup of rank 2 (Proposition 1.3) can be lifted to a set on which the semigroup acts appropriately. Since, by the previous theorem, the group of isometries of the plane satisfies the hypothesis of Proposition 1.9, the Sierpiński-Mazurkiewicz Paradox is an immediate corollary.

Proposition 1.9. *Suppose a group G acting on X contains τ, ρ such that, for some $x \in X$, any two words in τ, ρ, beginning with τ, ρ, respectively, disagree when applied to x. Then there is a nonempty G-paradoxical subset of X.*

Proof. Let S be the subsemigroup of G generated by τ and ρ, and let E be the S-orbit of x. Then $E \supseteq \tau(E), \rho(E)$, and the hypothesis on x implies that $\tau w_1(x) \neq \rho w_2(x)$ for any words $w_1, w_2 \in S$. Hence, $\tau(E) \cap \rho(E) = \varnothing$. Since $\tau^{-1}(\tau E) = E = \rho^{-1}(\rho E)$, this shows that E is G-paradoxical. $\square$

The proof of the Sierpiński-Mazurkiewicz Paradox is really quite constructive. Following it through, one sees that the set

$$E = \{a_0 + a_1 e^i + a_2 e^{2i} + \cdots + a_n e^{ni} : n \in \mathbf{N}, a_i \in \mathbf{N}\}$$

is paradoxical; each of the two subsets of E defined according to whether a_0 is positive or zero is congruent to all of E. The existence of such a set is by no means contradictory. After all, E is countable and so has measure zero; the fact that E is paradoxical implies only that $2 \cdot 0 = 0$. Still, this construction raises many questions about the sorts of planar sets that are paradoxical, and about the possibilities in other dimensions. To give some flavor of what is to come, we list some of these related results:

- No nonempty subset of $\mathbf{R}^1$ is paradoxical (12.12).
- There are uncountable paradoxical subsets of $\mathbf{R}^2$ (6.9).
- Any bounded subset of $\mathbf{R}^3$ (or $\mathbf{R}^n$, $n \geqslant 3$) with nonempty interior is paradoxical (3.11). This is a generalization of the Banach-Tarski Paradox.
- No bounded subset of $\mathbf{R}^2$ with nonempty interior is paradoxical (10.10).
- It is not known whether a nonempty bounded subset of $\mathbf{R}^2$ can be paradoxical, but none can be paradoxical using two pieces (12.14).
- The subgroups G of G_2 such that a nonempty G-paradoxical subset of the plane exists are precisely the subgroups having a free non-Abelian subsemigroup (12.19).

The ideas of the Sierpiński-Mazurkiewicz construction form the foundation of much of the early history of geometrical paradoxes, though more in the context of groups rather than semigroups. The analogue of 1.9 is that a paradoxical decomposition of a group is easily lifted to a set on which the group acts *without nontrivial fixed points* (by which is meant that no nonidentity element of the group fixes a point of the set). The conclusion of the following proposition is stronger than that of Proposition 1.9 since it states explicitly which set is paradoxical, namely the whole set upon which the group acts. On the other hand, unlike 1.9, the Axiom of Choice is required.

Proposition 1.10 (AC). *If G is paradoxical and acts on X without nontrivial fixed points, then X is G-paradoxical. Hence X is F-paradoxical whenever F, a free group of rank 2, acts on X with no nontrivial fixed points.*

Proof. Suppose A_i, $B_j \subseteq G$, g_i, $h_j \in G$ witness that G is paradoxical. By the Axiom of Choice, there is a set M containing exactly one element from each G-orbit in X. Then $\{g(M): g \in G\}$ is a partition of X; pairwise disjointness of this family is an easy consequence of the lack of nontrivial fixed points in G's action. Now, let $A_i^* = \bigcup\{g(M): g \in A_i\}$ and $B_j^* = \bigcup\{g(M): g \in B_j\}$. Then $\{A_i^*\} \cup \{B_j^*\}$ is a pairwise disjoint collection of subsets of X (because $\{A_i\} \cup \{B_j\}$ is pairwise disjoint) and the equations $X = \bigcup g_i A_i^* = \bigcup h_j B_j^*$ follow from the corresponding equations in $G: G = \bigcup g_i A_i = \bigcup h_j B_j^*$. The assertion about F follows from Theorem 1.2. Note that the number of sets used for X is the same as the number of A_i, B_j originally given for G.   □

If the action of G on X is transitive (as is the case if, as in 1.9, X is replaced by a single G-orbit in X), then the Axiom of Choice is not needed to define M. As an exercise, the reader can show that the converse of 1.10 is valid for all actions: if X is G-paradoxical then G is G-paradoxical (transfer the paradoxical decomposition of a single orbit to G). This result has an important measure-theoretic interpretation (see Theorem 10.3).

The main example of a paradoxical group is a free group of rank 2 (Theorem 1.2) and constructions such as the Banach-Tarski Paradox are based on the realization of such a group as a group of isometries of $\mathbf{R}^n$. But actions of isometries on $\mathbf{R}^n$ have, in general, many fixed points, and so the main applications of Proposition 1.10 involve figuring out some way to deal with them. Nonetheless, the idea of lifting a paradox from a group to a set upon which it acts is, by itself, sufficient to obtain interesting pseudo-paradoxical results that have important implications in measure theory (see Theorem 2.6).

PARADOXICAL GROUPS

Proposition 1.10 sheds some light on exactly which groups are paradoxical. Since a subgroup of a group acts by left translation on the whole group, without nontrivial fixed points, the following result is an immediate consequence of 1.10.

Corollary 1.11 (AC).   *A group with a paradoxical subgroup is paradoxical. Hence any group with a free subgroup of rank 2 (in particular, any non-Abelian free group) is paradoxical.*

Are the paradoxical groups given by Corollary 1.11 the only ones? More precisely, is it true that the only paradoxical groups are the ones having a free subgroup of rank 2? This tantalizing problem had been unsolved ever since von Neumann first formulated Corollary 1.11 in 1929. But progress on several fronts in the 1960s and 1970s—specifically, the negative resolution of Burnside's Problem on finitely generated groups of finite exponent (discussed at the end of this chapter) and the development of the theory of growth conditions in groups (see Chapter 12)—has led to a solution in the 1980s. Although a detailed proof has not yet appeared, Ol'shanskii [178] has announced the following result, showing that the class of paradoxical groups is larger than one might have expected.

Theorem 1.12 (AC).   *There is a paradoxical group that has no element of infinite order, and therefore no free subgroup of any rank.*

Before mentioning some of the ideas underlying this result, we show how this problem can be given a measure-theoretic formulation. In fact, nonparadoxical groups coincide with the groups that bear a left-invariant, finitely additive measure of total measure one that is defined on all subsets. Such groups are called *amenable*, and most of Part II is devoted to the study of amenable groups and their connection with the theory of paradoxical decompositions. Chapter 9 contains a general theorem of Tarski which yields that the amenable groups coincide with the nonparadoxical groups. To summarize some of the main results relating free groups and amenability, let AG denote the class of all amenable groups and let EG denote the class of elementary groups, that is, the smallest class containing all finite groups and all Abelian groups, and satisfying the following properties:

(i) if H is a subgroup of $G \in EG$, then $H \in EG$,

(ii) if H is a normal subgroup of $G \in EG$, then $G/H \in EG$,

(iii) if H is a normal subgroup of G, and both H and G/H are in EG, then $G \in EG$,

(iv) if $\{G_i : i \in I\}$ is a directed system with respect to the subgroup relation*, and each $G_i \in EG$, then the union of the G_i is a group in EG.

---

* This means that for all $i, j \in I$ there is some $k \in I$ such that G_i and G_j are subgroups of G_k. Note that in the presence of (i), condition (iv) is equivalent to: if all finitely generated subgroups of G are in EG, then $G \in EG$.

This class can be defined more explicitly using transfinite induction (see [30]). It is easy to see that no group in EG can have a free non-Abelian subgroup, and we shall show (10.4) that, in fact, no group in EG is paradoxical; more precisely, we shall show that all elementary groups are amenable.

Let NF consist of all groups without a free subgroup of rank 2. By 1.11 and the results just mentioned, $EG \subseteq AG \subseteq NF$, and Theorem 1.12 shows that the second inclusion is proper. Moreover, Grigorchuk's recent counter-example to the Milnor-Wolf Conjecture (Theorem 12.17) shows that the first inclusion is proper as well.

Even before the work of Grigorchuk and Ol'shanskii, it was known that $EG \neq NF$. This is related to the famous problem of Burnside: Is each group $B(m, n)$ finite? (The group $B(m, n)$ is $\langle x_1, x_2, \ldots, x_m : w^n = 1 \rangle$, where w ranges over all words in $x_1^{\pm 1}, \ldots, x_m^{\pm 1}$.) This was solved negatively in 1968 by Adian and Novikov (see [2]); indeed, Adian has shown that $B(2, 665)$ is infinite. Now, $B(2, 665)$ obviously has no free subgroup and it is not hard to see that $B(2, 665) \notin EG$. The latter assertion is a consequence of the fact (see [30, Thm. 2.3]) that there are no infinite groups that are elementary, finitely generated, and *periodic* (meaning, each group element has finite order). It is not yet known whether $B(2, 665)$ is paradoxical—Mycielski and Adian have conjectured that it is—but the groups used by Ol'shanskii and Grigorchuk are similar to $B(2, 665)$ in that they are periodic. It should be noted that the three classes EG, AG, and NF share many properties; for instance, they all satisfy conditions (i)–(iv) in the definition of elementary groups (for AG, this is proved in 10.4).

Theorem 1.12 notwithstanding, there is a wide class of groups in which the paradoxical groups do coincide with the groups having a free subgroup of rank 2. A relatively recent theorem of Tits [237] shows that, when restricted to groups of matrices, $EG = AG = NF$. In particular, his result applies to any group of Euclidean isometries. See Theorem 10.6 for a further discussion of Tits's work. To consider another direction, note that $B(2, 665)$ has two generators and infinitely many relations. Cohen [35] has conjectured that $AG = NF$ when restricted to finitely presented groups, that is, groups defined by a finite generating set together with finitely many relations.

Finally, we note that if a paradoxical decomposition of a group bears a closer resemblance to the sort caused by a free subgroup of rank 2 (i.e., uses the minimal number of pieces—four), then in fact the group contains such a free subgroup (see Corollary 4.9).

NOTES

Galileo's observations on the problems of infinity can be found in [74, p. 31]. For more information on nineteenth-century thoughts on these sorts of paradoxes, see [18]. The definition of G-paradoxical appears in [219], but has its roots in earlier work of Hausdorff, Banach and Tarski, and Sierpiński and Mazurkiewicz. Banach and Tarski [11] published their paradoxical construction in $\mathbf{R}^3$ in 1924, and many papers discussing or simplifying their work have appeared since. The initial discovery that free groups (actually, free products) bear paradoxes that can be given a geometric interpretation is due to Hausdorff (see Notes to Chapter 2).

The question of whether the assertion of Theorem 1.4(a) (or (b)) implies the Axiom of Choice was posed by Tarski [231] and solved by Sageev [200]. The classical construction of a non-Lebesgue measurable subset of the circle (Theorem 1.5) was given by Vitali [243] in 1905. The slightly more general Corollary 1.6 was pointed out by Hausdorff [90].

Sierpiński raised the possibility of a paradoxical subset of the plane (Theorem 1.7), and the existence of such a set was initially proved by Mazurkiewicz, whose proof was, in turn, simplified by Sierpiński (see [143]). The results in this chapter related to semigroups (Theorem 1.8 and Propositions 1.3 and 1.9) are derived from their work. The Sierpiński-Mazurkiewicz Paradox is discussed in [70, p. 277], [87, p. 26], and [144, p. 155]. These three books contain discussions of the Banach-Tarski Paradox as well. Generalizations of the choiceless Sierpiński-Mazurkiewicz Paradox are discussed in Chapter 6.

The problem of whether free groups of rank 2 are essentially the only paradoxical groups is stated in [43, p. 520] (in the context of measures, Corollary 9.2 shows the equivalence with the formulation in terms of paradoxical decompositions), but had been considered by Tarski and others in the 1930s. Day also mentioned the possibility that $EG = AG$. In the literature, the problem whether $AG = NF$ has been referred to as Day's Conjecture and as von Neumann's Conjecture. Mycielski [163] and Kesten [103] realized that a counterexample to a problem posed by Burnside—a finitely generated periodic group that is infinite—would necessarily lie in $NF \setminus EG$. Then Adian and Novikov (see [2]) showed that $B(2, 665)$ is such a group, and Adian [2, p. 254] conjectured that $B(2, 665)$ is not amenable, a possibility first raised by Mycielski [168]. On the other hand, Kesten [103] asked if the assertion that each $B(m, n)$ is amenable might be a viable weakening of Burnside's Conjecture that each $B(m, n)$ is finite. Although the status of the group $B(2, 665)$ is not yet resolved, Ol'shanskii [178] combined some of the ideas behind the solution of Burnside's Problem with a characterization of amenability in terms of growth conditions to construct a group in $NF \setminus AG$ (Theorem 1.12).

Chapter 2

The Hausdorff Paradox

It was shown in Chapter 1 (Proposition 1.10) how free non-Abelian groups can be used to generate paradoxes. The first opportunity for such a free group to appear among the Euclidean isometry groups is in 3-space. This is because G_1 and G_2 are solvable (Appendix A), and hence cannot contain any free non-Abelian subgroup. In this chapter we explicitly construct a free non-Abelian subgroup of G_3, and describe a paradoxical decomposition that this causes. A subset S of a group G will be called *independent* if S is a free set of generators of H, the subgroup of G generated by S; H is then a free group of rank $|S|$.

Theorem 2.1. *There are two independent rotations, ϕ and ρ, about axes through the origin in $\mathbf{R}^3$. Hence, if $n \geqslant 3$, SO_n has a free subgroup of rank 2.*

Proof. Let ϕ and ρ be counterclockwise rotations around the z-axis and x-axis, respectively, each through the angle $\arccos \frac{1}{3}$. Then $\phi^{\pm 1}$, $\rho^{\pm 1}$ are represented by matrices as follows:

$$\phi^{\pm 1} = \begin{pmatrix} \dfrac{1}{3} & \mp\dfrac{2\sqrt{2}}{3} & 0 \\[2mm] \pm\dfrac{2\sqrt{2}}{3} & \dfrac{1}{3} & 0 \\[2mm] 0 & 0 & 1 \end{pmatrix} \qquad \rho^{\pm 1} = \begin{pmatrix} 1 & 0 & 0 \\[2mm] 0 & \dfrac{1}{3} & \mp\dfrac{2\sqrt{2}}{3} \\[2mm] 0 & \pm\dfrac{2\sqrt{2}}{3} & \dfrac{1}{3} \end{pmatrix}.$$

We wish to show that no nontrivial reduced word in $\phi^{\pm 1}$, $\rho^{\pm 1}$ equals the identity. Since conjugation by ϕ does not affect whether or not a word vanishes, we may restrict ourselves to words ending (on the right) in $\phi^{\pm 1}$. Hence to get a contradiction, assume that w is such a word and w equals the identity.

We claim that $w(1,0,0)$ has the form $(a, b\sqrt{2}, c)/3^k$ where a, b, c are integers and b is not divisible by 3. This implies that $w(1,0,0) \neq (1,0,0)$, which is the required contradiction. The claim is proved by induction on the length of w. If w has length one, then $w = \phi^{\pm 1}$ and $w(1,0,0) = (1, \pm 2\sqrt{2}, 0)/3$. Suppose then that $w = \phi^{\pm 1} w'$ or $w = \rho^{\pm 1} w'$ where $w'(1,0,0) = (a', b'\sqrt{2}, c')/3^{k-1}$. A single application of the matrices above shows that $w(1,0,0) = (a, b\sqrt{2}, c)/3^k$ where $a = a' \mp 4b'$, $b = b' \pm 2a'$, $c = 3c'$, or $a = 3a'$, $b = b' \mp 2c'$, $c = c' \pm 4b'$ according as w begins with $\phi^{\pm 1}$ or $\rho^{\pm 1}$. It follows that a, b, c are always integers.

It remains only to show that b never becomes divisible by 3. Four cases arise according as w equals $\phi^{\pm 1}\rho^{\pm 1}v$, $\rho^{\pm 1}\phi^{\pm 1}v$, $\phi^{\pm 1}\phi^{\pm 1}v$, or $\rho^{\pm 1}\rho^{\pm 1}v$ where, possibly, v is the empty word. In the first two cases, using the notation and equations of the previous paragraph, $b = b' \mp 2c'$ where 3 divides c' or $b = b' \pm 2a'$ where 3 divides a'. Thus if b' is not divisible by 3, neither is b. For the other two cases, let a'', b'', c'' be the integers arising in $v(1,0,0)$. Then, in either case, $b = 2b' - 9b''$. For instance, in the third case, $b = b' \pm 2a' = b' \pm 2(a'' \mp 4b'') = b' + b'' \pm 2a'' - 9b'' = 2b' - 9b''$; an essentially identical proof works in the fourth case. Thus if b' is not divisible by 3, neither is b, completing the proof.   □

The proof given above is just one of many constructions of a free non-Abelian group of rotations in $\mathbf{R}^3$. Hausdorff gave the first such construction in 1914; he showed that if ϕ and ρ are rotations through 180°, 120°, respectively, about axes containing the origin, and if $\cos 2\theta$ is transcendental where θ is the angle between the axes, then ϕ and ρ are free generators of $\mathbf{Z}_2 * \mathbf{Z}_3$. Since $\mathbf{Z}_2 * \mathbf{Z}_3$ has a free subgroup of rank 2 ($\rho\phi\rho$, $\phi\rho\phi\rho\phi$ freely generate such a subgroup), Theorem 2.1 follows. Recently, Osofsky simplified Hausdorff's approach by showing that ϕ and ρ generate $\mathbf{Z}_2 * \mathbf{Z}_3$ even if $\theta = 45°$.

The approach we have used was developed by Świerczkowski, and is noteworthy because of its avoidance of transcendental numbers, its use of perpendicular axes, and its use of the same angle for each rotation. Świerczkowski was motivated by the following interesting question of Steinhaus, a negative answer to which can be derived from Theorem 2.1. Can a finite chain of neatly piled regular tetrahedra (meaning: two consecutive tetrahedra share exactly one face, and each tetrahedron is distinct from its predecessor's predecessor) end in a tetrahedron that is a translation of the first one? In Chapter 5 this question will be answered negatively using a group of reflections rather than rotations. More precisely, it will be shown (5.10) that the four reflections in the faces of a regular tetrahedron satisfy no relation except that the square of each is the identity; that is, these four reflections generate the free product $\mathbf{Z}_2 * \mathbf{Z}_2 * \mathbf{Z}_2 * \mathbf{Z}_2$.

There are other results that yield specific independent pairs of rotations. We omit the proof of the following theorem (see [46, 229]), since from the point of view of paradoxical decompositions, the additional information it provides is not necessary. (However, a special case of part (b) is implicit in the proof of Theorem 5.7.)

Theorem 2.2. *Suppose ϕ and ρ are rotations of S^2 with the same angle, θ, of rotation. Then ϕ and ρ are independent if either (a) the axes are perpendicular and $\cos\theta$ is a rational different from 0, $\pm\frac{1}{2}$, ± 1, or (b) the axes are distinct and the cosine of the angle formed by the axes is transcendental.*

While Theorems 2.1 and 2.2 give specific examples of free groups, it turns out that, from a topological point of view, almost any pair of rotations is independent. More precisely, if $SO_3 \times SO_3$ is given the product topology, then $\{(\phi,\rho) \in SO_3 \times SO_3 : \phi \text{ and } \rho \text{ are independent}\}$ is comeager (and hence dense); this is a consequence of the discussion following Corollary 6.6.

It was pointed out that the properties of being solvable and of containing a free non-Abelian group are mutually exclusive. Because of nonsolvable finite groups, the two conditions are not exhaustive, but one can ask whether every group either has a solvable subgroup of finite index, or has a free non-Abelian subgroup. Though not true for all groups, it is valid for a wide class of groups. This is a deep result proved by Tits in 1972 (see Theorem 10.6 for a fuller discussion). In particular, Tits showed the property to be valid for all subgroups of GL_n, the nonsingular linear transformations of $\mathbf{R}^n$, from which it follows for all groups of Euclidean isometries. It follows that the classes EG, AG, and NF coincide for such groups. Thus the fact that as soon as SO_n loses its solvability, it gains a free non-Abelian subgroup, is a special case of a far-reaching result in the theory of matrix groups.

Later on we shall see that the existence of larger free groups has implications for paradoxical decompositions. Thus, despite the power of Tits's Theorem, we shall investigate (in Chapters 4, 5, and 6) other sorts of free groups of isometries. Of particular interest will be the fact that there is an independent set of rotations in SO_3 with the same cardinality as the continuum; in fact the free product of any sequence of continuum many cyclic groups is representable in SO_3. Also, there are continuum many independent isometries of $\mathbf{R}^3$ such that the group they generate acts on $\mathbf{R}^3$ without nontrivial fixed points (see Theorems 5.8 and 6.5).

Each element of the free group of rotations (call it F) constructed in Theorem 2.1 fixes all points on some line in $\mathbf{R}^3$, and so Proposition 1.10 cannot yet be applied. A naive approach to this difficulty turns out to be fruitful. Each nonidentity rotation in F has two fixed points on S^2, the unit sphere, namely the intersection of the rotation's axis with the sphere. Let D be the collection of all such points; since F is countable, so is D. Now, if $P \in S^2 \backslash D$ and $g \in F$, then $g(P)$ lies in $S^2 \backslash D$ as well: if h fixed $g(P)$, then P would be a fixed point of $g^{-1}hg$. Hence F acts on $S^2 \backslash D$ without nontrivial fixed points, and Proposition 1.10 may be applied to this action to obtain the following result.

Theorem 2.3 (Hausdorff Paradox) (AC).   *There is a countable subset D of S^2 such that $S^2 \backslash D$ is SO_3-paradoxical.*

A countable subset of the sphere can be dense, and so the paradoxical nature of 2.3 is not immediately evident. Still, countable sets are very small in size compared to the whole (uncountable) sphere. We shall see in the next chapter how the smallness of D allows it to be eliminated completely, yielding the Banach-Tarski Paradox: S^2 is SO_3-paradoxical. But even without eliminating D, Hausdorff's Paradox has an important measure-theoretic consequence.

The fact (1.6) that there is no countably additive, isometry-invariant measure on $\mathscr{P}(\mathbf{R}^n)$ normalizing the unit cube naturally leads to the question of whether measures exist satisfying a weaker set of conditions. The usual approach is to allow the measure to assign values to a smaller collection of sets, that is use Lebesgue measure and live with the fact that some sets are not Lebesgue measurable. Finitely additive measures had been studied prior to Lebesgue, and it is natural to ask whether there might be a finitely additive, isometry-invariant measure defined for all subsets of $\mathbf{R}^n$. It was this question that motivated Hausdorff to carry out his construction on S^2, because he was able to use it to provide an answer in all dimensions except one and two. First we give a definition and proposition that state precisely the fundamental connection between paradoxical decompositions and the nonexistence of finitely additive measures. Suppose a group G acts on a set X, and $E \subseteq X$.

Definition 2.4.   *E is called G-negligible if $\mu(E) = 0$ whenever μ is a finitely additive, G-invariant measure on $\mathscr{P}(X)$ with $\mu(E) < \infty$.*

Proposition 2.5.   *If E is G-paradoxical, then E is G-negligible.*

Proof.   Suppose μ is a finitely additive, G-invariant measure on $\mathscr{P}(X)$ and $\mu(E) < \infty$. Let the fact that E is G-paradoxical be witnessed by A_i, g_i, B_j, h_j.   Then $\mu(E) \geqslant \sum \mu(A_i) + \sum \mu(B_j) = \sum \mu(g_i A_i) + \sum \mu(h_j B_j) \geqslant \mu(\bigcup g_i A_i) + \mu(\bigcup h_j B_j) = \mu(E) + \mu(E) = 2\mu(E)$. Since $\mu(E) < \infty$, this means $\mu(E) = 0$. $\square$

One of the more noteworthy results of the theory of finitely additive measures is Tarski's Theorem that the converse of Proposition 2.5 is valid: If E is not G-paradoxical, then a finitely additive, G-invariant measure on $\mathscr{P}(X)$ normalizing E must exist. This will be proved in Chapter 9.

The next theorem deduces the SO_3-negligibility of S^2 from the Hausdorff Paradox by proving that countable sets are negligible with respect to finite measures on $\mathscr{P}(S^2)$.

Theorem 2.6 (AC). *S^2 is SO_3-negligible. Hence there is no finitely additive, rotation-invariant measure on $\mathscr{P}(S^2)$ having total measure one. Moreover, for any $n \geqslant 3$ there is no finitely additive, isometry-invariant measure on $\mathscr{P}(\mathbf{R}^n)$ normalizing the unit cube.*

Proof. Suppose μ is a finitely additive, SO_3-invariant measure on $\mathscr{P}(S^2)$ with $\mu(S^2) < \infty$. If D is the countable set in the Hausdorff Paradox then, by Proposition 2.5, $\mu(S^2 \backslash D) = 0$. Hence it suffices to show that $\mu(D) = 0$. Let ℓ be a line through the origin that is disjoint from D. For each point P in D let $A(P)$ be the set of angles θ such that the rotation of P around ℓ through θ takes P to another point in D. The countability of D implies that $A(P)$ is countable, and hence that $A = \bigcup \{A(P) : P \in D\}$ is countable too. If ρ is chosen to be a rotation around ℓ through one of the uncountably many angles not in A, then ρ has the property that $\rho(D)$ is disjoint from D. It follows that $\mu(S^2) \geqslant \mu(D \cup \rho(D)) = \mu(D) + \mu(\rho(D)) = \mu(D) + \mu(D)$, so $\mu(D) \leqslant \mu(S^2)/2$. This same argument with D replaced by the countable set $D \cup \rho(D)$ yields a rotation ρ' such that $D \cup \rho(D)$ is disjoint from $\rho'(D \cup \rho(D))$. Hence $\mu(S^2) \geqslant \mu(D \cup \rho(D)) + \mu(\rho'(D) \cup \rho'\rho(D)) = 4\mu(D)$, and $\mu(D) \leqslant \mu(S^2)/4$. Continuing in this way, one sees that $\mu(D) \leqslant \mu(S^2)/2^n$ for each n, so $\mu(D) = 0$ as desired.

To prove the assertion about $\mathbf{R}^n$, it suffices to consider $n = 3$, since a measure in a higher dimension induces one in $\mathbf{R}^3$ as described in the proof of 1.6. Now, if μ is an isometry-invariant measure on $\mathbf{R}^3$ normalizing the unit cube, then μ must vanish on singletons. This is because any two singletons are congruent, and so receive the same measure; hence, if a singleton's measure were positive, then the measure of the unit cube would be infinite. Moreover, translation-invariance implies that any cube has finite, nonzero measure, and it follows that $0 < \mu(B) < \infty$, where B denotes the unit ball. Define a measure v on $\mathscr{P}(S^2)$ by the adjunction of radii, that is, $v(A) = \mu\{\alpha P : P \in A, 0 < \alpha \leqslant 1\}$. Since $\mu(\{\mathbf{0}\}) = 0$, $v(S^2) = \mu(B)$. Moreover, v is finitely additive and SO_3-invariant because μ is. This contradicts the SO_3-negligibility of S^2. □

As observed at the beginning of this chapter, there are no free non-Abelian subgroups in the lower-dimensional Euclidean isometry groups. This is why the ideas of the Hausdorff Paradox cannot be used to decide the existence of isometry-invariant, finitely additive measures defined on $\mathscr{P}(\mathbf{R}^1)$ or $\mathscr{P}(\mathbf{R}^2)$. In Chapter 10 we shall show that invariant measures always exist with respect to groups satisfying certain abstract conditions. Since these conditions include solvability, it will follow that isometry-invariant, finitely additive measures defined on $\mathscr{P}(\mathbf{R})$ and $\mathscr{P}(\mathbf{R}^2)$ exist.

NOTES

Hausdorff's original embedding of $Z_2 * Z_3$ in a rotation group appears in [90; 89, p. 469]; for a modern treatment see [227]. Osofsky's simpler construction appears as a problem in the *American Mathematical Monthly* [179, 180] (see also [A6]). The fact that $Z_2 * Z_3$, and hence SO_3, contains a free subgroup of rank 2 is first stated in [246, p. 80, n. 22]. In fact, $G * H$ always has a free subgroup of rank 2 unless one of G or H is trivial, or each of G, H has order 2 (see [135, p. 195, Ex. 19]). In an addendum to [180], Lyndon shows, using quaternions, that the three rotations through $120°$ around the three orthogonal axes in $\mathbf{R}^3$ freely generate $Z_3 * Z_3 * Z_3$.

It is curious that Hausdorff's embedding of a free product in SO_3 appeared in the same year as the Sierpiński-Mazurkiewicz Paradox (1.7) that involves the embedding of a free semigroup in G_2. Perhaps Hausdorff was motivated by Felix Klein's representation of $Z_2 * Z_3$ as linear fractional transformations and hence as isometries of the hyperbolic plane [108]. (See the discussion of the hyperbolic plane in Chapter 5.)

The proof of Theorem 2.1 given here is due to Świerczkowski [229], who used it to answer a problem of Steinhaus on tetrahedra, which is considered here at the end of Chapter 5. Many papers on the representation of free groups using rotations or isometries ([6, 23, 46, 47, 52, 54, 170,]) appeared in Holland and Poland in the 1950s, largely inspired by work of Sierpiński and de Groot, and many results were obtained independently in both countries. Thus, part (b) of Theorem 2.2, originally conjectured by de Groot [46], was proved by Dekker [54], while a similar result was proved by Balcerzyk [6]. Part (a) of Theorem 2.2 is stated in [229]. The existence of a free group of rotations of rank $2^{\aleph_0}$ was essentially proved by Sierpiński [212] in 1945, and rediscovered by de Groot and Dekker [46, 47] in 1954. De Groot [46] conjectured that any free product of continuum many subgroups of SO_3, each of size strictly less than $2^{\aleph_0}$, is also a subgroup of SO_3. This was proved by Balcerzyk and Mycielski [6], although Dekker [54] independently proved the special case where all the groups are cyclic. The existence of free groups of isometries acting on $\mathbf{R}^3$ without nontrivial fixed points was proved independently by Mycielski and Świerczkowski [170] and Dekker [52]. Dekker also studied the isometry groups of higher dimensional Euclidean and non-Euclidean spaces (see Chapter 5).

The result of Tits on matrix groups, which answered a conjecture of Bass and Serre, appears in [237].

The Hausdorff Paradox (2.3) is, of course, due to Hausdorff [90; 89, p. 469], who used it to prove Theorem 2.6. Hausdorff's formulation of 2.3 was slightly different though. He showed that, modulo a countable set of points, a "half" of a sphere could be congruent to a "third" of a sphere, rather than that a sphere could be duplicated (Corollary 3.10). From the point of view of disproving the existence of measures, this is all that is required.

Chapter 3

The Banach-Tarski Paradox: Duplicating Spheres and Balls

The idea of cutting a figure into pieces and rearranging them to form another figure goes back at least to Greek geometry, where this method was used to derive area formulas for regions such as parallelograms. When forming such rearrangements, one totally ignores the boundaries of the pieces. The consideration of a notion of dissection in which every single point is taken into account, that is, a set-theoretic generalization of the classical geometric definition, leads to an interesting, and very general, equivalence relation. By studying the abstract properties of this new relation, Banach and Tarski were able to improve on Hausdorff's Paradox (2.3) by eliminating the need to exclude a countable subset of the sphere. Since geometric rearrangements will be useful too, we start with the classical definition in the plane.

Definition 3.1.   *Two polygons in the plane are* congruent by dissection *if one of them can be decomposed into finitely many polygonal pieces that can be rearranged using isometries (and ignoring boundaries) to form the other polygon.*

It is clear that polygons that are congruent by dissection have the same area. The converse was proved in the early nineteenth century, and a simple proof can be given by efficiently making use of the fact that congruence by dissection is an equivalence relation (transitivity is easily proved by superposition, using the fact that the intersection of two polygons is a polygon (see Boltianskii [17, p. 50] or Eves [70, p. 233])).

Theorem 3.2 (Bolyai-Gerwien Theorem). *Two polygons are congruent by dissection if and only if they have the same area.*

Proof. In order to prove the reverse direction, it suffices, because of transitivity, to show that any polygon is congruent by dissection to a square (necessarily of the same area). We do this first for a triangle. Figure 3.1(a) shows that any triangle is congruent by dissection to a rectangle. Figure 3.1(b) shows how a rectangle whose length is at most four times its width can be transformed to a square: the triangles to be moved are clearly similar to their images, and the fact that the area of the square equals that of the rectangle implies that they are, in fact, congruent. The figure does not correctly describe the situation when the length is greater than four times the width, but any such unbalanced rectangle can be transformed to one of the desired type by repeated halving and stacking as illustrated in Figure 3.1(c). Hence any triangle is congruent by dissection with a square.

The proof is concluded by observing that the Pythagorean Theorem can be proved in a way that can be used to transform two (or more) squares into one by dissection. Consider Figure 3.1(d), which proves that $c^2 = a^2 + b^2$ by showing how the squares on a and b can be transformed by dissection into

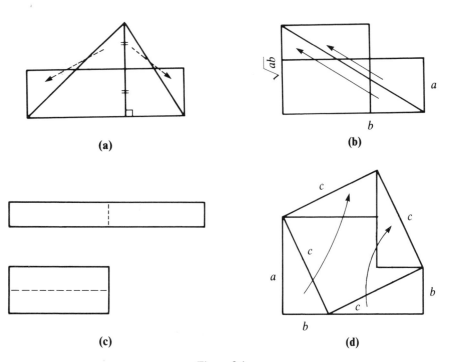

Figure 3.1

one on c. Used repeatedly, this construction shows how any finite set of squares can be transformed by dissection to a single square. Now, any polygon can be split into triangles (see [70, p. 238]). Squaring the triangles and combining the squares as just indicated yields a square that is congruent by dissection to the original polygon.   □

This efficient proof does not lead to particularly efficient or beautiful dissections. For several interesting examples, such as a four-piece squaring of an equilateral triangle, see Chapter 5 of [70].

The theory of geometrical dissections in higher dimensions, or other geometries, is not at all as simple as in the plane. In fact, the third problem on Hilbert's famous list asks whether a regular tetrahedron in $\mathbf{R}^3$ is congruent by dissection (into polyhedra) with a cube. All proofs of the volume formula for a tetrahedron were based on a limiting process of one sort or another, such as the devil's staircase [93, p. 390] or Cavalieri's Principle [149, chap. 23]. Hopes for an elegant dissection proof were dashed when Dehn proved, in 1900, that a regular tetrahedron is not congruent by dissection with any cube (see [17] for a proof). But it is possible that for a suitable generalization of dissection where a wider class of pieces is allowed, a regular tetrahedron is piecewise congruent to a cube. Indeed, one consequence of the Banach-Tarski Paradox is that a regular tetrahedron can be cubed if arbitrary sets are allowed as pieces (see Theorems 3.10 and 9.18, and Questions 3.14 and 8.15).

The set-theoretic version of congruence by dissection may be stated in the context of an arbitrary group action.

Definition 3.3. *Suppose G acts on X and $A, B \subseteq X$. Then A and B are G-equidecomposable (sometimes called* finitely G-equidecomposable *or piecewise G-congruent) if A and B can each be partitioned into the same finite number of respectively G-congruent pieces. Formally,*

$$A = \bigcup_{i=1}^{n} A_i, \qquad B = \bigcup_{i=1}^{n} B_i,$$

$A_i \cap A_j = \varnothing = B_i \cap B_j$ *if $i < j \leqslant n$, and there are $g_1, \ldots, g_n \in G$ such that, for each $i \leqslant n$, $g_i(A_i) = B_i$.*

The notation $A \sim_G B$ will be used to denote the equidecomposability relation, but the G will be suppressed if X is a metric space and G is the full isometry group of X, or if it is obvious which group G is meant. Thus, for sets in $\mathbf{R}^n$, equidecomposability means G_n-equidecomposability. We say A is G-equidecomposable with B using n pieces (denoted $A \sim_n B$) if the disassembly can be effected with n pieces.

It is straightforward to verify that $\sim_G$ is an equivalence relation. Transitivity of $\sim_G$ is proved in the same way as for congruence by dissection, yielding that if $A \sim_m B$ and $B \sim_n C$ then $A \sim C$, using at most mn pieces. Not surprisingly then, the relation $\sim_n$ is not transitive. A simple counterexample is: $A = \{1, 2, 3, 4\}$, $B = \{1, 2, 5, 6\}$, and $C = \{1, 5, 9, 13\}$; $A \sim_2 B \sim_2 C$, but $A \sim_4 C$ and the 4 cannot be lowered. We may now rephrase more succinctly the notion of a set being G-paradoxical. E is G-paradoxical if and only if E contains disjoint sets A, B such that $A \sim_G E$ and $B \sim_G E$. One can then obtain immediately the following easy but useful fact, which shows that the property of being G-paradoxical is really a property of the $\sim_G$-equivalence classes in $\mathscr{P}(X)$.

Proposition 3.4. *Suppose G acts on X and E, E' are G-equidecomposable subsets of X. If E is G-paradoxical, so is E'.*

It is not immediately apparent that there is any connection between equidecomposability and congruence by dissection. Indeed, they differ in a most fundamental way. Since there is no restriction on the subsets that may be used to verify that $A \sim B$, there is no guarantee that A and B have the same area (or n-dimensional Lebesgue measure, if A, $B \subseteq \mathbf{R}^n$). For if the pieces are non-Lebesgue measurable, then the straightforward proof that works in the case of congruence by dissection cannot be used, since it involves summing the areas of the pieces. Thus it is conceivable that all polygons in the plane are equidecomposable, or that all polyhedra in $\mathbf{R}^3$ are equidecomposable. In fact, the latter assertion is true (see 3.10) while the former is not! The preservation of a given (G-invariant) measure under G-equidecomposability is related to the existence of a finitely additive, G-invariant extension of the measure to all subsets of X, and this, in turn, depends on abstract properties of the group G (and of G's action on X). The second part of this book deals extensively with the question of which groups have the property that invariant measures on an algebra of subsets of X may be extended to invariant, finitely additive measures on all subsets of X.

In a different vein, one can ask whether polygons that are congruent by dissection are necessarily equidecomposable. The problem is that, somehow, the boundaries of the pieces in a geometrical dissection must be accounted for in a precise way. In a typical dissection, such as those in Figure 3.1, the boundaries do double duty and so cannot simply be assigned to one of the pieces. This problem can be solved though, and the main tool is a very important property of the equivalence relation $\sim_G$.

Whenever one has an equivalence relation on the collection of subsets of a set, one may define another relation, $\preccurlyeq$, by $A \preccurlyeq B$ if and only if A is equivalent to a subset of B. Then $\preccurlyeq$ is really a relation on the equivalence classes and, in fact, is reflexive and transitive. The Schröder-Bernstein Theorem of classical set theory states that if the cardinality relation is used— A and B are equivalent if there is a bijection from A to B—then $\preccurlyeq$ is antisymmetric as well; that is, if $A \preccurlyeq B$ and $B \preccurlyeq A$, then A and B have the same cardinality. Thus $\preccurlyeq$ is a partial order on the equivalence classes. (Under the

Axiom of Choice, every set is equivalent to an ordinal, so $\leqslant$ is a well-ordering.) Banach realized that the proof of the Schröder-Bernstein Theorem could be applied to any equivalence relation satisfying two abstract properties; in particular, it applies to G-equidecomposability. From now on we use the notation $A \leqslant B$ only in the context of equidecomposability: $A \leqslant B$ means A is G-equidecomposable with a subset of B.

Theorem 3.5 (Banach-Schröder-Bernstein Theorem). *Suppose G acts on X and $A, B \subseteq X$. If $A \leqslant B$ and $B \leqslant A$, then $A \sim_G B$. Thus $\leqslant$ is a partial ordering of the $\sim_G$-classes in $\mathscr{P}(X)$.*

Proof. The relation $\sim_G$ is easily seen to satisfy the following two conditions:

(a) if $A \sim B$ then there is a bijection $g: A \to B$ such that $C \sim g(C)$ whenever $C \subseteq A$, and
(b) if $A_1 \cap A_2 = \varnothing = B_1 \cap B_2$, and if $A_1 \sim B_1$ and $A_2 \sim B_2$, then $A_1 \cup A_2 \sim B_1 \cup B_2$.

The rest of the proof assumes only that $\sim$ is an equivalence relation on $\mathscr{P}(X)$ satisfying (a) and (b).

Let $f: A \to B_1$, $g: A_1 \to B$, where $B_1 \subseteq B$ and $A_1 \subseteq A$, be bijections as guaranteed by (a). Let $C_0 = A \backslash A_1$ and, by induction, define C_{n+1} to be $g^{-1}f(C_n)$; let $C = \bigcup_{n=0}^{\infty} C_n$. Then it is easy to check that $g(A \backslash C) = B \backslash f(C)$, and hence the choice of g implies that $A \backslash C \sim B \backslash f(C)$. But, by the choice of f, $C \sim f(C)$ and property (b) yields $(A \backslash C) \cup C \sim (B \backslash f(C)) \cup f(C)$, or $A \sim B$ as desired. $\square$

It is clear from the proof that $m + n$ pieces suffice for the final decomposition if m, n, respectively, suffice for the hypothesized decompositions. This proof serves as a proof of the classical Schröder-Bernstein Theorem as well, since the cardinality relation satisfies properties (a) and (b).

This theorem eases dramatically the verification of equidecomposability. As an illustration, suppose a subset E of X is G-paradoxical, say A, B are disjoint subsets of E with $A \sim E \sim B$. Then $E \sim B \subseteq E \backslash A \subseteq E$, so the Banach-Schröder-Bernstein Theorem implies that $E \backslash A \sim E$. This proves the following result.

Corollary 3.6. *A subset E of X is G-paradoxical if and only if there are disjoint sets $A, B \subseteq E$ with $A \cup B = E$ and $A \sim E \sim B$.*

Another application of the theorem is the following counterintuitive result, the underlying idea of which will appear later (3.11 and 7.8). It yields, for example, that a given circle and a given square may each be split into two Borel sets, such that corresponding pieces are similar; as an exercise, the reader can construct explicitly the four pieces in this special case. Recall that the group of all similarities of $\mathbf{R}^n$ is the group generated by all isometries and all magnifications from the origin, where the latter refers to a function of the form $f(P) = \alpha P$, where $\alpha \neq 0$.

Corollary 3.7. *Any two subsets, X, Y, of $\mathbf{R}^n$, each of which is bounded and has nonempty interior, may be partitioned as follows: $X = X_1 \cup X_2$, $Y = Y_1 \cup Y_2$, where X_1 is similar to Y_1 and X_2 is similar to Y_2. If X and Y are Borel sets, so are the four sets in the decomposition.*

Proof. The hypotheses on X and Y guarantee the existence of similarities g_1, g_2 such that $g_1(X) \subseteq Y$ and $g_2(Y) \subseteq X$; simply shrink X so that it fits into Y, and vice versa. Therefore $X \preccurlyeq Y$ and $Y \preccurlyeq X$ with respect to equidecomposability using similarities, and since $m + n = 1 + 1 = 2$, the result now follows from Theorem 3.5. If X and Y are Borel, then the sets introduced by the proof of 3.5 will also be Borel. $\quad\square$

The following application of Theorem 3.5 is important in that it shows that polygons that are congruent by (geometric) dissection are also equidecomposable, that is, congruent by set-theoretic dissection.

Theorem 3.8. *If the polygons P_1 and P_2 are congruent by dissection, then they are equidecomposable.*

Proof. Let Q_1, Q_2 be the open sets obtained by forming the union of all the interiors of the polygonal subsets of P_1, P_2, respectively, arising from the hypothesized dissection. Then $Q_1 \sim Q_2$ and so the proof will be complete once it is shown that $P_1 \sim Q_1$ and $P_2 \sim Q_2$, that is, that the boundary segments can be absorbed. This follows from the following fact (by setting $A = P_1$ and $T = P_1 \backslash Q_1$): If A is a bounded set in the plane with nonempty interior and T is a set, disjoint from A, consisting of finitely many (bounded) line segments, then $A \sim A \cup T$.

To prove this fact, let D be a disc contained in A, and let r be its radius. By subdividing the segments in T, we may assume that each one has length less than r. Let θ be any rotation of D about its center having infinite order, let R be any radius of D (excluding the center of D), and let $\bar{R} = R \cup \theta(R) \cup \theta^2(R) \cup \dots$. Now, if $s \in T$, then $D \cup s \preccurlyeq D$. This is because $\theta(\bar{R})$ is disjoint from R and $D \backslash \bar{R}$, so $D \cup s = (D \backslash \bar{R}) \cup \bar{R} \cup s \sim (D \backslash \bar{R}) \cup \theta(\bar{R}) \cup \sigma(s) \subseteq D$, where σ is any isometry taking s to a subset of R. Since, obviously, $D \preccurlyeq D \cup s$, the Banach-Schröder-Bernstein Theorem implies that $D \sim D \cup s$. Since each of the segments in T may thus be absorbed, one at a time, into D, we have that $D \sim D \cup T$. Adding $A \backslash D$ to both sides yields $A \sim A \cup T$, as required. $\quad\square$

Because of the Bolyai-Gerwien Theorem (3.2), the preceding theorem implies that any two polygons of the same area are equidecomposable. The converse is true, though much harder to prove, since it follows from the existence of a finitely additive, G_2-invariant measure on all subsets of $\mathbf{R}^2$ that agrees with area on polygons (see Corollary 10.8).

The proof of 3.8 might be called a *proof by absorption*, since it shows how a troublesome set (the boundary segments) can be absorbed in a way that, essentially, renders it irrelevant. Now, we have seen that free groups of rank 2 cause paradoxes when they act without fixed points, and so situations where the fixed points can be absorbed will be especially important. The following proof is typical of the absorption proofs, and immediately yields the Banach-Tarski Paradox.

Theorem 3.9. *If D is a countable subset of S^2, then S^2 and $S^2 \setminus D$ are SO_3-equidecomposable (using two pieces).*

Proof. We seek a rotation, ρ, of the sphere such that the sets D, $\rho(D)$, $\rho^2(D),\ldots$ are pairwise disjoint. This suffices, since then $S^2 = \bar{D} \cup (S^2 \setminus \bar{D}) \sim \rho(\bar{D}) \cup (S^2 \setminus \bar{D}) = S^2 \setminus D$, where $\bar{D} = \bigcup \{\rho^n(D): n = 0, 1, 2, \ldots\}$. The construction of ρ is similar to the proof of 2.6. Let ℓ be a line through the origin that misses the countable set D. Let A be the set of angles θ such that for some $n > 0$ and some $P \in D$, $\rho(P)$ is also in D where ρ is the rotation about ℓ through $n\theta$ radians. Then A is countable, so we may choose an angle θ not in A; let ρ be the corresponding rotation about ℓ. Then $\rho^n(D) \cap D = \varnothing$ if $n > 0$, from which it follows that whenever $0 \leqslant m < n$, then $\rho^m(D) \cap \rho^n(D) = \varnothing$ (consider $\rho^{n-m}(D) \cap D$); therefore ρ is as required. $\square$

Corollary 3.10 (*The Banach-Tarski Paradox*) (*AC*). S^2 *is SO_3-paradoxical, as is any sphere centered at the origin. Moreover, any solid ball in $\mathbf{R}^3$ is G_3-paradoxical and $\mathbf{R}^3$ itself is paradoxical.*

Proof. The Hausdorff Paradox (2.3) states that $S^2 \setminus D$ is SO_3-paradoxical for some countable set D (of fixed points of rotations). Combining this with the previous theorem and Proposition 3.4 yields that S^2 is SO_3-paradoxical. Since none of the previous results depends on the size of the sphere, spheres of any radius admit paradoxical decompositions.

It suffices to consider balls centered at $\mathbf{0}$, since G_3 contains all translations. For definiteness, we consider the unit ball B, but the same proof works for balls of any size. The decomposition of S^2 yields one for $B \setminus \{\mathbf{0}\}$ if we use the radial correspondence: $P \to \{\alpha P: 0 < \alpha \leqslant 1\}$. Hence it suffices to show that B is G_3-equidecomposable with $B \setminus \{\mathbf{0}\}$, that is, that a point can be absorbed. Let $P = (0, 0, \frac{1}{2})$ and let ρ be a rotation of infinite order about an axis through P but missing the origin. Then, as usual, the set $D = \{\rho^n(\mathbf{0}): n \geqslant 0\}$ may be used to absorb $\mathbf{0}$: $\rho(D) = D \setminus \{\mathbf{0}\}$, so $B \sim B \setminus \{\mathbf{0}\}$.

If, instead, the radial correspondence of S^2 with all of $\mathbf{R}^3 \backslash \{0\}$ is used, one gets a paradoxical decomposition of $\mathbf{R}^3 \backslash \{0\}$ using rotations. Since, exactly as for the ball, $\mathbf{R}^3 \backslash \{0\} \sim_{G_3} \mathbf{R}^3$, $\mathbf{R}^3$ is paradoxical via isometries. $\qquad\square$

Because of its use of Theorem 3.9, this proof of the Banach-Tarski Paradox seems to depend on having uncountably many rotations available. But, as a consequence of a more general approach in the next chapter, we shall see (Theorem 11.22) that for subgroups G of SO_3, S^2 is G-paradoxical if and only if G has a free subgroup of rank 2. In fact, it follows from Theorem 4.5 that S^2 is paradoxical with only the rotations ϕ and ρ of Theorem 2.1 being used to move the pieces of the decomposition.

The version of the Banach-Tarski Paradox in 3.10 does not add anything to our knowledge of finitely additive measures not already derivable from the Hausdorff Paradox (see 2.6), but the result is much more striking, indeed more bizarre, than Hausdorff's. A ball, which has a definite volume, may be taken apart into finitely many pieces that may be rearranged via rotations of $\mathbf{R}^3$ to form two, or even 1,000,000 balls, each identical to the original one! Or, more whimsically, the unit ball may be decomposed into finitely many pieces, forming a three-dimensional jig-saw puzzle with the following property. For each $n \leq 1,000,000$ the pieces of the puzzle may be arranged using rotations to form n disjoint unit balls. Of course, because the Axiom of Choice is used to produce the pieces, the jig-saw would have to be inconceivably sharp!

Rotations preserve volume, and this is why the result has come to be known as a paradox. A resolution is that there may not be a volume for the rotations to preserve; the pieces in the decomposition may be (indeed, will have to be) non-Lebesgue measurable. In fact, we have already seen (Theorem 2.6) that S^2 is SO_3-negligible, that is, there is no rotation-invariant, finitely additive measure defined for all subsets of S^2 (or of the unit ball).

The proof of the Banach-Tarski Paradox, like all proofs of the existence of a nonmeasurable set, is nonconstructive in that it appeals to the Axiom of Choice. It has been argued that the result is so counterintuitive, so patently false in the real world, that one of the underlying assumptions must be incorrect; the Axiom of Choice is usually selected as the culprit. We shall discuss this argument more fully in Chapter 13, where the role that Choice plays in the foundations of measure theory will be examined in detail. For now, let us mention two points. First, strange results that do not require the Axiom of Choice abound in all of mathematics. In particular, there is the Sierpiński-Mazurkiewicz Paradox (1.7); while not as startling as the duplication of the sphere, it runs counter to intuition despite its constructive proof. Second, the Axiom of Choice is consistent with the other axioms of set theory; as shown by Gödel, the axiom is true in the "constructible universe." Hence, independent of its truth in any individual's view of the set-theoretic universe, the Banach-Tarski Paradox is, at the least, consistent.

Though volume is not one of them, there are properties that are preserved by equidecomposability in $\mathbf{R}^3$. If A is bounded, then so is any set equidecomposable with A; the same is true if A has nonempty interior. Banach and Tarski were able to show that any two sets, each having these two properties, are equidecomposable. Thus they generalized their already surprising result so that it applies to solids of any shape. It is a consequence of this strengthening that any bounded subset of $\mathbf{R}^3$ with nonempty interior is paradoxical; in fact, the unit ball is equidecomposable with any other ball, no matter how large or small. Another consequence is that *any* two polyhedra in $\mathbf{R}^3$ are equidecomposable; this should be contrasted with the result (Theorem 3.8 and remarks following) that two polygons are equidecomposable if and only if they have the same area.

Theorem 3.11 (Banach-Tarski Paradox, Strong Form) (AC). *If A and B are any two bounded subsets of $\mathbf{R}^3$, each having nonempty interior, then A and B are equidecomposable.*

Proof. It suffices to show that $A \leqslant B$, for then by the same argument, $B \leqslant A$ and Theorem 3.5 yields $A \sim B$. Choose solid balls K and L such that $A \subseteq K$ and $L \subseteq B$, and let n be large enough that K may be covered by n (overlapping) copies of L. Now, if S is a set of n disjoint copies of L, then using the Banach-Tarski Paradox to repeatedly duplicate L, and using translations to move the copies so obtained, yields that $L \geqslant S$. Therefore $A \subseteq K \leqslant S \leqslant L \subseteq B$, so $A \leqslant B$. $\square$

This remarkable result usually is viewed negatively since it so forcefully illustrates Hausdorff's Theorem (2.6) that certain measures do not exist. Not all consequences of the Banach-Tarski Paradox are negative, however. Tarski used it to prove a result on finitely additive measures on the algebra of Lebesgue measurable subsets of $\mathbf{R}^3$, and this result was used in recent work on the uniqueness of Lebesgue measure (see Lemma 9.7 and remarks following Theorem 11.11).

There are some interesting problems of a topological nature regarding the Banach-Tarski Paradox. As pointed out, the pieces into which the sphere S^2 is decomposed in order to be duplicated cannot be Lebesgue measurable, and hence cannot be Borel sets. There is another family of subsets of $\mathbf{R}^n$ or S^n that shares many of the properties of the measurable sets. A set is said to have the *Property of Baire* if it differs from a Borel set by a meager set; that is, the symmetric difference $X \vartriangle B \left(=(X\setminus B) \cup (B\setminus X)\right)$ is meager for some Borel set B. (Recall that a set is meager—often said to be of first category—if it is a countable union of nowhere dense sets.) In fact (see [181, p. 20]), the set B in the definition above can be taken to be open. Not all sets have the Property of Baire: using the representation using open sets just mentioned, it is not hard to see that the set M of the proof of

Theorem 1.5 (whose construction used the Axiom of Choice) fails to have
the Property of Baire. See [181] for more on this family of sets.

There are many similarities between the σ-algebras of measurable
sets and sets with the Property of Baire; for instance, a set is Lebesgue measurable if and only if it differs from a Borel set (in fact an F_σ) by a null
(i.e., Lebesgue measure zero) set. But the two notions do not coincide. Every
subset of $\mathbf{R}^n$ may be split into a null set and a meager set (see [181, p. 5]),
and if a nonmeasurable set is so split, the meager part cannot be measurable. While the measurable subsets of S^2 carry a finitely additive (in fact,
countably additive) SO_3-invariant measure of total measure one—namely
Lebesgue measure itself—it is not known whether such a finitely additive
measure exists on the sets having the Property of Baire. The existence of
such a measure would imply a negative answer to the following question.

Question 3.12 (Marczewski's Problem).   Does S^2 admit a paradoxical decomposition (with respect to rotations) in which each piece has the
Property of Baire?

Although little is known about this question, it seems reasonable to
conjecture that the answer is no. First of all, it is unlikely that the proof
of the Banach-Tarski Paradox, which uses the Axiom of Choice to define
the pieces, could be modified so that the pieces have the Property of Baire.
Furthermore, if such a decomposition of S^2 exists, then by the proofs of
3.10 and 3.11, a paradoxical decomposition, using pieces with the Property of
Baire, of the unit cube in $\mathbf{R}^3$ exists. This has the following seemingly im-
plausible consequence (see (2) $\Rightarrow$ (8) of Theorem 9.5): Any cube, no matter
how small, contains pairwise disjoint open subsets $U_1, \ldots, U_m$ such that, for
suitable isometries ρ_i, $\bigcup \rho_i U_i$ is dense in the unit cube. In the case of a ball
in $\mathbf{R}^3$, Marczewski's Problem is equivalent to a statement about packing
(Borel) measure zero sets into arbitrarily small cubes using finitely many
pieces with the Property of Baire (see page 168).

Another old problem on the topology of pieces concerns more general metric spaces. No paradoxical decomposition of the sphere can use
only Borel pieces, as previously pointed out. How general is this fact? If $\mathbf{N}$
is given the discrete metric, then any permutation is an isometry, and since
$\mathbf{N}$ is countable, all subsets are Borel; hence (see 1.4) $\mathbf{N}$ is Borel paradoxical.
Because of this, and other examples, we restrict our attention to compact
metric spaces. Now, any compact metric space bears a countably additive,
G-invariant Borel measure of total measure one, where G is the group of
isometries of the space. (To prove this, put a metric on G by defining the
distance between g_1 and g_2 to be the supremum of the distance between $g_1(x)$
and $g_2(x)$, for $x \in X$. One can check that this turns G into a compact
topological group, and therefore G bears a left-invariant Borel measure, ν,

of total measure one, namely Haar measure (see [36]). We may use v to define the desired Borel measure, μ, on X by fixing some $x \in X$ and setting $\mu(A) = v(\{g \in G : g(x) \in A\})$.) The existence of such a Borel measure, even if it were only finitely additive, means that a compact metric space is never paradoxical using Borel pieces. But a rather different problem arises if we relax the definition of isometry by considering partial isometries, rather than global isometries. Let us reserve the term *congruent* for two subsets A and B of a metric space to mean that there is a distance-preserving bijection from A to B, that is, $\sigma(A) = B$ for some partial isometry σ. For $\mathbf{R}^n$ this leads to nothing new since partial isometries extend to isometries of $\mathbf{R}^n$, but in more general spaces such extensions may not exist. In compact metric spaces, however, it is easy to see that any isometry on A extends uniquely to one on $\bar{A}$, the closure of A (see [124, p. 215]). This congruence relation does not arise from a group action, but the definition of paradoxical decomposition is easily modified to apply. Simply replace the sets $g_i(A_i)$ and $h_j(B_j)$ by sets congruent to A_i, B_j, respectively. Note that if A is Borel and congruent to B, it is not obvious that B also is Borel; in general, continuous images of even closed sets need not be Borel. But for complete, separable spaces (sometimes called Polish spaces; compact spaces are Polish), a one-to-one, continuous image of a Borel set is Borel (see [117, p. 487]) and hence the Borel sets are, indeed, closed under congruence.

Question 3.13. Is it true that no compact metric space is paradoxical (with respect to congruence) using Borel pieces?

Of course, an affirmative answer would follow from the existence of a finitely additive, congruence-invariant, Borel measure on the space, having total measure one. This problem, in the (possibly stronger) measure-theoretic form just given, is an old one; it was posed by Banach and Ulam in 1935 as the second problem in the famous collection of primarily Polish problems known as *The Scottish Book* [140]. There are three interesting partial results. Mycielski [165] showed that any compact metric space admits a countably additive, Borel measure of total measure one that assigns congruent open sets the same measure; hence there are no paradoxical decompositions using just open sets. Because Mycielski's construction yields a countably additive measure, the existence of countable, compact metric spaces (e.g., $\{0, 1, \frac{1}{2}, \frac{1}{3}, \ldots\}$) shows that it cannot be modified to work for closed sets. Nevertheless, Davies and Ostaszewski [41] have shown how to construct finitely additive, congruence-invariant measures of total measure one for all countable compact metric spaces. Moreover, Bandt and Baraki [A5] have recently shown that under the additional assumption that the metric space is *locally homogeneous* (any two points have congruent neighborhoods), Mycielski's measure does indeed assign congruent Borel sets the same measure.

The strong form of the Banach-Tarski Paradox implies that a regular tetrahedron in $\mathbf{R}^3$ is equidecomposable with a cube. This is not really in the spirit of the sort of dissection Hilbert asked for in his third problem, since any cube will do; the cube need not have the same volume as the tetrahedron. But this raises the possibility of cubing a regular tetrahedron using pieces restricted to some nice collection of sets more general than polyhedra (the context of Hilbert's Third Problem), but not so general that paradoxes exist and volume is not preserved. A natural choice is to restrict the pieces to be Lebesgue measurable, for then equidecomposable sets must have the same measure.

Question 3.14.   Is a regular tetrahedron in $\mathbf{R}^3$ equidecomposable with a cube using isometries,* and pieces that are Lebesgue measurable?

The reason that this question, and Hilbert's Third Problem, is posed for a regular tetrahedron is that nonregular tetrahedra congruent by dissection with a cube are known to exist (see [17]). There are related questions in $\mathbf{R}^1$ and $\mathbf{R}^2$(see 7.5 and 8.15).

Finally, a different, rather more bizarre, question arises if one asks whether the pieces in a paradoxical duplication can be moved physically to form the two new copies of, say, the sphere. Since the pieces do not exist physically, this calls for some clarification. Consider the equidecomposability of a unit ball with two disjoint unit balls. Can the motions of the pieces, assuming the latter somehow to be given, be actually carried out in $\mathbf{R}^3$ in such a way that the pieces never overlap? This can be formalized as follows.

Question 3.15.   Let B, B_1, B_2 be pairwise disjoint unit balls in $\mathbf{R}^3$. Can B be partitioned into $A_1,\ldots,A_m$, $A_{m+1},\ldots,A_{m+n}$ such that for each $i = 1,\ldots,m + n$ and $t \in [0, 1]$ there is an isometry σ_t^i of $\mathbf{R}^3$ satisfying:

(i) σ_0^i is the identity; $\sigma_1^i(A_i)$, $i = 1,\ldots,n$, partitions B_1; $\sigma_1^i(A_i)$, $i = n + 1,\ldots,n + m$, partitions B_2.

(ii) For each i, the isometries σ_t^i depend continuously on t.

(iii) For each $t \in [0, 1]$, the sets $\sigma_t^i(A_i)$, $i = 1,\ldots,m + n$ are pairwise disjoint.

As pointed out by Klee [87, p. 51], the isometries used in the Banach-Tarski Paradox are used to transform pieces instantaneously to a congruent copy. The preceding question asks whether a continuous movement of the pieces can be realized in $\mathbf{R}^3$.

In the rest of Part I we shall show how versions of the Banach-Tarski Paradox can be constructed in other spaces, such as higher-dimensional spheres and non-Euclidean spaces. As pointed out at the end of Chapter 2, the Banach-Tarski Paradox does not exist in $\mathbf{R}^1$ or $\mathbf{R}^2$, but

---

* See note on page 101.

other constructions are possible that have similar measure-theoretic implications. And we shall also study refinements of the original construction in $\mathbf{R}^3$, showing, for example, how to duplicate a ball using the smallest possible number of pieces.

NOTES

The use of dissection to derive area formulas dates at least to Greek geometry, but the converse idea (Theorem 3.2) was not considered until the nineteenth century. F. Bolyai and P. Gerwien discovered Theorem 3.2 independently around 1832 (see [75]), but apparently William Wallace of England had proved the result already in 1807 ([185]; see also [65, p. 225] and [97]). Gerwien also proved [76] that Theorem 3.2 is valid for spherical polyhedra. A complete treatment of the Bolyai-Gerwien Theorem, including a discussion of the smallest possible group of isometries with respect to which it is true, may be found in [17]. For a treatment of the Bolyai-Gerwien Theorem in both Euclidean and hyperbolic geometry see [145, §10.4].

For an exposition of the solution, due to Dehn, of Hilbert's Third Problem, see [17]. More advanced topics related to this problem may be found in [201].

The definition of G-equidecomposability is due to Banach and Tarski ([11, 230]), and [11] contains many elementary properties of $\sim_G$. The example showing that $\sim_2$ is not transitive is taken from Sierpiński's monograph [219], which contains a fairly complete discussion of equidecomposability, including the Hausdorff and Banach-Tarski Paradoxes.

Banach's version of the Schröder-Bernstein Theorem (3.5) is proved in [9] (see also [117, p. 190]). Theorem 3.8 on the equidecomposability of polygons is due to Tarski [230], and appears in [11] as well. The application of Theorem 3.5 to similarities (Corollary 3.7) is due to Klee [107, p. 139].

Theorem 3.9 and its use in deriving the Banach-Tarski Paradox (3.10) from the Hausdorff Paradox is due to Sierpiński [215; 219, pp. 42, 92]. The original derivation that appeared twenty years earlier in [11] is slightly different. Theorem 3.11, the strong form of the Banach-Tarski Paradox, appears in the original paper [11].

For the reader interested in pursuing the foundations of set theory, modern proofs of Gödel's famous theorem that the Axiom of Choice is consistent with Zermelo-Fraenkel Set Theory may be found in [113] and [115].

Marczewski's Problem has been discussed often by Mycielski [163, 166, 168, 169], who found many interesting equivalent questions, such as the one mentioned after 3.12. Question 3.13, in the form asking about invariant, finitely additive, Borel measures in compact metric spaces, is Problem 2 of *The Scottish Book* [140] and is due to Banach and Ulam. It is also mentioned in [12] and [238, p. 43]. Question 3.15 is due to de Groot (see [51, p. 25]).

Chapter 4

Locally Commutative Actions: Minimizing the Number of Pieces in a Paradoxical Decomposition

If one analyzed carefully the proof presented in Chapter 3 that a sphere may be duplicated using rotations, one would find that that proof used ten pieces. More precisely, $S^2 = A \cup B$ where $A \cap B = \emptyset$ and $A \sim_4 S^2$ and $B \sim_6 S^2$. (By 3.6, 2.6, and 1.10, $S^2 \setminus D$ splits into A' and B' with $A' \sim_2 S^2 \setminus D \sim_3 B'$; 3.9 shows that $S^2 \sim_2 S^2 \setminus D$, whence A' and B' yield A, $B \subseteq S^2$ with the properties claimed.) It is easy to see that at least four pieces are necessary whenever a set X, acted upon by a group G, is G-paradoxical. For if X contains disjoint A, B with $A \sim_m X \sim_n B$ and $m + n < 4$, then one of m or n equals 1. If, say, $m = 1$, then $X = g(A)$ for some $g \in G$ whence $A = g^{-1}(X) = X$ and $B = \emptyset$, a contradiction. It turns out that an interesting feature of the rotation group's action on the sphere allows the minimal number of pieces to be realized: there are disjoint sets A, $B \subseteq S^2$ such that $A \sim_2 S^2 \sim_2 B$. Moreover, the techniques used to cut the number of pieces to a minimum lead to significant new ideas on how to deal with the fixed points of an action of a free group, adding to our ability to recognize when a group's action is paradoxical. First, we state precisely the way in which the pieces will be counted.

Definition 4.1. *Suppose G acts on X and $E \subseteq X$. Then E is G-paradoxical using r pieces if there are disjoint A, B with $E = A \cup B$, $A \sim_m E \sim_n B$, and $m + n = r$.*

Note that this definition is stronger than merely adding to Definition 1.1 the condition that $m + n = 4$, because this definition requires $E = A \cup B$ rather than just $E \supseteq A, B$. In fact, if r pieces are used in Definition 1.1, then by virtue of Corollary 3.6 and its proof, E is paradoxical using $r + 1$ pieces, in the sense of Definition 4.1. This distinction is strikingly illustrated by the free group, F, of rank two. The simple proof of Theorem 1.2 uses four pieces, but since the identity of F does not appear in any of the pieces, it yields only that F is paradoxical using five pieces. In order to do the same with four pieces, let σ and τ be the free generators of F and let $A_1 = W(\tau)$, $A_2 = W(\tau^{-1})$, $A_3 = W(\sigma) \cup \{1, \sigma^{-1}, \sigma^{-2}, \ldots\}$, and $A_4 = W(\sigma^{-1}) \setminus \{\sigma^{-1}, \sigma^{-2}, \ldots\}$ (see Figure 4.1). Then $A_1 \cup \tau(A_2) = F = A_3 \cup \sigma(A_4)$. The following theorem shows how this construction can be modified to yield an important additional condition that is satisfied by the decomposition.

Theorem 4.2.   *F, the free group generated by σ and τ, can be partitioned into A_1, A_2, A_3, and A_4 such that $\sigma(A_2) = A_2 \cup A_3 \cup A_4$ and $\tau(A_4) = A_1 \cup A_2 \cup A_4$; therefore F is paradoxical using four pieces. Moreover, for any fixed $w \in F$, the partition can be chosen so that w is in the same piece as the identity of F.*

Proof.   The sets of Figure 4.1 prove the first part of the theorem, provided we let A_1 and A_2 be the sets A_3 and A_4 of Figure 4.1, and vice versa.

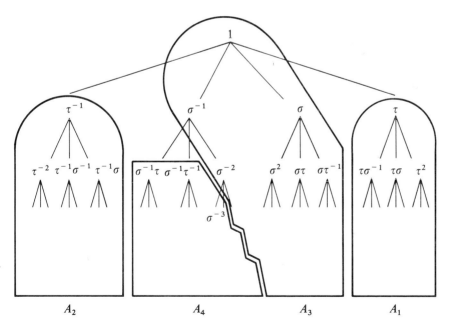

Figure 4.1.

For the second part, note that the two desired equations are equivalent to four containment relations:

$$\sigma(A_2) \subseteq A_2 \cup A_3 \cup A_4$$

$$\sigma^{-1}(A_2 \cup A_3 \cup A_4) \subseteq A_2$$

$$\tau(A_4) \subseteq A_1 \cup A_2 \cup A_4$$

$$\tau^{-1}(A_1 \cup A_2 \cup A_4) \subseteq A_4.$$

We shall use "domain of σ" to denote A_2, "range of σ" to denote $A_2 \cup A_3 \cup A_4$, and so on.

Suppose that $w = \rho_n \cdots \rho_1$ is given, where each ρ_i is one of $\sigma^{\pm 1}, \tau^{\pm 1}$. Before assigning arbitrary words to the four sets, we show how to successfully place $1, \rho_1, \rho_2\rho_1, \ldots, w$. Place 1 and w in A_2, A_1, A_4 or A_3 according as $\rho_1 = \sigma, \sigma^{-1}, \tau$, or τ^{-1}. To see how to place the other end-segments of w assume, for definiteness, that $\rho_1 = \sigma$. Consider first $\rho_{n-1} \cdots \rho_1$, which equals $\rho_n^{-1}w$. If w is in the domain of ρ_n^{-1}, $\rho_{n-1} \cdots \rho_1$ is placed in the range of ρ_n^{-1}; if w is not in the domain of ρ_n^{-1} place $\rho_{n-1} \cdots \rho_1$ in one of the A_i disjoint from ρ_n^{-1}'s range. Then $\rho_{n-2} \cdots \rho_1$ is considered as $\rho_{n-1}^{-1}(\rho_{n-1} \cdots \rho_1)$ and placed appropriately, and so on until only $\rho_1 = \sigma$ remains to be dealt with. Because $\sigma = \sigma \cdot 1$ and $1 \in A_2$, σ must be placed in $A_2 \cup A_3 \cup A_4$, and because $\sigma = \rho_1 = \rho_2^{-1}(\rho_2\rho_1)$, σ must also be placed to satisfy the equation involving ρ_2^{-1} with respect to the location of $\rho_2\rho_1$. But, for each of the three possibilities, $\sigma^{-1}, \tau, \tau^{-1}$, for ρ_2^{-1}, $A_2 \cup A_3 \cup A_4$ intersects both the range of ρ_2^{-1} and its complement, so a successful placement of σ is possible. The other cases, $\rho_1 = \sigma^{-1}$ or $\tau^{\pm 1}$, proceed in an identical manner.

Now, an arbitrary word $u \in F$ may be written uniquely as vt, where $t \in \{1, \rho_1, \rho_2\rho_1, \ldots, w\}$. Such words can now be assigned by induction on the length of v. If $v = 1$, then u is an end-segment of w and has already been assigned. If v does not begin with σ^{-1}, put σvt in $A_2 \cup A_3 \cup A_4$ or A_1 according as vt is, or is not, in A_2; this ensures that vt satisfies the first containment. The other noncancelling extensions of vt are dealt with similarly. $\square$

Corollary 4.3 (AC).   *If F, a free group of rank 2, acts on X without nontrivial fixed points, then X is F-paradoxical using four pieces. Hence any group having a free non-Abelian subgroup is paradoxical using four pieces.*

Proof.   Use Theorem 4.2 to get $F = A \cup B$ witnessing that F is paradoxical using four pieces. Let M be a choice set for the orbits of X with respect to F, and define $A^*, B^* \subseteq X$ by $A^* = \bigcup \{g(M) : g \in A\}$, $B^* = \bigcup \{g(M) : g \in B\}$. Since the lack of fixed points means that for each $y \in X$ there is a unique $g \in F$ such that $y \in g(M)$, $A^* \cup B^* = X$, and $A^* \cap B^* = \varnothing$. Moreover, $A^* \sim_2 X \sim_2 B^*$ follows directly from

$A \sim_2 F \sim_2 B$. A free non-Abelian group has a free subgroup, F, of rank 2, so any group containing a free non-Abelian subgroup is acted upon, via left translation, by F. This action has no nontrivial fixed points, so this situation is a special case of the first part of the corollary. □

Although from a geometrical point of view, it is nicer to have decompositions into as few pieces as possible, such efficient decompositions do not have any new implications for the existence of finitely additive measures. If X is G-paradoxical, then X is G-negligible (Definition 2.4) no matter how many pieces are needed. But Theorem 4.2, in its entirety, can be used to provide a way of dealing with fixed points of action of free groups that is entirely different from the method used in Chapter 3. This, in turn, yields the G-negligibility of X for a wider class of actions than we have yet identified.

If G acts on X and $x \in X$, let $Stab(x)$, the stabilizer of x, denote $\{\sigma \in G : \sigma(x) = x\}$; $Stab(x)$ is a subgroup of G.

Definition 4.4.   *An action of a group G on X is called* locally commutative *if $Stab(x)$ is commutative for every $x \in X$; equivalently, if two elements of G have a common fixed point, then they commute.*

Of course, any action without any nontrivial fixed points is locally commutative. A more interesting example is the action of the rotation group SO_3, on the sphere S^2. If two rotations of the sphere share a fixed point, they must have the same axis, and therefore they commute; in this case each $Stab(x)$ is isomorphic to the Abelian group SO_2.

The technique used earlier (1.10, 4.3) for transferring a decomposition from a group to a set is based on having a unique representation of each $y \in X$ as $g(x)$, where x is some representative of the orbit of y. If the action has fixed points, then such representations are not unique, but in the case of a locally commutative action, there is an intricate way around this difficulty.

Theorem 4.5 (AC).   *If the action of F on X is locally commutative, where F is freely generated by σ and τ, then X is F-paradoxical using four pieces.*

Proof.   We shall use Theorem 4.2 to partition X into A_1^*, A_2^*, A_3^*, and A_4^* satisfying $\sigma(A_2^*) = A_2^* \cup A_3^* \cup A_4^*$ and $\tau(A_4^*) = A_1^* \cup A_2^* \cup A_4^*$, which suffices to prove the theorem. However, we shall not use a single partition of F into A_1, A_2, A_3, A_4, but rather many such partitions. More precisely, divide X up into orbits with respect to F; then the sets A_i^* will be defined orbit by orbit, appealing to possibly different partitions of F in each orbit.

Each orbit consists entirely of nontrivial fixed points, or contains no such points (if $w(x) = x$ and $u \in F$, then $u(x)$ is fixed by uwu^{-1}). For an orbit containing no nontrivial fixed points, the assignment of points to the A_i^* is straightforward, and the same partition of F into A_1, A_2, A_3, and A_4 may be used for all such orbits, namely the one illustrated in Figure 4.1. For each orbit without fixed points, choose any representative point x in the orbit. Then any y in the orbit may be written uniquely as $v(x)$, and y is placed in A_i^* if A_i is the subset of F containing v. Then the equations for the A_i^* (as far as they have yet been defined) follow from the corresponding equations for the A_i.

For the assignment of points in an orbit, $\mathcal{O}$, of fixed points, the choice of x cannot be as arbitrary. Choose a nontrivial word w in F of shortest length fixing an element of $\mathcal{O}$, and let x be one of w's fixed points in $\mathcal{O}$. Let ρ denote whichever of $\sigma^{\pm 1}, \tau^{\pm 1}$ that is the leftmost term of w. Note that w does not end in ρ^{-1}, for if it did, $\rho^{-1} w \rho$, which fixes $\rho^{-1}(x)$, would be shorter than w. Now, use Theorem 4.2 to partition F into A_1, A_2, A_3, A_4, satisfying the two equations and having the additional property that 1 and w lie in the same piece of the partition.

We claim that every point y in $\mathcal{O}$ may be written uniquely as $v(x)$, where v does not end in w, and does not end in ρ^{-1}. The existence of such a representation is proved by considering a word v of minimal length such that $y = v(x)$. Then v does not end in w or w^{-1}, so in case v does end in ρ^{-1}, we may replace v by vw. The uniqueness of this representation depends on the following important aspect of local commutativity. The only elements of F that fix x are the powers of w. To prove this, note that if u fixes x, then local commutativity implies that $uw = wu$, whence by an elementary property of free groups (see [135, p. 42]), $u = t^j$ and $w = t^k$ for some $t \in F$ and $j, k \in \mathbf{Z}$. But then the minimality of w implies that $|j| > |k|$, so j may be written as $\ell k + r$, where $\ell, r \in \mathbf{Z}$, and $0 \leqslant r < |k|$. Then $x = u(x) = t^r(t^k)^\ell(x) = t^r(x)$ and, again using the minimality of w, this means r must be 0. Therefore k divides j, and u is a power of w, as claimed.

Now, if $y = v(x) = u(x)$ are two representations of y in the desired form, then $u^{-1} v(x) = x$, so one of $u^{-1} v$ or $v^{-1} u$ is a positive power of w. Assume it is $u^{-1} v$ (the other case is similar); then either u^{-1} begins with a ρ, contradicting the fact that u does not end in ρ^{-1}, or all of u^{-1} cancels with v, implying that v ends in a w, again a contradiction.

With this representation in hand, we may assign points in $\mathcal{O}$ to the sets A_i^*. Place y in A_i^* if A_i is the subset of F containing v, where $v(x)$ is the unique representation of y defined above. To show that this assignment works, consider first the relation $\sigma(A_2^*) \subseteq A_2^* \cup A_3^* \cup A_4^*$. Suppose $y \in A_2^*$— therefore y has the representation $v(x)$, with $v \in A_2$—and consider $\sigma(y)$. If $\sigma v(x)$ is the correct representation of $\sigma(y)$, then since $v \in A_2$ implies $\sigma v \in A_2 \cup A_3 \cup A_4$, $\sigma(y)$ is properly placed in $A_2^* \cup A_3^* \cup A_4^*$. But it is possible that σv ends in w or ρ^{-1}. In the former case, since v does not end in w, σv must equal w, so $\sigma(y) = w(x) = x$. Since σv lies in $A_2 \cup A_3 \cup A_4$, so does w, and by our choice of a partition of F, so does 1. And since $1(x)$ is the unique representation of x, this implies that x, and hence $\sigma(y)$, is in $A_2^* \cup A_3^* \cup A_4^*$. If

σv ends in ρ^{-1}, then since v does not, v must equal 1, and so $y = x$, $\sigma v = \sigma$, $\rho^{-1} = \sigma$, and w begins with σ^{-1}. Now, $1 = v \in A_2$, so w is in A_2 and hence $\sigma w \in A_2 \cup A_3 \cup A_4$. But $\sigma w(x)$ is the unique representation of $\sigma(x)$, so $\sigma(y) = \sigma(x) \in A_2^* \cup A_3^* \cup A_4^*$. An identical treatment works for the other three containments involving $\sigma^{-1}(A_2^* \cup A_3^* \cup A_4^*)$, $\tau(A_4^*)$, and $\tau^{-1}(A_1^* \cup A_2^* \cup A_4^*)$, completing the proof. □

Regardless of the number of pieces used, the fact that any locally commutative action of a non-Abelian free group is paradoxical is interesting by itself, because it implies that no invariant, finitely additive measure of total measure one exists; that is, X is F-negligible. There is in fact a simpler proof of this result that, though not efficient in its use of pieces, avoids the rather involved reasonings of 4.2 and 4.5 (see Corollary 8.6).

Corollary 4.6 (AC).   *S^2 is SO_3-paradoxical using four pieces, and the four cannot be improved.*

Proof.   Since the action of the rotations on the sphere is locally commutative, and since SO_3 has a free subgroup of rank 2 (Theorem 2.1), this follows from Theorem 4.5. It was explained at the beginning of this chapter why four is the best possible number of pieces. □

There are other naturally occurring, locally commutative actions to which Theorem 4.5 applies. For instance, SL_2, the group of area-preserving and orientation-preserving linear transformations of $\mathbf{R}^2$, is locally commutative in its action on $\mathbf{R}^2 \backslash \{0\}$. If σ_1 and σ_2 fix P, then, choosing a basis that contains P, σ_i may be represented by $\left[\begin{smallmatrix} 1 & a_i \\ 0 & b_i \end{smallmatrix}\right]$. Since $\det(\sigma_i) = 1$, b_i must equal 1, so $\sigma_1 \sigma_2 = \sigma_2 \sigma_1 = \left[\begin{smallmatrix} 1 & a_1 + a_2 \\ 0 & 1 \end{smallmatrix}\right]$. Moreover, SL_2 (even $SL_2(\mathbf{Z})$) has two independent elements. We shall prove (Proposition 7.1) that $\left[\begin{smallmatrix} 1 & r \\ 0 & 1 \end{smallmatrix}\right]$ and its transpose are independent whenever $r \geqslant 2$. By 4.5 then, $\mathbf{R}^2 \backslash \{0\}$ is $SL_2(\mathbf{Z})$-paradoxical using four pieces. Actually, with more work one can produce two independent elements of $SL_2(\mathbf{Z})$ such that the group they generate has no nontrivial fixed points in $\mathbf{R}^2 \backslash \{0\}$ (see remarks following Proposition 7.1).

A similar situation arises with the isometries of the hyperbolic plane, H^2. If we take the upper half-plane of the complex numbers as our model of the hyperbolic plane, then the orientation-preserving isometries correspond to linear fractional transformations $z \mapsto (az + b)/(cz + d)$, where $ad - bc \neq 0$ and $a, b, c, d \in \mathbf{R}$ (see [79], [122], or [145]). Since a transformation is unchanged if each entry is divided by the same constant, we may assume $ad - bc = 1$; therefore the group is isomorphic to $PSL_2 = SL_2/\{\pm I\}$. (The orientation-reversing motions are given by $z \mapsto (a\bar{z} + b)/(c\bar{z} + d)$, where $ad - bc = -1$.) The orientation-preserving transformations are locally commutative on H^2 (see the discussion of H^2 in Chapter 5). Although this fact is useful (see Corollary 6.6(b)), it is less important here because of the fact (proved in Chapter 5) that the independent pairs of isometries corresponding to the independent pairs of matrices, $\begin{bmatrix} 1 & r \\ 0 & 1 \end{bmatrix}$ and $\begin{bmatrix} 1 & 0 \\ r & 1 \end{bmatrix}$ ($r \geqslant 2$), generate a free group that acts on H^2 without nontrivial fixed points. Hence H^2 is paradoxical using four pieces. But much more is true. Because this free group is discrete, there is a polygonal fundamental region for its action and this yields (see Fig. 5.4) a paradoxical decomposition of H^2 using polygonal (and hence measurable) pieces! There is no contradiction in this because the whole hyperbolic plane has infinite measure, and a paradoxical decomposition yields only the harmless equation: $2 \cdot \infty = \infty$. Nevertheless, this paradox is surprising, especially as similar paradoxical decompositions of $\mathbf{R}^n$ using measurable sets do not exist (see 11.16).

A MINIMAL DECOMPOSITION OF A BALL

The preceding work on the sphere may also be applied to decompositions of a solid ball with respect to G_3, the isometry group of $\mathbf{R}^3$. This situation differs from the sphere in two ways: (1) G_3's action is not locally commutative (consider two noncommuting rotations about axes containing $\mathbf{0}$), and (2) G_3 does not act on B, the unit ball in $\mathbf{R}^3$. We have already seen (3.9) that despite the lack of local commutativity, B is G_3-paradoxical. But the second point means that it is not clear what the minimal number of pieces is, as the necessity of four pieces depends on having a group of transformations from a set to itself. For instance, in the Sierpiński-Mazurkiewicz Paradox (1.7) a set is constructed that is paradoxical using two pieces. Nevertheless, a geometric argument can be used to show that five pieces are necessary for the ball, while the finer analysis available for the sphere allows a five-piece paradoxical decomposition of a ball to be constructed.

> **Theorem 4.7 (AC).** *A solid ball in $\mathbf{R}^3$ does not admit a paradoxical decomposition using fewer than five pieces, and five-piece decompositions of any ball exist.*

> *Proof.* The theorem will be proved for B, the unit ball of $\mathbf{R}^3$ centered at $\mathbf{0}$, but the proof applies to any ball. Let S denote the unit sphere that is the surface of B. To see that at least five pieces are necessary, suppose $B = B_1 \cup B_2 \cup B_3 \cup B_4$ where the B_i are pairwise disjoint and $\sigma_1 B_1 \cup \sigma_2 B_2 = B = \sigma_3 B_3 \cup \sigma_4 B_4$ for isometries σ_i; this is the only possibility for a decomposition using fewer than five pieces. Not all of the σ_i can fix $\mathbf{0}$, otherwise one copy of B would be missing the origin, so suppose $\sigma_4(\mathbf{0}) \neq \mathbf{0}$. Then $\sigma_4(B)$ is a unit ball different from B, and it follows that there is a closed hemispherical surface $H \subseteq S$ that is disjoint from $\sigma_4(B)$. (Let H be the hemisphere of S symmetric about the point of intersection of S with the

extension of the directed line segment from $\sigma_4(\mathbf{0})$ to $\mathbf{0}$.) Since $\sigma_3(B_3)$ must contain H, B_3 must contain $\sigma_3^{-1}(H)$, a closed hemisphere of S. This means that $(B_1 \cup B_2) \cap S$ is contained in an open hemisphere of S, namely the complement of $\sigma_3^{-1}(H)$. Since, therefore, neither B_1 nor B_2 contains a closed hemisphere, the argument used on $\sigma_4(B)$ yields that each of σ_1, σ_2 fix $\mathbf{0}$ and hence map S to S. But then $\bigl(\sigma_1(B_1) \cup \sigma_2(B_2)\bigr) \cap S = \sigma_1(B_1 \cap S) \cup \sigma_2(B_2 \cap S)$, which is contained in the union of two open hemispheres and hence is a proper subset of S. This contradicts the fact that $\sigma_1(B_1) \cup \sigma_2(B_2) = B$.

To construct a five-piece decomposition, we shall work separately on each S_r, the sphere of radius r, where $0 < r \leqslant 1$. Use Theorem 2.1 to select σ and τ, two independent rotations of B, and let F be the free group they generate. By Theorem 4.5, we may partition each S_r, for $0 < r < 1$, into A_1^r, A_2^r, A_3^r, and A_4^r satisfying $\sigma(A_2^r) = A_2^r \cup A_3^r \cup A_4^r$ and $\tau(A_4^r) = A_1^r \cup A_2^r \cup A_4^r$. For S, however, we need a partition into five sets, $A_1^1, A_2^1, A_3^1, A_4^1$, and a single point $\{P\}$, satisfying $\sigma(A_2^1) = A_2^1 \cup A_3^1 \cup A_4^1 \cup \{P\}$, $\tau(A_4^1) = A_1^1 \cup A_2^1 \cup A_4^1 \cup \{P\}$. Such a partition can be constructed by examining the proof of Theorem 4.5 and selecting a single orbit, $\mathcal{O}$, of nonfixed points. There must be such an orbit because F has only countably many fixed points on S, so only countably many points lie in orbits of fixed points. We may now select P to be any point in $\mathcal{O}$, and assign other points Q in $\mathcal{O}$ to A_1^1, A_2^1, A_3^1, A_4^1 according as w begins with a $\sigma, \sigma^{-1}, \tau$, or τ^{-1}, where w is the unique word in $F \backslash \{1\}$ such that $Q = w(P)$. Since $\sigma W(\sigma^{-1}) = F \backslash W(\sigma)$ and $\tau W(\tau^{-1}) = F \backslash W(\tau)$ (see proof of Theorem 1.2), this partition is as desired. Now, B may be partitioned into B_1, B_2, B_3, B_4, and $\{P\}$, where $B_1 = \{\mathbf{0}\} \cup \bigcup \{A_1^r : 0 < r \leqslant 1\}$ and for $i = 2, 3, 4$, $B_i = \bigcup \{A_i^r : 0 < r \leqslant 1\}$. Letting ρ be the translation taking P to $\mathbf{0}$, we see that this partition works since $B_1 \cup \sigma(B_2) = B$ and $B_3 \cup \tau(B_4) \cup \rho(\{P\}) = B$. $\qquad\square$

By partitioning spheres of arbitrary positive radius as in the previous proof, and stealing a point from S in the same way, we can get a paradoxical decomposition of all of $\mathbf{R}^3$ using five pieces. But this decomposition is not the simplest possible. Later we shall construct two independent isometries of $\mathbf{R}^3$ such that the free group that they generate acts on $\mathbf{R}^3$ without nontrivial fixed points (Theorem 5.7). Corollary 4.3 then yields a (best possible) four-piece paradoxical decomposition of $\mathbf{R}^3$.

Although it took over 20 years (from 1924 to 1947), it is remarkable that the original Banach-Tarski duplication of the ball, which was somewhat more complicated than the approach presented in Chapter 3, could be modified and refined to yield a duplication requiring only five pieces, the minimal number possible. In general, optimization along these lines is quite difficult. For instance, if we apply, with minor modification, Theorem 3.8 to the simple two-piece geometric rearrangement of a square into an isosceles right triangle (cut along a diagonal), then five pieces are used to prove these polygons equidecomposable. No lower bound (except the trivial one, 2) for this problem is known.

There is as yet no criterion that can be used to tell, in general, whether or not an action is paradoxical. But Theorem 4.5 allows such a criterion to be given for the special case of actions of a group of rotations on the sphere. Because the entire action of SO_3 on S^2 is locally commutative, it follows from 4.5 that S^2 is G-paradoxical whenever G is a subgroup of SO_3 containing two independent rotations. We shall see in Part II (Theorem 11.15) that these are the only subgroups for which the sphere is paradoxical: if G is a subgroup of SO_3 not containing a free subgroup of rank 2, then a finitely additive, G-invariant measure on $\mathscr{P}(S^2)$ of total measure 1 exists, whence S^2 is not G-paradoxical.

Another situation in which paradoxical actions may be characterized arises if we restrict consideration to decompositions using four pieces. This is because, as we now prove, the converse of Theorem 4.5 is valid, so if G acts on X, then X is G-paradoxical using four pieces if and only if G has a free subgroup of rank 2 whose action on X is locally commutative.

Theorem 4.8. *If G acts on X and X is G-paradoxical using four pieces, then G has two independent elements σ, τ such that the action of F, the group they generate, on X is locally commutative.*

Proof. A four-piece paradoxical decomposition can only arise from a partition of X into A_1, A_2, A_3, A_4 such that for suitable $g_i \in G$, $g_1(A_1)$ and $g_2(A_2)$ partition X, as do $g_3(A_3)$ and $g_4(A_4)$. Let $\sigma = g_1^{-1}g_2$, $\tau = g_3^{-1}g_4$; then the pairs $A_1, \sigma(A_2)$ and $A_3, \tau(A_4)$ each partition X, yielding the four familiar equations:

$$\sigma(A_2) = X\setminus A_1, \quad \sigma^{-1}(A_1) = X\setminus A_2, \quad \tau(A_4) = X\setminus A_3, \quad \tau^{-1}(A_3) = X\setminus A_4.$$

We shall use the terms "domain of σ," "range of τ^{-1}," and so on, with respect to these equations as in the proof of 4.2.

To prove that σ and τ are independent, suppose w is a nontrivial word in $\sigma, \sigma^{-1}, \tau, \tau^{-1}$; say $w = \rho_n \cdots \rho_1$ where each ρ_i is one of $\sigma^{\pm 1}, \tau^{\pm 1}$. Choose $x \in X$ so that x is not in the domain of ρ_1, but is in the range of ρ_n. These two conditions eliminate at most two of the A_i, so a suitable x exists. But if x is not in the domain of ρ_1, then $\rho_1(x)$ is not in the range of ρ_1, and hence because $\rho_2 \neq \rho_1^{-1}$, $\rho_1(x)$ is not in the domain of ρ_2. Continuing through w in this way yields that $w(x)$ is not in the range of ρ_n. By the choice of x then, $w(x) \neq x$ so w is not equal to the identity and the independence of σ, τ is proved.

Suppose the action of F, the free subgroup of G generated by σ and τ, on X is not locally commutative. First we prove the claim that whenever $u, v \in F\setminus\{1\}$ have a common fixed point x in X, then one of the rightmost terms of u or u^{-1} equals one of the same for v or v^{-1}. For suppose $u = \rho_n \cdots \rho_1$ and $v = \phi_m \cdots \phi_1$ where each ρ_i, ϕ_j is one of $\sigma^{\pm 1}, \tau^{\pm 1}$. If $\rho_1 \neq \phi_1$, then x

cannot be in the domains of both ρ_1 and ϕ_1. Suppose x is not in the domain of ρ_1. Then, as proved in the first part of this theorem, $u(x) = x$ is not in the range of ρ_n. If $\rho_n \neq \phi_m$, then since $v(x) = x = u(x)$, $v(x)$ cannot be in the complement of ϕ_m's range too, and so as before, x must be in the domain of ϕ_1. But the domain of ϕ_1 is contained in the range of ρ_n, unless $\rho_n = \phi_1^{-1}$, as sought. Finally, if x is not in the domain of ϕ_1, then the same proof yields $\phi_m = \rho_1^{-1}$.

Now, let u, v be a noncommuting pair in F that shares a fixed point, such that the sum of the lengths of u and v is as small as possible with respect to all other noncommuting pairs that share a fixed point. Suppose $u = \rho_n \cdots \rho_1$ and $v = \phi_m \cdots \phi_1$ ($\rho_i, \phi_j \in \{\sigma^{\pm 1}, \tau^{\pm 1}\}$), and let $x \in X$ be one of the fixed points of both u and v. Then $\rho_n \neq \rho_1^{-1}$; otherwise, by the claim of the previous paragraph, the pair $\rho_1 u \rho_1^{-1}$, $\rho_1 v \rho_1^{-1}$, which fixes $\rho_1(x)$ and does not commute, would have total length smaller, by 2, than that of u, v. Similarly, $\phi_m \neq \phi_1^{-1}$. Now, by the same claim, we may assume $\phi_1 = \rho_1$; otherwise replace one, or both, of u, v by its inverse. Consider uv^{-1} and $v^{-1}u$, both of which fix x. The minimality of u, v implies that there cannot be so much cancellation in uv^{-1} or $v^{-1}u$ as to affect any of the endterms. For if, say, the u of uv^{-1} was completely absorbed, then the pair uv^{-1}, u^{-1}, which fixes x and does not commute, would have a smaller total length than u, v. But because $\rho_1^{-1} \neq \rho_n$, $\phi_1^{-1} \neq \phi_m$ and $\rho_1 = \phi_1$, the endterms of uv^{-1} and $v^{-1}u$ contradict the claim of the previous paragraph. $\qquad\square$

The characterization provided by the theorem just proved and its converse, Theorem 4.5, is especially succinct for actions without nontrivial fixed points: if G acts on X in such a way, then X is G-paradoxical using four pieces if and only if G has a free non-Abelian subgroup. In particular, this applies to the action of a group on itself by left translation, yielding the following corollary.

Corollary 4.9 (AC).   *A group G is paradoxical using four pieces if and only if G has a free subgroup of rank 2.*

Proof.   By Corollary 4.3 and Theorem 4.8. $\qquad\square$

This result shows that the equality $AG = NF$—which, by Theorem 1.12, is false—is equivalent to the assertion that all paradoxical groups are paradoxical using four pieces. Phrased this way, the conjecture that $AG = NF$ is less plausible, and it is not surprising that it has turned out to be false.

GENERAL SYSTEMS OF CONGRUENCES

Theorem 4.5, and the group theory result upon which its proof is based, Theorem 4.2, are really special cases of a much more general result on partitioning a set or group into subsets satisfying a given system of congruences. Since the derivation of 4.5 from 4.2 is independent of the system of congruences, we seek to characterize the sort of systems for which Theorem 4.2, including the assertion about a fixed word w, remains valid. An example for which 4.2 is false is the single congruence $\sigma(A) = B$, with respect to a partition of F into A and B. For if w is taken to be σ, then if 1 is in A, w is in B, and if 1 is not in A, w is not in B, whence 1 and w cannot be placed together. In fact, this example is essentially the only one for which the sought-after generalization fails, for as we now show, Theorems 4.2 and 4.5 are valid for any system of congruences except those that, explicitly or implicitly, yield that a set is congruent to its complement. If $D \subseteq \{1,\ldots,r\}$ then D^c denotes $\{1,\ldots,r\}\setminus D$.

Definition 4.10. *Consider an abstract system of m relations involving set-variables $A_1,\ldots,A_r$, each relation having the form:*

$$\bigcup\{A_j : j \in L_i\} \cong \bigcup\{A_j : j \in R_i\},$$

where L_i and R_i are subsets of $\{1,\ldots,r\}$. Because it is intended that the relations be witnessed by isometries, each relation in the system is called a congruence. *Only congruences where L_i and R_i are nonempty proper subsets of $\{1,\ldots,r\}$ are of interest, and a system where each congruence has that form will be called a* proper system. *Suppose that when we add to a proper system the m complementary congruences, that is, the one referring to L_i^c and R_i^c instead of L_i and R_i, and then consider all congruences obtainable from these $2m$ ones by transitivity, we do not obtain a congruence of the form $\bigcup\{A_j : j \in D\} \cong \bigcup\{A_j : j \in D^c\}$. Then the original system of congruences is called a* weak system.

The system that we considered in 4.2 and 4.5 is certainly weak, for if we close $\{A_2 \cong A_2 \cup A_3 \cup A_4, A_4 \cong A_1 \cup A_2 \cup A_4\}$ under complementation and transitivity (and ignore the ones of the form $A_i \cong A_i$), only the congruences $A_1 \cong A_1 \cup A_3 \cup A_4$ and $A_3 \cong A_1 \cup A_2 \cup A_3$ are added. Another interesting example is the set of $r - 1$ congruences involving $A_1,\ldots,A_r$ given by $A_1 \cong A_2, A_1 \cong A_3,\ldots,A_1 \cong A_r$. If $r = 2$, then $A_1 \cong A_2$ already shows that the system is not weak. But if $r > 2$, then the closure of this system consists of all congruences of the form $A_i \cong A_j$ and its complement,

$$A_1 \cup \cdots \cup A_{i-1} \cup A_{i+1} \cup \cdots \cup A_r \cong A_1 \cup \cdots \cup A_{j-1} \cup A_{j+1} \cup \cdots \cup A_r.$$

Thus, provided $r > 2$, this system, which is satisfied by a partition of a set into r congruent pieces, is a weak system. As a final example, consider a partition into three sets satisfying $A_1 \cong A_2, A_1 \cong A_3, A_1 \cong A_1 \cup A_2$. Such a partition is called a *Hausdorff decomposition*, since it was the sort considered by Hausdorff in his seminal investigations into paradoxical decompositions and finitely additive measures (see Notes for Chapter 2). Clearly this system is not weak, since it implies $A_3 \cong A_1 \cup A_2$. We now generalize Theorem 4.2 by proving a purely group theoretic result about the satisfiability of proper and weak systems of congruences in free groups.

Theorem 4.11.   Suppose $\bigcup\{A_j : j \in L_i\} \cong \bigcup\{A_j : j \in R_i\}$, $i = 1,\ldots,m$, is a proper system of congruences involving $A_1,\ldots,A_r$, and F is the free group generated by $\sigma_1,\ldots,\sigma_m$. Then F may be partitioned into $A_1,\ldots,A_r$ satisfying the system, with σ_i witnessing the ith congruence. If, in addition, the congruences form a weak system, then it can be guaranteed that any fixed word $w \in F$ is placed in the same piece of the partition as 1, the identity of F.

Proof.   We shall show how the proof of Theorem 4.2 must be modified to yield this generalization. As in 4.2, we replace the m congruences by the equivalent set of $2m$ containments:

$$\sigma_i(\bigcup\{A_j : j \in L_i\}) \subseteq \bigcup\{A_j : j \in R_i\},$$

$$\sigma_i^{-1}(\bigcup\{A_j : j \in R_i\}) \subseteq \bigcup\{A_j : j \in L_i\},$$

$i = 1,\ldots,m$. Because each L_i, R_i is nonempty and proper, a straightforward induction exactly as in the last part of 4.2's proof (taking $w = 1$ there) yields the partition of F for an arbitrary system. The stronger result for weak systems is, as in 4.2, a bit more complicated.

Suppose $w = \rho_n \cdots \rho_1$, with each $\rho_k \in \{\sigma_i^{\pm 1} : i = 1,\ldots,m\}$. As in 4.2, it is sufficient to assign the end-segments $1, \rho_1, \rho_2\rho_1, \ldots, w$ successfully to sets A_j of the partition, for then the rest of F may be placed by induction. Note that any of the $2m$ containments is equivalent to the one obtained by replacing L_i, R_i by L_i^c, R_i^c, respectively. We use the terms "domain" or "range" of $\sigma_i^{\pm 1}$ as in the proof of 4.2.

Case 1.   For some $k = 1,\ldots,n$, the range of ρ_k (or its complement) intersects both the range of ρ_{k+1}^{-1} and its complement (arithmetic on the index k is assumed to be modulo n; thus $n + 1$ represents 1, etc.).

This case is handled in essentially the same way as the proof of 4.2. By replacing the containment for ρ_k by its complementary one if necessary, we may assume that the hypothesis of this case applies to the range of ρ_k rather than its complement. Now, place $\rho_{k-1} \cdots \rho_1$ in the domain of ρ_k, and then place the segments $\rho_{k-2} \cdots \rho_1, \ldots, \rho_1, 1, w, \rho_{n-1} \cdots \rho_1, \ldots, \rho_{k+1} \cdots \rho_1$ in order to satisfy the appropriate containments, and with w in the same piece as 1. This leaves the last end-segment, $\rho_k \cdots \rho_1$, which must be placed in the range of ρ_k and in either the range of ρ_{k+1}^{-1} or its complement. By the hypothesis on k, a successful placement of $\rho_k \cdots \rho_1$ is possible.

Case 2.   The hypothesis of Case 1 fails.

The failure of Case 1's hypothesis implies that for each $k = 1,\ldots,n$, the range of ρ_k equals the range of ρ_{k+1}^{-1} or its complement. But the range of ρ^{-1} equals the domain of ρ, so we have that the range of ρ_k equals the domain of ρ_{k+1} or its complement. Now, place 1 into the domain of ρ_1, place ρ_1 into the range of ρ_1, place $\rho_2\rho_1$ into the range of ρ_2 or its complement according as the range of ρ_1 equals the domain of ρ_2 or its complement, and continue in this way until w must be placed. Then w is forced to be in either the range of ρ_n or its complement, and hence either in the domain of ρ_1 or its complement. But the latter cannot occur, for otherwise the condition

$$w(\text{domain of } \rho_1) = \text{complement of the domain of } \rho_1$$

is a consequence of the system of congruences, contradicting the weakness of the system. Hence w can be placed in the domain of ρ_1 and in the same set containing the identity, completing the proof of this case, and of the theorem.   □

Corollary 4.12 (AC).   *Suppose G acts on X and a proper system of m congruences involving $A_1, \ldots, A_r$ is given. If G has m independent elements such that the group F that they generate acts on X without nontrivial fixed points, then X may be partitioned into sets A_i satisfying the system. If the action of F is merely locally commutative, then the partition is possible provided the congruences form a weak system.*

Proof.   This follows from the previous theorem in exactly the same way that Corollary 4.3 and Theorem 4.5 follow from Theorem 4.2.   □

Now, a free group of rank 2 contains m independent elements for any $m < \aleph_0$; indeed, if σ and τ are independent, then $\{\tau, \sigma\tau\sigma^{-1}, \sigma^2\tau\sigma^{-2}, \ldots\}$ is a set of $\aleph_0$ independent elements, as is $\{\sigma\tau, \sigma^2\tau^2, \sigma^3\tau^3, \ldots\}$ (see [135, p. 43]). Hence, by Theorem 2.1, SO_3 contains a free locally commutative subgroup of any finite rank, and therefore by Corollary 4.12, any weak system of congruences is solvable, using rotations, by a partition of S^2. Infinite, even uncountable systems of congruences will be discussed in the next chapter.

The general theory of systems of congruences has some interesting, nonparadoxical, geometric consequences concerning partitions of a set into congruent pieces.

Definition 4.13.   *A set X is m-*divisible *with respect to a group G acting on X if X splits into m pairwise disjoint, pairwise G-congruent subsets. Each of the m subsets is then called an mth* part *of X. If no group is mentioned, it is understood that the isometry group of X is being used.*

As examples, note that for $m < \infty$ a half-open interval on the real line is m-divisible with respect to translations. Similarly, the circle S^1 is m-divisible. Also, each $\mathbf{R}^n$ is m-divisible: an mth part of $\mathbf{R}^1$ is given by $\bigcup \{[mk, mk + 1) : k \in \mathbf{Z}\}$, and this extends easily to $\mathbf{R}^n$. On the other hand, S^2 is not 2-divisible using rotations, since any rotation of a sphere has a fixed point on the sphere. Of course, the antipodal map $(x \mapsto -x)$ yields that S^2

is 2-divisible if arbitrary isometries are allowed: one half of S^2 is given by the open northern hemisphere together with one half of the equator.

The m-divisibility of X is equivalent to the existence of a partition of X satisfying the system $\{A_1 \cong A_j : 2 \leqslant j \leqslant m\}$. This system is weak if (and only if) $m \geqslant 3$, so Corollary 4.12 yields divisibility results when locally commutative free groups of rank 2 (and hence rank $\aleph_0$) are present. Applying this to the locally commutative action of SO_3 on S^2 yields the following result.

Corollary 4.14 (AC).   *For any m with $3 \leqslant m < \infty$, S^2 is m-divisible.*

Unlike the decompositions of the Banach-Tarski Paradox, it is not clear whether the pieces in a three-piece splitting of the sphere as in Corollary 4.14 can be Lebesgue measurable. This leads to Question 4.15. Since it is consistent with Zermelo-Fraenkel set theory that all sets are Lebesgue measurable (see 13.1), a negative answer to part (a) implies an affirmative answer to part (b).

Question 4.15.   (a) Can S^2 be split into three Lebesgue measurable pieces, any two of which are congruent by a rotation? (b) Is the Axiom of Choice necessary in Corollary 4.14?

If in part (a) of this question, we ask instead for pieces having the Property of Baire, we obtain a question similar to Marczewski's Problem (3.11). In both cases, 3-divisibility and paradoxical decompositions, a system of congruences is solvable using arbitrary sets, but the restriction to sets with the Property of Baire yields an open question. The 3-divisibility case is different in that it is not even known that a splitting into Borel sets is impossible.

In fact, a much stronger result than Corollary 4.14—one that necessitates nonmeasurable pieces—is possible. By considering a more complicated system of congruences we shall prove (see remarks following Corollary 6.9) that there is a subset of S^2 that is *simultaneously* a third, a quarter, a fifth, and so on, part of S^2; such a set must be nonmeasurable, otherwise its measure would be equal to both $\frac{1}{3}$ and $\frac{1}{4}$. Chapter 6 will also contain a discussion of simultaneous m-divisibility for $\mathbf{R}^n$ and for hyperbolic spaces. (See also the discussion of Hausdorff decompositions at the end of this chapter.)

There are other scattered results on m-divisibility where only one m at a time is considered, but by no means a general theory. Corollary 4.14 can be extended to show that S^2 (and higher-dimensional spheres) is m-divisible using rotations for any m with $3 \leqslant m \leqslant 2^{\aleph_0}$ (see Corollary 6.9); indeed, simultaneous m-divisibility is possible here too (see p. 88). As pointed out, S^1 is m-divisible for any finite m using half-open arcs; moreover, Vitali's classical example of a non-Lebesgue measurable subset of the circle yields the $\aleph_0$-divisibility of S^1

(see the proof of Theorem 1.5), and this can be extended to all larger cardinals $m \leqslant 2^{\aleph_0}$ (see Corollary 6.9). In short, if arbitrary isometries are allowed rather than just rotations, then all spheres are m-divisible for all possibilities for $m: 2 \leqslant m \leqslant 2^{\aleph_0}$.

The $\aleph_0$-divisibility of S^1 yields in a straightforward manner the existence of $\aleph_0$ pairwise disjoint subsets of a half-open interval, any two of which are equidecomposable (via translations) using two pieces. The extra piece is needed to account for the problem of addition mod 1, if $[0, 1)$ is the interval. Steinhaus (see Mazurkiewicz [142, p. 8]) raised the question whether an interval is, in fact, $\aleph_0$-divisible, and this was settled by von Neumann [245] who proved that all intervals—half-open, open, or closed—are $\aleph_0$-divisible via translations. This result too has been generalized to all m with $\aleph_0 \leqslant m \leqslant 2^{\aleph_0}$. Since the finite case for a half-open interval is obvious, such intervals are m-divisible via translations for all possible m. On the other hand, it is known that open or closed intervals are not m-divisible using isometries for any finite $m \geqslant 2$. For higher dimensional balls, all that is known is that the n-dimensional ball is not m-divisible if $2 \leqslant m \leqslant n$. Thus the simplest open case is whether the disc in the plane is 3-divisible. There are examples of Banach spaces in which the unit ball is 2-divisible*—for example, the space of all real sequences converging to zero, with the sup norm—but for a wide class of Banach spaces (reflexive and strictly convex) the unit ball is not 2-divisible (see [64]).

It should be noted that the idea of splitting a set into congruent pieces is, like equidecomposability, a generalization of an older geometric problem: given a figure (usually in the plane), divide it using finitely many cuts (straight or along curves) into m congruent pieces, where the boundaries of the pieces are ignored. Unlike geometric equidecomposability, for which the Bolyai-Gerwien Theorem (3.2) controls what can be done with polygons, this problem is usually attacked on an ad hoc basis. Here is one attractive unsolved problem in this area: if p is an odd prime, is there any way of cutting a square into p congruent pieces other than the obvious way into horizontal (or vertical) rectangles? Note that a polygon is a countable part (in the geometric sense) of $\mathbf{R}^2$ if and only if $\mathbf{R}^2$ can be tiled using $\aleph_0$ copies of the polygon. For a survey of results and questions about such tilings see [83].

Theorem 4.8 and Corollary 4.12 yield necessary and sufficient conditions for the solvability of all weak systems of congruences, namely, the group G acting on X must have a free, locally commutative subgroup of rank 2. The fixed-point condition of the first part of 4.12, however, is sufficient but not necessary for the solvability of all proper systems. For consider the sphere, S^2, acted upon by its full group, O_3, of isometries, including orientation-

---

* Here the definition is widened to allow distance-preserving functions from one subset of the Banach space to another, rather than only bijections from the Banach space to itself that preserve distance.

reversing ones. Any of the latter, when squared, yields an orientation-preserving isometry that can only be a rotation; therefore any subgroup of O_3 has nonidentity elements with fixed points on S^2. Nevertheless, a judicious use of the antipodal map shows that all proper systems of congruences are solvable with respect to O_3.

Theorem 4.16 (AC). *A proper system $\{\bigcup\{A_j : j \in L_i\} \cong \bigcup\{A_j : j \in R_i\}$: $i = 1, \ldots, m\}$ of congruences involving $A_1, \ldots, A_r$ is solvable via a partition of S^2 provided arbitrary isometries of the sphere may be used to witness the congruences.*

Proof. Let $\zeta : S^2 \to S^2$ denote the antipodal map, $\zeta(x) = -x$, and choose $\sigma_1, \ldots, \sigma_m$ to be independent rotations of S^2. Choose any two-coloring of the subsets of $\{1, \ldots, r\}$ such that the color of any set differs from the color of its complement. Then define τ_i to be either σ_i or $\sigma_i \zeta$, according as L_i and R_i do, or do not, have the same color.

Note that because each rotation of S^2 is also a linear transformation of $\mathbf{R}^3$, ζ commutes with each σ_i. It follows from this and the fact that $\zeta^2 = 1$ that the τ_i are independent, for if a word involving the τ_i equals the identity, then the square of the corresponding word in the σ_i does too, contradicting the independence of the σ_i. Note also that for any rotation θ, $\zeta\theta$ has a fixed point on S^2 if and only if θ is a rotation of order 2. Since there are no such rotations in the free group generated by the σ_i, the local commutativity of that group implies the same for F, the group generated by the τ_i.

We may now proceed as we did for weak systems in Corollary 4.12, that is, use the method of proof of Theorem 4.5, except that a problem will arise when it comes to partitioning an orbit of fixed points. If w is the chosen short word fixing a point in the orbit, then because the given system of congruences is not necessarily weak, Theorem 4.11 as stated does not yield the required partition of F with 1 and w in the same piece. Nevertheless, the only spot where 4.11's proof appeals to the weakness of the system is in Case 2, where the range of each ρ_k equals the domain of ρ_{k+1} or its complement. But consider the sequence of colors assigned to the sequence of sets:

$$\text{domain}(\rho_1) \quad \text{range}(\rho_1) \quad \text{domain}(\rho_2) \quad \cdots \quad \text{range}(\rho_n).$$

Because w has a fixed point, w cannot equal $\theta\zeta$ for a rotation θ, and so w must contain an even number of ζ's, that is, an even number of τ_i's that equal $\sigma_i\zeta$. This means that there is an even number of color switches in the steps from domain to range in the sequence of sets just displayed. Now, if there is an odd number of switches in the range to domain steps, then the placement of 1 in domain (ρ_1) forces w to be in the complement of range (ρ_n), and since the total number of color switches is odd, the complement of ρ_n's range must be the domain of ρ_1. Therefore w may be placed in the same piece as 1. If there is an even number of switches in the range to domain steps, then w is forced to be in the range of ρ_n, which has the same color as, and therefore equals, the domain of ρ_1. Again, this allows a successful placement of w. Thus there is a partition of F that can be used to assign the points of the orbit to the appropriate sets. This can be done for each orbit of fixed points, yielding the desired partition of S^2.   □

    The solution of all proper systems involving $A_1, \ldots, A_r$ is equivalent to the solution of the single, all-encompassing system of $\binom{2^r-2}{2}$ congruences: $\bigcup \{A_j : j \in L\} \cong \bigcup \{A_j : j \in R\}$, where L, R vary over all pairs of nonempty, proper subsets of $\{1, \ldots, r\}$. A partition of S^2 satisfying this latter system, which exists by the preceding theorem, is surely a striking object. For S^2 is divided into r pieces such that if someone gathers a nonempty, proper collection of these pieces into one pile, and then gathers another such collection into another pile (a piece may appear in both piles), then the union of the pieces in one pile is congruent to the union of the other pile. In particular, each of the r sets is, simultaneously, a half, a third, $\ldots$, an rth part of S^2.

    This theorem has one application that is interesting from a historical point of view. Hausdorff's original decomposition of the sphere (see Chapter 2 Notes) showed that except for a countable set, the sphere could be partitioned into A_1, A_2, and A_3, satisfying $A_1 \cong A_2 \cong A_3 \cong A_1 \cup A_2$, where the congruences are realized by rotations. These congruences do not form a weak system, and because rotations have fixed points, it is impossible to eliminate the countable subset of the sphere excluded at the outset. The Banach-Tarski Paradox eliminates this countable set, but constructs a partition satisfying a different paradoxical system of congruences. It is a consequence of Theorem 4.16 that if all isometries of the sphere are allowed, then a Hausdorff decomposition of the entire sphere is possible. For a version of the Hausdorff Paradox that retains the restriction to rotations but weakens the assumption that a countable set of points on the sphere must be deleted, see the appendix to [192]. For a constructive geometric realization of the Hausdorff Paradox, that is, one that does not require the Axiom of Choice, see Figure 5.2.

NOTES

Von Neumann was apparently the first to consider the problem of counting the number of pieces in a paradoxical decomposition, and in [246, p. 77] he states, without proof, that nine pieces suffice for the duplication of a solid ball. Sierpiński [211] was able to improve this to eight, but it was Raphael Robinson [192] who had the idea of first trying to optimize the situation for the sphere and then extending the analysis to the ball.

    In fact, most of the main ideas of this chapter stem from Robinson's paper, which contains the proof that five pieces are necessary and sufficient for the ball, and the consideration of arbitrary weak systems of congruences and the application in Corollary 4.14. Question 4.15 was raised by Mycielski [158, 159].

The generalization of Robinson's work from the special context of a sphere to general locally commutative actions is due to Dekker [48, 51], who was responsible for the definition of local commutativity and its applications (4.5 and 4.12). A special case of these aspects of local commutativity was rediscovered by Akemann [4, Prop. 4]. Dekker also considered the converse problem, proving Theorem 4.8, and investigated higher-dimensional Euclidean and non-Euclidean spaces [49, 50]; these generalizations are treated in Chapters 5 and 6. The part of Theorem 4.8 that yields the independence of σ and τ is similar to a result known as MacBeath's Lemma (see [130, 131]).

Actually, the special case of Dekker's converse that deals solely with groups acting on themselves, Corollary 4.9, had been considered earlier. While a student of Tarski in the 1940s, B. Jonsson proved 4.9, but since the result did not lead to any progress on the question whether paradoxical groups necessarily had free subgroups of rank 2 (see the end of Chapter 1), it was never published.

The idea of weaving the antipodal map into strings of rotations to solve all proper systems of congruences on S^2 is due to Adams [1], who proved Theorem 4.16 and derived the Hausdorff Paradox without the exclusion of a countable set.

The m-divisibility of S^2 with respect to rotations for m satisfying $\aleph_0 \leqslant m \leqslant 2^{\aleph_0}$ is due to Mycielski [155], who also proved simultaneous m-divisibility (see remarks following Corollary 6.9). The m-divisibility of S^1 for $\aleph_0 < m \leqslant 2^{\aleph_0}$ is due to Ruziewicz [198]. Von Neumann's proof [245] of the $\aleph_0$-divisibility of intervals is simplified and extended to larger cardinals by Mycielski [158]. The nondivisibility of an open or closed interval into finitely many congruent pieces is due to Gustin [84] although special cases of this result have been rediscovered by Sierpiński [219, p. 63], Schinzel (see [158]), and Cater [26]. Van der Waerden [240] asked whether a disc in $\mathbf{R}^2$ is 2-divisible, and a negative answer was provided by Gysin [85]. Puppe (see [87, pp. 27, 81]) extended this to apply to any bounded, closed, convex subset of $\mathbf{R}^2$. R. M. Robinson (see [250]) showed that a ball in $\mathbf{R}^n$ is not m-divisible for any m with $2 \leqslant m \leqslant n$. The results on divisibility in Banach spaces are due to Edelstein [64].

Chapter 5

Higher Dimensions and Non-Euclidean Spaces

Since paradoxical decompositions depend on free groups, and since the group of rotations of S^2 is contained in higher-dimensional rotation groups, it comes as no surprise that paradoxical decompositions exist for higher-dimensional spaces. This generalization is not completely obvious, though, since the fixed point set of an isometry does expand when the isometry is extended to a higher dimension by fixing additional coordinates. Nevertheless, the basic results from Chapter 3 on the existence of paradoxical decompositions do extend without requiring any new techniques (see Theorem 5.1). For example, we have already seen that the unit ball in $\mathbf{R}^n$ is G_n-negligible if $n \geq 3$ (see proof of 2.6), and by the theorem of Tarski alluded to just prior to Theorem 2.6, it follows that such balls are paradoxical. But it is useful to see how the decompositions in higher dimensions may be obtained quite directly from the construction on S^2, as is done in Theorem 5.1.

The expansion of the fixed point set is a crucial impasse to generalizing the finer analysis of Chapter 4, however. This is because new fixed points completely destroy the local commutativity of a group when it is viewed as acting on a higher-dimensional space. Nonetheless, locally commutative free groups of isometries (and, where possible, free groups without fixed points) do exist; hence there are minimal paradoxical decompositions in all higher dimensions (see 5.5).

Theorem 5.1 (AC).   *Assume $n \geqslant 3$.*

(a) *Any sphere in $\mathbf{R}^n$ is paradoxical with respect to its group of rotations.*
(b) *Any solid ball in $\mathbf{R}^n$ is G_n-paradoxical, as is $\mathbf{R}^n$ itself.*
(c) *Any two bounded subsets of $\mathbf{R}^n$ with nonempty interior are equidecomposable.*

Proof.   (a) The result is true for $n = 3$ (Corollary 3.10) and we proceed from there by induction. We consider S^n, the unit sphere in $\mathbf{R}^{n+1}$ centered at $\mathbf{0}$, but the same proof applies to all spheres. Now, suppose $A_i, B_j \subseteq S^{n-1}$ and σ_i, $\tau_j \in SO_n$ witness the fact that S^{n-1} is paradoxical. Define A_i^*, B_j^* to partition S^n, excluding the two poles $(0, \ldots, 0, \pm 1)$, by putting $(x_1, \ldots, x_n, z)$ in A_i^* or B_j^* according to which of the A_i, B_j contains $(x_1, \ldots, x_n)/|(x_1, \ldots, x_n)|$. Extend σ_i, τ_j to $\sigma_i^*, \tau_j^* \in SO_{n+1}$ by fixing the new axis. In matrix form:

$$
\sigma_i^* = \begin{bmatrix} & & & 0 \\ & \sigma_i & & \vdots \\ & & & 0 \\ 0 \cdots 0 & & & 1 \end{bmatrix}.
$$

Then A_i^*, B_j^*, σ_i^*, τ_j^* provide a paradoxical decomposition of $S^n \backslash \{(0, \ldots, 0, \pm 1)\}$. But any two-dimensional rotation of infinite order, viewed as rotating the last two coordinates and fixing the first $n - 1$ coordinates, can be used as usual (see proof of 3.10) to show that $S^n \backslash \{(0, \ldots, 0, \pm 1)\} \sim_2 S^n$; by 3.4 this proves part (a). The rest of the theorem follows from (a) exactly as it does in $\mathbf{R}^3$ (see 3.10 and 3.11). $\square$

As pointed out above, we must make a bit of a fresh start in order to solve arbitrary weak systems of congruences, and hence obtain minimal paradoxical decompositions, for S^3 and beyond. We first consider S^3, where a free non-Abelian group of rotations with no nontrivial fixed points can be constructed directly. Then this group will be combined with the free subgroup of SO_3 constructed in Chapter 2 to obtain free subgroups in all higher-dimensional rotation groups except SO_5, which requires special treatment.

Theorem 5.2.   *The group SO_4 has a free subgroup of rank 2 whose action on the sphere S^3 is without nontrivial fixed points.*

Proof.   Choose θ to be an angle whose cosine is transcendental (e.g., $\theta = 1$ radian), and let σ, τ be the rotations in SO_4 given, respectively, by

$$
\begin{bmatrix} \cos\theta & -\sin\theta & 0 & 0 \\ \sin\theta & \cos\theta & 0 & 0 \\ 0 & 0 & \cos\theta & -\sin\theta \\ 0 & 0 & \sin\theta & \cos\theta \end{bmatrix}, \quad \begin{bmatrix} \cos\theta & 0 & 0 & -\sin\theta \\ 0 & \cos\theta & -\sin\theta & 0 \\ 0 & \sin\theta & \cos\theta & 0 \\ \sin\theta & 0 & 0 & \cos\theta \end{bmatrix}.
$$

We shall prove that no nontrivial word in $\sigma^{\pm 1}, \tau^{\pm 1}$ has any fixed points on S^3, which implies both the independence of σ, τ, and the lack of fixed points for elements of the group they generate.

A nontrivial word w has one of the four forms: $\sigma^{\pm 1} \cdots \tau^{\pm 1}, \tau^{\pm 1} \cdots \sigma^{\pm 1}$, $\sigma^{\pm 1} \cdots \sigma^{\pm 1}$, or $\tau^{\pm 1} \cdots \tau^{\pm 1}$. The second form reduces to the first by considering w^{-1}, and repeated conjugation reduces the last two cases to one of the first two, unless w is simply a power of σ or τ. Because $\cos \theta$ is not algebraic, θ is not a rational multiple of π, and therefore powers of the rotation $\left[\begin{smallmatrix} \cos \theta & -\sin \theta \\ \sin \theta & \cos \theta \end{smallmatrix}\right]$, and hence powers of σ or of τ, do not have any fixed points. Thus it remains to prove that the same is true for a word, w, of the form $\sigma^{\pm 1} \cdots \tau^{\pm 1}$. This will be done by showing that 1 is not an eigenvalue of the matrix corresponding to such a word w.

Each of $\sigma^{\pm 1}, \tau^{\pm 1}$ has the form

$$
\begin{bmatrix}
P & -Q & -R & -S \\
Q & P & -S & R \\
R & S & P & -Q \\
S & -R & Q & P
\end{bmatrix}
$$

where P and R are polynomials in $\cos \theta$ (with integer coefficients), and Q and S are the products of such polynomials with $\sin \theta$. An obvious induction yields that the same is true for any word w in $\sigma^{\pm 1}, \tau^{\pm 1}$. Computing the determinant of $w - \lambda I$ then yields that the characteristic equation for w is

$$
\lambda^4 - 4P\lambda^3 + \left(6P^2 + 2(Q^2 + R^2 + S^2)\right)\lambda \\
- 4P(P^2 + Q^2 + R^2 + S^2)\lambda + (P^2 + Q^2 + R^2 + S^2) = 0.
$$

But w is orthogonal, so $P^2 + Q^2 + R^2 + S^2 = 1$ and the characteristic equation is $\lambda^4 - 4P\lambda^3 + (4P^2 + 2)\lambda^2 - 4P\lambda + 1 = 0$. If 1 is an eigenvalue of w, then $4P^2 - 8P + 4 = 0$. Since $\cos \theta$ is transcendental, this will be a contradiction once it is shown that P, a polynomial in $\cos \theta$, is not just a constant.

Since σ comes from two two-dimensional rotations, for positive integers m we have that

$$
\sigma^{\pm m} = \begin{bmatrix}
\cos m\theta & \mp \sin m\theta & 0 & 0 \\
\pm \sin m\theta & \cos m\theta & 0 & 0 \\
0 & 0 & \cos m\theta & \mp \sin m\theta \\
0 & 0 & \pm \sin m\theta & \cos m\theta
\end{bmatrix}.
$$

Because of the identities

$$
\cos m\theta = 2^{m-1} \cos^m \theta + \text{terms of lower degree in } \cos \theta,
$$

and

$$
\sin m\theta = \sin \theta (2^{m-1} \cos^{m-1} \theta + \text{terms of lower degree in } \cos \theta),
$$

$\sigma^{\pm m}$ has the following form, where $\doteq$ denotes that only the term of highest degree in $\cos\theta$ is retained in each entry:

$$\sigma^{\pm m} \doteq 2^{m-1}\cos^{m-1}\theta \begin{bmatrix} \cos\theta & \mp\sin\theta & 0 & 0 \\ \pm\sin\theta & \cos\theta & 0 & 0 \\ 0 & 0 & \cos\theta & \mp\sin\theta \\ 0 & 0 & \pm\sin\theta & \cos\theta \end{bmatrix}.$$

There is a similar representation for $\tau^{\pm k}$. Multiplying the two representations, and substituting $1 - \cos^2\theta$ for $\sin^2\theta$, yields the following (where $\varepsilon, \delta = \pm 1$):

$$\sigma^{\varepsilon m}\tau^{\delta k} \doteq 2^{m+k-2}\cos^{m+k-1}\theta \begin{bmatrix} \cos\theta & -\varepsilon\sin\theta & -\varepsilon\delta\cos\theta & -\delta\sin\theta \\ \varepsilon\sin\theta & \cos\theta & -\delta\sin\theta & \varepsilon\delta\cos\theta \\ \varepsilon\delta\cos\theta & \delta\sin\theta & \cos\theta & -\varepsilon\sin\theta \\ \delta\sin\theta & -\varepsilon\delta\cos\theta & \varepsilon\sin\theta & \cos\theta \end{bmatrix}.$$

Now, we claim that if $w = \sigma^{\varepsilon_n m_n}\tau^{\delta_n k_n}\cdots\sigma^{\varepsilon_1 m_1}\tau^{\delta_1 k_1}$ and Σ denotes $|k_1| + |m_1| + \cdots + |k_n| + |m_n|$, then

$$w \doteq 2^{\Sigma-n-1}\cos^{\Sigma-1}\theta \begin{bmatrix} \xi\cos\theta & -\mu\sin\theta & -\mu\cos\theta & -v\sin\theta \\ \mu\sin\theta & \xi\cos\theta & -v\sin\theta & \zeta\cos\theta \\ \zeta\cos\theta & v\sin\theta & \xi\cos\theta & -\mu\sin\theta \\ v\sin\theta & -\zeta\cos\theta & \mu\sin\theta & \xi\cos\theta \end{bmatrix},$$

where $\xi, \mu, v, \zeta = \pm 1$. To prove this claim, assume inductively that w has the correct form (with ξ_n, μ_n, ζ_n, and v_n), and consider $(\sigma^{\varepsilon m}\tau^{\delta k})w$. Multiplying yields the desired form, with

$$\xi_{n+1} = \xi_n + \varepsilon\mu_n + \delta v_n - \varepsilon\delta\zeta_n$$

$$\mu_{n+1} = \mu_n + \varepsilon\xi_n + \delta\zeta_n - \varepsilon\delta v_n$$

$$v_{n+1} = v_n - \varepsilon\zeta_n + \delta\xi_n + \varepsilon\delta\mu_n$$

$$\zeta_{n+1} = \zeta_n - \varepsilon v_n + \delta\mu_n + \varepsilon\delta\xi_n,$$

but with one less power of two than desired. It follows easily from these equations that the equation $\mu v = \xi\zeta$, which is true in $\sigma^{\varepsilon m}\tau^{\delta k}$, remains true. Hence

$$\xi_{n+1} = \xi_n + \varepsilon\mu_n + \delta v_n - \frac{\varepsilon\delta\mu_n v_n}{\xi_n} = \xi_n + \varepsilon\mu_n + \delta v_n - \xi_n(\varepsilon\mu_n)(\delta v_n),$$

which, since all terms are ± 1, equals one of ± 2. The same holds true for μ_{n+1}, v_{n+1}, ζ_{n+1}. Factoring out a 2 yields the correct power of 2 and the correct form, ± 1, for the coefficients, completing the proof of the claim.

This claim shows that P is a polynomial of degree Σ in $\cos\theta$ and, as explained, it follows that 1 is not an eigenvalue of w. □

Corollary 5.3. *If $n \geqslant 3$ and $n \neq 5$, then SO_n has a free subgroup of rank 2 that is locally commutative in its action on S^{n-1}. If n is a multiple of 4, then SO_n has a free subgroup of rank 2 without any nontrivial fixed points on S^{n-1}.*

Proof. Any $n \geqslant 3$, except $n = 5$, may be written as $3k + 4\ell$ for some positive integers k, ℓ. Letting σ_3, τ_3, and σ_4, τ_4 denote, respectively, pairs of independent elements in SO_3 and SO_4, as constructed in Theorems 2.1 and 5.2, we can define σ, $\tau \in SO_n$ by using these pairs on the three- and four-dimensional subspaces of $\mathbf{R}^n$ as specified by the decomposition of n into $3k + 4\ell$. Thus the matrix of σ is block diagonal, with k 3×3 blocks (σ_3's) and ℓ 4×4 blocks (σ_4's). Assume the 3×3 blocks precede the 4×4 ones. It is clear that σ and τ are independent. If $k = 0$, then the lack of fixed points in the group generated by σ_4 and τ_4 implies the same for the group generated by σ and τ, proving the assertion of the corollary regarding multiples of 4.

Now, suppose $k > 0$ and w, u are two words in $\sigma^{\pm 1}$, $\tau^{\pm 1}$ that share a fixed point. Note that w and u are ordinary three-dimensional rotations when restricted to a three-dimensional subspace on which they act as if they were the corresponding words in $\sigma_3^{\pm 1}$, $\tau_3^{\pm 1}$. Since they share a fixed point on S^{n-1}, they must share a nonzero fixed point in one of these k three-dimensional subspaces. But then u and w commute when considered as words in $\sigma_3^{\pm 1}$, $\tau_3^{\pm 1}$. By the independence of σ_3 and τ_3, this implies that u and w are commuting words in $\sigma^{\pm 1}$, $\tau^{\pm 1}$. $\qquad\qquad\square$

This corollary leaves two possibilities outstanding: SO_{4n+2} and SO_5. In SO_m, m odd, every element has $+1$ as an eigenvalue, and hence has a fixed point on S^{m-1}. Dekker, who discovered the groups of 5.2 and 5.3, conjectured [49] that free groups of rank 2 without nontrivial fixed points exist in the other half of the possible cases, that is, in SO_{4n+2}, $n \geqslant 1$. He also conjectured that SO_5, which did not yield to his techniques, has a locally commutative free subgroup of rank 2. Quite recently, Deligne and Sullivan [57] settled the SO_{4n+2} case, using relatively deep techniques of algebraic number theory, as well as the theorem of Tits (Theorem 10.5). Their work was significantly extended by A. Borel [20], who showed how to get locally commutative, non-Abelian free subgroups of each SO_n, $n \geqslant 3$; in particular, such subgroups of SO_5 exist, as conjectured by Dekker. In fact, Borel's pair of independent rotations of S^4 is similar to the earlier examples in S^2 and S^3. The rotations are defined using angles with a transcendental cosine, and each is built from two two-dimensional rotations. Borel also showed how fixed-point free, rank 2 free subgroups of SO_{4n+2} could be constructed without using Tits's Theorem. These results, whose proofs use methods from the theory of Lie groups, provide a complete solution to the problem of the existence of these types of free rotation groups: free non-Abelian subgroups of SO_n that act on S^{n-1} without nontrivial fixed points, or are locally commutative, exist in all cases except those excluded for algebraic reasons (solvability or eigenvalues). Moreover, we shall see in Chapter 6 that the set of pairs from SO_n that serve as generators of these groups is dense in SO_n^2.

Theorem 5.4. *For any even $n \geqslant 4$, SO_n has a free subgroup of rank 2 having no nontrivial fixed points on S^{n-1}. For any $n \geqslant 3$, SO_n has a free subgroup of rank 2 whose action on S^{n-1} is locally commutative.*

Corollary 5.5 (AC). *If $n \geqslant 2$, then S^n (and $\mathbf{R}^{n+1} \backslash \{0\}$) may be partitioned to satisfy any weak system of congruences with respect to the rotation group SO_{n+1}. In particular, S^n is paradoxical using four pieces. If n is odd, $n \geqslant 3$, then all proper systems of congruences may be solved with respect to the action of SO_{n+1} on S^n. The same is true for even $n \geqslant 2$ if we enlarge the group witnessing the congruences from SO_{n+1} to O_{n+1}.*

Proof. Since a free group of rank 2 contains, for any m, m independent elements, the parts of this corollary that deal with SO_{n+1} follow from Corollary 4.12 and the free groups of 5.3 and 5.4. The result about O_{n+1} follows by closely analyzing the technique of Theorem 4.16, which showed how the presence of the antipodal map could be combined with a locally commutative free group to solve all proper systems of congruences. The essential feature of the locally commutative group used in that proof is that no group element sends any point P on S^n to its antipode, $-P$; that is, no group element has -1 as an eigenvalue. This is easily seen to be the case for any free subgroup of SO_3 and for the subgroup of SO_4 constructed in Theorem 5.2; for the latter note that if $w(P) = -P$, then w^2 would fix P. Hence this property holds for all the locally commutative groups of Corollary 5.3, that is, all relevant cases except SO_5.

For SO_5 (and this is also true for SO_3 and SO_4) one can show that, in fact, *any* locally commutative free subgroup of rank 2 has no element with -1 as an eigenvalue. In particular, then, this is true of Borel's group mentioned prior to Theorem 5.4, completing the proof of Corollary 5.5. To prove this fact about SO_5, suppose G is a locally commutative subgroup of SO_5 freely generated by σ, τ, and some word $u \in G$ has -1 as an eigenvalue. It follows from the orthogonality of u that its real eigenvalues include $+1$, -1, -1, whence u^2 is the identity on a three-dimensional subspace of $\mathbf{R}^5$. Suppose, without loss of generality, that u, and hence u^2, begins with σ. Then u^2 and $w = \tau u^2 \tau^{-1}$ have different leftmost terms, and so are not powers of a common word in G. It follows from G's freeness that u^2 and w do not commute (see proof of 4.5), and hence by local commutativity, that they do not share a fixed point on S^4. But this contradicts the fact that u^2 and w each fix (pointwise) a three-dimensional subspace of $\mathbf{R}^5$ (because w is a conjugate of u^2), and two such subspaces must have a point in common on S^4. $\qquad\square$

The proof in the last paragraph breaks down in SO_6 and beyond, and it is not known whether a locally commutative, non-Abelian free subgroup of SO_n $(n \geqslant 6)$ can contain an element having -1 as an eigenvalue. This gap did not affect the proof of Corollary 5.5, since this eigenvalue does not appear among the specific locally commutative groups of Corollary 5.3.

Note that, for n odd, the antipodal map has determinant $+1$ and so lies in SO_{n+1}. Hence the technique of the preceding proof, which uses the antipodal map and Theorem 4.16, yields another proof of the part of Corollary 5.5 that deals with n odd, $n \geqslant 3$. This approach does not require the fixed-point free groups of Theorem 5.4, but uses only the simpler locally commutative groups of Corollary 5.3.

Corollary 5.5 gives a complete answer to the question of m-divisibility of spheres for finite m. But there is a simpler approach that yields a more constructive solution, at least for spheres of odd dimension. For now we consider only finite m, but in the next chapter we shall consider infinite m, showing that any S^n (except S^1) has a subset that is, simultaneously, a half, a third,..., a $2^{\aleph_0}$th part of S^n (see Corollary 6.9 and remarks following).

Theorem 5.6.   *Assume m is an integer and $m \geqslant 2$.*

(a) *If n is odd, then S^n is m-divisible with respect to SO_{n+1}.*
(b) *(AC) If n is even, then S^n is m-divisible with respect to SO_{n+1} if and only if $m \geqslant 3$.*
(c) *(AC) For any n, S^n is m-divisible with respect to O_{n+1}.*

Proof.   First, observe that the m-divisibility of S^n can be deduced from that of S^k and S^ℓ if $k + 1 + \ell + 1 = n + 1$. For we may partition S^n by considering a point $P = (x_1,\ldots,x_{n+1}) \in S^n$ and seeing how the m-division of S^k treats $(x_1,\ldots,x_{k+1})/|(x_1,\ldots,x_{k+1})|$; if $x_1 = \cdots = x_{k+1} = 0$, consider instead $(x_{k+2},\ldots,x_{k+2+\ell})/|(x_{k+2},\ldots,x_{k+2+\ell})|$, which is in S^ℓ. Now, for (a) simply write $n + 1$ as $2k$ and use the above approach on the k pairs of coordinates, considering a pair as a point on S^1, which is m-divisible for all m. For (b) write $n + 1$ as $2k + 3$ and use the m-divisibility of S^1 and S^2; recall that the latter uses the Axiom of Choice (see 4.14). The failure of 2-divisibility arises because every rotation in SO_{n+1} has a fixed point when $n + 1$ is odd. Since the antipodal map, which yields 2-divisibility, is in O_{n+1}, (c) follows from (a) and (b).   □

Because orthogonal maps all fix the origin, the detailed study of such maps leaves unanswered the question of solving congruences by partitions of $\mathbf{R}^n$ and using the isometry group, G_n. For instance, it was shown following Theorem 4.7 how to get a five-piece paradoxical decomposition of $\mathbf{R}^3$ from the four-piece one of S^2 (and the same exists for $\mathbf{R}^n$ if $n \geqslant 3$ by the four-piece decomposition of the corresponding sphere that exists by

5.5). But, in fact, four-piece decompositions are possible, since it follows from Theorem 5.7 that any proper system of congruences can be solved by a partition of $\mathbf{R}^n$, provided $n \geqslant 3$. Note that the particular system asserting that $\mathbf{R}^n$ is m-divisible is trivial to solve for all m, n (see remarks following Definition 4.13).

Theorem 5.7. *There are two independent isometries in G_3 such that the group, F, that they generate has no nontrivial fixed points in R^3. The same is true for G_n and $\mathbf{R}^n$ if $n \geqslant 3$.*

Proof. Let ϕ and ψ be the rotations in SO_3 given by

$$\begin{bmatrix} \cos\theta & -\sin\theta & 0 \\ \sin\theta & \cos\theta & 0 \\ 0 & 0 & 1 \end{bmatrix} \quad \text{and} \quad \begin{bmatrix} 1 & 0 & 0 \\ 0 & \cos\theta & -\sin\theta \\ 0 & \sin\theta & \cos\theta \end{bmatrix},$$

respectively, where the common rotation angle θ is chosen so that $\cos\theta$ is transcendental. Letting $T_{\vec{v}}$ denote the translation of $\mathbf{R}^3$ by the vector $\vec{v}$, and $\vec{i}, \vec{k}$ the vectors $(1,0,0)$, $(0,0,1)$, respectively, define σ to be $T_{\vec{k}}\phi$ and τ to be $T_{\vec{i}}\psi$; σ and τ will be the independent isometries we seek. These two isometries are each screw motions (or glide rotations) of $\mathbf{R}^3$, that is, a rotation followed by a translation in the direction of the rotation's axis. Note that if the translation component of a screw-motion is not the identity, then the motion has no fixed points.

We will need to know that the matrix of a word $\phi^{n_1}\psi^{m_1}\cdots\phi^{n_s}\psi^{m_s}$ has the form

$$\begin{bmatrix} P\cos\theta & -\mathrm{sgn}(n_1)Q\sin\theta & -\mathrm{sgn}(n_1 m_s)Q\cos\theta \\ P\sin\theta & Q\cos\theta & -\mathrm{sgn}(m_s)Q\sin\theta \\ P\cos\theta & P\sin\theta & P\cos\theta \end{bmatrix}$$

where Q stands for a polynomial in $\cos\theta$ (possibly different in each entry) of degree $d = \Sigma - 1$ $(\Sigma = |n_1| + |m_1| + |n_2| + \cdots + |m_s|)$ with leading coefficient $2^{\Sigma-2s}$, and P represents (possibly different) polynomials in $\cos\theta$ of degree strictly less than d. We omit the details, as this may be easily proved by induction in exactly the same manner as the analogous fact proved in Theorem 5.2. (It follows from this representation that ϕ and ψ are independent; see Theorem 2.2(b).)

Now, it must be shown that no nontrivial word in $\sigma^{\pm 1}$, $\tau^{\pm 1}$ has any fixed points in $\mathbf{R}^3$; the independence of σ, τ follows. Since the existence of a fixed point is invariant under conjugation and inversion, it follows as in the proof of 5.2 that we need only consider pure powers of σ or τ and words of the form $w = \sigma^{n_1}\tau^{m_1}\cdots\sigma^{n_s}\tau^{m_s}$, with each exponent a nonzero integer. Since $\sigma^n = T_{n\vec{k}}\phi^n$ and $\tau^m = T_{m\vec{i}}\psi^m$, the pure powers of σ and τ have no fixed points.

It follows easily from the fact that rotations are linear transformations that for any rotation ρ, $\rho T_{\vec{v}} = T_{\rho(\vec{v})}\rho$, and hence $w = T_{\vec{t}}\hat{w}$ where $\hat{w}$ is the result of replacing σ, τ in w by ϕ, ψ, respectively, and $\vec{t}$ will be computed momentarily. For $1 \leqslant r \leqslant s$, let $\hat{w}_r$ represent the left segment of $\hat{w}$: $\sigma^{n_1}\tau^{m_1} \cdots \sigma^{n_r}\tau^{m_r}$. Then using the rule for moving translations left across a rotation, and using the fact that $\psi(\vec{i}) = \vec{i}$, it is easy to see that

$$\vec{t} = n_1\vec{k} + n_2\hat{w}_1(\vec{k}) + \cdots + n_s\hat{w}_{s-1}(\vec{k}) + m_1\hat{w}_1(\vec{i}) + m_2\hat{w}_2(\vec{i}) + \cdots + m_s\hat{w}_s(\vec{i}).$$

Let $\vec{a}$ be a unit vector in the direction of the axis of $\hat{w}$, oriented so that the rotation obeys the right-hand rule with respect to $\vec{a}$. Now, $\vec{t}$ is not necessarily in the same direction as $\vec{a}$, but as long as $\vec{t}$ is not perpendicular to $\hat{w}$'s axis, $w = T_{\vec{t}}\hat{w}$ will have no fixed points. Thus the proof will be complete once it is shown that $\vec{t} \cdot \vec{a} \neq 0$; in fact this condition is equivalent to the nonexistence of a fixed point of a screw motion.

Consider the expansion of $\vec{t} \cdot \vec{a}$ using the formula for $\vec{t}$ just given. There are two sorts of terms in this expansion. The first sort is $n_j\hat{w}_{j-1}(\vec{k}) \cdot \vec{a}$. By the invariance of inner product under orthogonal transformations (see Appendix A) each such term equals $n_j\vec{k} \cdot \hat{w}_{j-1}^{-1}(\vec{a})$. Since the rotation of ρ_1's axis by ρ_2 (ρ_i are rotations) yields the axis of $\rho_2\rho_1\rho_2^{-1}$, this last expression is the dot product of $n_j\vec{k}$ with the appropriately oriented unit vector along the axis of the rotation $\hat{w}_{j-1}^{-1}\hat{w}\hat{w}_{j-1} = \sigma^{n_j}\tau^{m_j} \cdots \tau^{m_s}\sigma^{n_1} \cdots \sigma^{n_{j-1}}\tau^{m_{j-1}}$. Recall that if (A_{ij}) represents a rotation of $\mathbf{R}^3$ with unit axis vector $\vec{b}$ and rotation angle ξ, then $2\vec{b}\sin\xi = (A_{32} - A_{23}, A_{13} - A_{31}, A_{21} - A_{12})$. This is proved in Appendix A. Also proved in that Appendix is the fact that the angle of a rotation is the same as the angle of a conjugate of that rotation by another rotation. Hence, if ξ is the angle of rotation of $\hat{w}$, then the term of $\vec{t} \cdot \vec{a}$ being considered equals $n_j(A_{21} - A_{12})/2\sin\xi$ where (A_{ij}) is the matrix of $\hat{w}_{j-1}^{-1}\hat{w}\hat{w}_{j-1}$. Using the representation of words in ϕ, ψ given at the beginning of the proof, this equals $(n_j/2\sin\xi)(P\sin\theta + \text{sgn}(n_j)Q\sin\theta)$. A similar analysis of the other sort of term in $\vec{t} \cdot \vec{a}$, $m_j\hat{w}_j(\vec{i}) \cdot \vec{a}$, yields a contribution having the form $(m_j/2\sin\xi)(P\sin\theta + \text{sgn}(m_j)Q\sin\theta)$. It follows that $(\sin\xi/\sin\theta)(\vec{t} \cdot \vec{a})$ is a polynomial in $\cos\theta$ with leading coefficient $2^{d-2s}(|n_1| + |m_1| + \cdots + |m_s|)$. Therefore, because of the transcendence of $\cos\theta$, $\vec{t} \cdot \vec{a}$ cannot vanish.

Extending the isometries σ and τ to higher dimensions by simply fixing the additional coordinates yields the desired independent isometries of $\mathbf{R}^n$ for any $n \geqslant 3$. $\square$

Corollary 5.8 (AC).   *If $n \geqslant 3$, then any proper system of m congruences involving $A_1, \ldots, A_r$ may be solved by a partition of $\mathbf{R}^n$, where isometries are used to realize the congruences. In particular, for $n \geqslant 3$, $\mathbf{R}^n$ is paradoxical using four pieces.*

Proof.   Since a free group of rank 2 contains one of rank m (see remarks preceding 4.14), this corollary is an immediate consequence of the preceding theorem and Corollary 4.12. Note that none of the machinery for dealing with weak systems is needed, only the proof of the first part of Theorem 4.11, which is relatively straightforward.   □

NON-EUCLIDEAN SPACES

The theory so far has been applied only to spheres and Euclidean spaces and their respective isometry (or rotation) groups, but the matrix computations of this chapter can be applied to elliptic and hyperbolic spaces as well. If we take RP^n (real projective n-space: S^n with antipodal points identified) as our model of L^n, elliptic n-space, with the distance from P to Q defined to be the smaller of the two spherical distances P to Q and $-P$ to Q (equivalently, $\arccos |P \cdot Q|$), then the isometry group is just the group of rotations of S^n, viewed as acting on RP^n. Since the only orthogonal transformation that collapses to the identity when viewed as acting on RP^n is the antipodal map, the isometry group of L^n coincides with SO_{n+1} if n is even, and is $SO_{n+1}/\{\pm I\}$ if n is odd. It follows that two independent rotations ϕ, ρ of S^n remain independent when viewed as rotations of L^n. Moreover, local commutativity (or lack of nontrivial fixed points) of the group generated by ϕ, ρ is preserved, for if u and w both fix P in L^n, then $u(P) = \pm P$ and $w(P) = \pm P$, so u^2 and w^2 share a fixed point on S^n. Hence u^2 and w^2 commute, and this implies the same for u and w (see [135, p. 41]). Combining these remarks with Theorem 5.4, we see that Corollary 5.5 holds in elliptic space. For each $n \geqslant 2$, all weak systems of congruences are satisfiable by a partition of L^n, and for odd $n \geqslant 3$, all proper systems are satisfiable. Note that the antipodal map on S^n becomes the identity in L^n, so the method of proof of the last part of Corollary 5.5 cannot be used; rather, we use Theorem 5.4's free groups without fixed points directly.

   For the hyperbolic plane, H^2, it has already been mentioned that the upper half-plane model allows the group of orientation-preserving isometries to be identified via linear fractional transformations with PSL_2. A nonidentity element of PSL_2 is called *elliptic*, *parabolic*, or *hyperbolic* according as the absolute value of its trace is $<2, = 2$, or >2. Of these, only the elliptic ones fix a point in H^2 (see [123, §I.1]). While a free subgroup of PSL_2 might contain an elliptic element, such elements can never appear in a free subgroup of $PSL_2(\mathbf{Z})$. This is because, by an easy computation, any elliptic element of $PSL_2(\mathbf{Z})$ has finite order: if $tr(\sigma) = 0$, then $\sigma = \begin{bmatrix} a & b \\ c & -a \end{bmatrix}$ and $\sigma^2 = \begin{bmatrix} -1 & 0 \\ 0 & -1 \end{bmatrix}$ which is the identity of PSL_2, and if $tr(\sigma) = \pm 1$, then $\sigma = \begin{bmatrix} a & b \\ c & \pm 1 - a \end{bmatrix}$ and σ has order 3. This shows that any independent pair in $SL_2(\mathbf{Z})$, such as $\begin{bmatrix} 1 & 2 \\ 0 & 1 \end{bmatrix}$ and its transpose (see 7.1), induces an independent pair of isometries of H^2 such that the group they generate has no nontrivial fixed points in H^2. Hence by 4.12 any proper system of congruences is solvable by a partition of the hyperbolic plane, using isometries.

   But in this case a much stronger result is true: the Axiom of Choice can be avoided and the sets of the partition can be taken to be Borel sets!

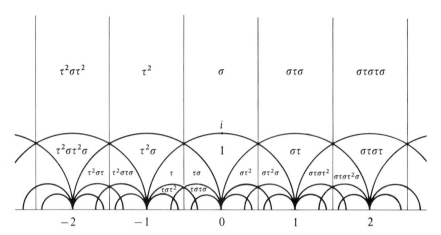

Figure 5.1. The tesselation of the upper half-plane with respect to the modular group, which is generated by $\sigma(z) = 1/z$ and $\tau(z) = -1/(z + 1)$.

Whenever G is a discrete subgroup* of PSL_2 there is a fundamental polygon for the action of G on H^2, that is, a hyperbolic (open) polygon P such that $\{\rho(P) : \rho \in G\}$ are pairwise disjoint and $\bigcup \{\overline{\rho(P)} : \rho \in G\} = H^2$ (see [123, §I.4]). In other words, H^2 can be tiled by pairwise interior-disjoint copies of the fundamental polygon, one copy for each transformation in G. The classic example of such a tiling is due to Klein and Fricke [108] (see [123, p. 29] or [133, p. 174]) and comes from letting $G = PSL_2(\mathbf{Z})$ (the modular group). This group is isomorphic to $\langle \sigma, \tau : \sigma^2 = \tau^3 = 1 \rangle$, and hence to $\mathbf{Z}_2 * \mathbf{Z}_3$, where $\sigma(z) = -1/z$ and $\tau(z) = -1/(z + 1)$; see [119, App. B] or [122, pp. 140, 234] for a proof. The tiling of H^2 corresponding to the action of the modular group is illustrated in Figure 5.1, where the triangle with vertices at 0 and $\pm 1/2 + (\sqrt{3}/2)i$ is the fundamental polygon and several copies of this triangle are labelled with the appropriate transformation from $PSL_2(\mathbf{Z})$.

Now, as observed by Hausdorff (see Chapter 2 Notes), the abstract group $\mathbf{Z}_2 * \mathbf{Z}_3$ is paradoxical; indeed there is a Hausdorff decomposition into $A \cup B \cup C$ such that $\tau A = B$, $\tau^2(A) = C$, and $\sigma A = B \cup C$, where σ and τ are the generators of $\mathbf{Z}_2 * \mathbf{Z}_3$. This decomposition can be obtained explicitly by assigning the identity to A and proceeding inductively according to the three desired equations (see [219] or [227]). Because of the isomorphism of $\mathbf{Z}_2 * \mathbf{Z}_3$ with $PSL_2(\mathbf{Z})$ and the one-to-one correspondence between transformations in G and tiles in the tiling of Figure 5.1, this immediately yields subsets (also called A, B, and C) of H^2 that provide a Hausdorff decomposition of $H^2 \setminus N$, where N is the (measure zero and nowhere dense) set consisting of the boundaries of all the polygons. The subset A of H^2 is just the union of the

---

* This means that the set of matrices corresponding to elements of G contains no convergent sequence of matrices.

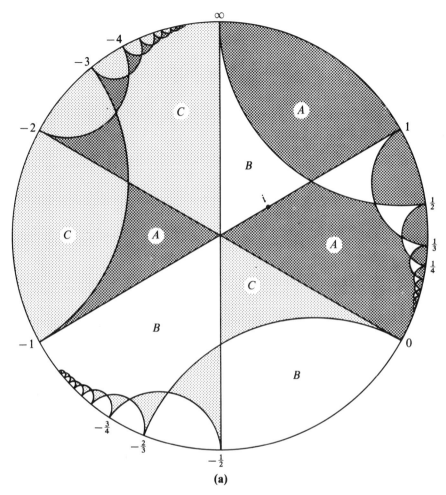

Figure 5.2(a).   *A* is congruent to *B* via τ and to *C* via τ^2.

images of the fundamental triangle using transformations in the subset *A* of *G*. This yields the three sets illustrated in Figure 5.2(a), where for the sake of greater (Euclidean!) symmetry, the half-plane model has been projected onto an open disc. The sets *A*, *B*, and *C* in Figure 5.2(a) are related by $\tau(A) = B$, $\tau^2(A) = C$, and $\sigma(A) = B \cup C$. Indeed, τ corresponds to a Euclidean rotation of 120° around $-1/2 + (\sqrt{3}/2)i$, the center of the circle, so it is quite evident that *A* (and *B* and *C*) are each a third of $H^2 \backslash N$ (here *N* is only a subset of the set of boundary segments of the tiles, as some of these segments have been assigned to *A*, *B*, or *C*). The transformation σ is a hyperbolic 180° rotation about *i*, and it is easiest to visualize by considering a different projection of the half-plane onto the disc, one that sends *i* to the center of the circle. This

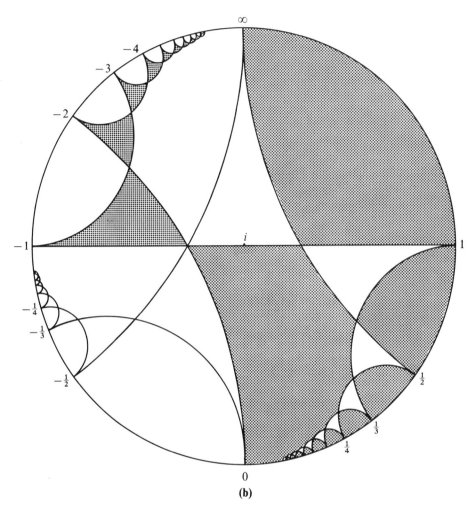

Figure 5.2(b).   The grey region is A and the white region is $B \cup C$; the regions are congruent via σ.

projection is illustrated in Figure 5.2(b); σ corresponds to a Euclidean half-turn about the center and the relation $\sigma(A) = B \cup C$ is clearly visible to the Euclidean observer. Thus A is also a half of $H^2 \backslash N$.

Although the decomposition described above has a pleasing degree of symmetry, it is flawed in the sense that the set A is a half, and a third, of $H^2 \backslash N$, rather than simply of H^2. To remedy this flaw, which is caused by the fact that σ and τ are elliptic and hence have a fixed point, we consider the nonelliptic subgroup of $PSL_2(\mathbf{Z})$ mentioned earlier; namely, let $\sigma(z) = z/(2z + 1)$ and $\tau(z) = z + 2$, and let F be the group generated by σ and τ. F consists of all matrices that are congruent to the identity modulo 2, and is called the principal congruence subgroup of the modular group of level 2 (see [123, p. 60]). As already shown, F has no elliptic elements, and because the matrices of σ and τ are $\begin{bmatrix} 1 & 0 \\ 2 & 1 \end{bmatrix}$ and its transpose, which by Proposition 7.1 are independent, F is a free group of rank 2. Actually, the independence of σ and τ can also be proved geometrically as follows. Use induction on the length of a nontrivial word w to show that either $\mathrm{Re}(w(i)) < -1$, $\mathrm{Re}(w(i)) > 1$, $|w(i) + \frac{1}{2}| < \frac{1}{2}$, or $|w(i) - \frac{1}{2}| < \frac{1}{2}$ according as w's leftmost term is τ^{-1}, τ, σ^{-1}, or σ. This implies that $w(i) \neq i$, so $w \neq 1$.

Since F is a subgroup of $PSL_2(\mathbf{Z})$, F is discrete and a fundamental polygon for F's action on H^2 exists. In order to get a choice set for the orbits of F's action on H^2—without using the Axiom of Choice—we use the fact that the boundary of the fundamental polygon consists of a countable number of sides (open hyperbolic segments) and vertices, and F maps vertices to vertices and sides to sides (see [123, §§I.4E,F]). It follows that there is a choice set M for the F-orbits that consists of the interior of the fundamental polygon together with some of its vertices and some of the sides. Clearly, M is a Borel set. To summarize, F is a rank 2 free group of hyperbolic isometries acting on H^2 without nontrivial fixed points, and there is a Borel choice set for the orbits of this action. This immediately yields the following decomposition result, by appealing to Theorem 4.11 and its proof and the fact that F contains free subgroups of any finite rank to which all the above remarks apply.

Theorem 5.9. *For any proper system of congruences involving r set-variables there is a partition of H^2 into Borel sets $A_1, \ldots, A_r$ that satisfy (using isometries) the given system. In particular, H^2 is paradoxical using Borel sets. Moreover, there is a Hausdorff decomposition of H^2 using Borel sets; that is, there is a Borel set in H^2 that is, simultaneously, a half and a third of H^2. These results all hold in H^n as well, if $n \geqslant 2$.*

Proof. Only the assertion about H^n remains to be proved. Using the upper half-space $\{(x_1, \ldots, x_{n-1}, t) : t, x_i \in \mathbf{R}, t > 0\}$ as a model for H^n, an isometry of H^2 may be extended to H^n by letting it act on (x_1, t), leaving the other coordinates fixed. Moreover, any Borel subset A of H^2 extends to one, A^*, of H^n by putting a point in A^* if and only if (x_1, t) is in A. It follows that a partition of H^2 that satisfies a given system of congruences induces one of H^n satisfying the same system. Note that it also follows that the isometry group of H^n contains a free subgroup of rank 2 that acts without nontrivial fixed points. $\square$

As an example, we explicitly construct a paradoxical decomposition of H^2 by using the free group F generated by σ and τ introduced prior to the preceding theorem. The (labelled) tiling corresponding to F is shown in Figure 5.3. Note that the geometric arrangement of words is quite similar to the tree

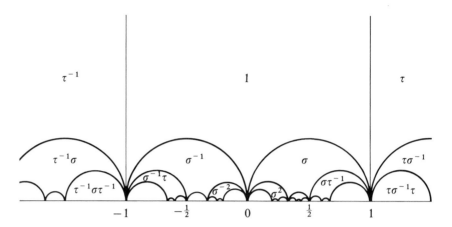

Figure 5.3. The tesselation of the upper-half plane with respect to a free subgroup of the modular group, which is freely generated by $\sigma(z) = z/(2z + 1)$ and $\tau(z) = z + 2$.

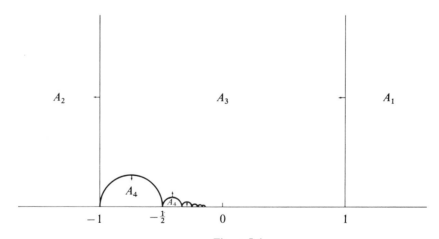

Figure 5.4.

representation of F given in Figure 4.1. Now, if the paradoxical decomposition of the group is transferred to H^2 via the labelling of tiles, one gets the four sets A_1, A_2, A_3, A_4 in Figure 5.4. The arrows indicate which sets get the boundary segments. Thus H^2 is divided into four sets such that $H^2 = A_1 \cup \tau(A_2) = A_3 \cup \sigma(A_4)$; that is, H^2 is paradoxical using four Borel sets. Such Borel sets in $\mathbf{R}^n$ do not exist, although this is not obvious since the fact that $\mathbf{R}^n$ (and H^n) has infinite Lebesgue (or hyperbolic) measure means that a Borel paradox would yield only the harmless equation $\infty = 2\infty$. The Euclidean cases and the measure-theoretic consequences of the hyperbolic paradox will be discussed in Chapter 11.

X	Spheres (S^n)	Spheres (S^n)	Elliptic spaces (L^n)	Hyperbolic spaces (H^n)	Euclidean spaces ($\mathbf{R}^n$)
G	Orientation-preserving isometries (SO_{n+1})	Isometries (O_{n+1})	Isometries	Isometries	Isometries (G_n)
X is G-paradoxical using four pieces $\Leftrightarrow$ G has a free locally commutative subgroup of rank $2 \Leftrightarrow$ all countable weak systems of congruences are solvable.	$2, 3, 4, \ldots$	$2, 3, 4, \ldots$	$2, 3, 4, \ldots$	$2, 3, 4, \ldots$	$3, 4, 5, \ldots$
G has a free subgroup of rank 2 without nontrivial fixed points on X.	$3, 5, 7, \ldots$	$3, 5, 7, \ldots$	$3, 5, 7, \ldots$	$2, 3, 4, \ldots$	$3, 4, 5, \ldots$
All countable proper systems of congruences are solvable.	$3, 5, 7, \ldots$	$2, 3, 4, \ldots$	$3, 5, 7, \ldots$	$2, 3, 4, \ldots$	$3, 4, 5, \ldots$

Figure 5.5. The table is complete in that the nonappearance of a certain dimension means that the phenomenon is known not to exist.

The hyperbolic case differs from the spherical and Euclidean cases in another way. We have pointed out that for S^n ($n \geqslant 3$, n odd) and $\mathbf{R}^3$ the set of independent pairs of isometries that generate a free group without nontrivial fixed points is dense in the square of SO_n, G_3, respectively. But in H^2 this is false. The elliptics, which correspond to matrices whose trace is strictly between -2 and 2, form an open subset of PSL_2, and therefore the set of independent pairs generating a fixed-point free group is certainly disjoint from the open set of pairs whose first coordinate is elliptic. Therefore the set of pairs is not dense. As we shall see in Chapter 6, H^3 does have this denseness property.

We conclude this discussion of the hyperbolic plane by pointing out that the action of all of PSL_2 on H^2 is locally commutative. For if σ_1 and σ_2 both fix $z \in H^2$, they must fix $\bar{z}$ as well. Choose σ to be a linear fractional transformation of $\mathbf{C} \cup \{\infty\}$ that takes z to 0 and $\bar{z}$ to ∞; then $\sigma\sigma_i\sigma^{-1}$ has $\{0, \infty\}$ as its fixed point set in $\mathbf{C} \cup \{\infty\}$, and so there must be $a_1, a_2 \in \mathbf{R}$ such that $\sigma\sigma_i\sigma^{-1}(z) = a_i z$. But then $\sigma\sigma_1\sigma^{-1}, \sigma\sigma_2\sigma^{-1}$ commute, so σ_1, σ_2 do also. Hence any free non-Abelian group of isometries of H^2 is rich enough to solve all weak systems of congruences, despite the possible presence of elliptic elements.

A summary of results about free groups and paradoxical decompositions in spheres and in Euclidean and non-Euclidean spaces appears in Figure 5.5. In the next chapter these results will be extended still further, to free groups of larger rank and to infinite systems of congruences. With the exception of the hyperbolic plane, these generalizations will resemble closely the results just proved about finite systems of congruences.

TETRAHEDRAL SNAKES

To conclude this chapter we discuss a geometric problem raised by Steinhaus that has nothing to do with systems of congruences, but its solution uses free groups and the sort of matrix computations that we have been considering. We shall show that the four reflections in the faces of a regular tetrahedron in $\mathbf{R}^3$ are free generators of $\mathbf{Z}_2 * \mathbf{Z}_2 * \mathbf{Z}_2 * \mathbf{Z}_2$; that is, the reflections satisfy no relation except those derivable from the fact that each has order 2. This is false for an equilateral triangle in the plane, since $\phi\psi\phi = \psi\phi\psi$ for two reflections ϕ, ψ in the triangle's sides, but is true for higher dimensional simplices.

This property of tetrahedral reflections can be used to solve the following geometric problem. Let us call a sequence of (at least two) congruent regular tetrahedra in $\mathbf{R}^3$ a *Steinhaus snake* if two consecutive tetrahedra share exactly one face, and each tetrahedron is distinct from its predecessor's predecessor. The terminology is derived from a popular toy currently on the market called The Magic Snake that uses, instead of tetrahedra, congruent triangular prisms made from an isosceles right triangle, with the height of the

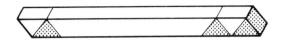

Figure 5.6.

prism equal to the length of a leg of the triangle. Thus each prism has five faces: two triangles, one rectangle, and two squares. Two consecutive prisms are joined on a square face, and the whole string of 24 prisms can be arranged, by 90° twists of the squares between prisms, into a variety of interesting shapes (see [7]), the simplest of which is a straight line (Figure 5.6).

The Magic Snake was invented in the 1970s by Ernö Rubik, but Steinhaus had considered the tetrahedral version in 1956, posing the question: can the last tetrahedron in a Steinhaus snake be a translation of the first one? Of course, the answer is yes for The Magic Snake, as it is for equilateral triangles in the plane: △▽. A negative answer to the Steinhaus problem can be derived from the result on free products of reflections, which we now prove. The problem was originally solved by Świerczkowski [229], who used the rotation group of Theorem 2.1; the approach using reflections given here is due to Dekker [53].

Theorem 5.10. *Let* $\phi_1, \phi_2, \phi_3, \phi_4$ *be the four reflections in the faces of a regular tetrahedron T in* $\mathbf{R}^3$. *Then no word of the form* $\phi_{i_1} \phi_{i_2} \ldots \phi_{i_s}$, *where* $s \geqslant 1$ *and adjacent terms are distinct, equals the identity; that is, the* ϕ_i *are free generators of* $\mathbf{Z}_2 * \mathbf{Z}_2 * \mathbf{Z}_2 * \mathbf{Z}_2$. *This same result, with* $n + 1$ *copies of* $\mathbf{Z}_2$, *holds for the* $n + 1$ *reflections in the faces of a regular n-dimensional simplex, provided* $n \geqslant 3$.

Proof. Any point P in $\mathbf{R}^3$ may be represented uniquely as $x_1 V_1 + x_2 V_2 + x_3 V_3 + x_4 V_4$ where the V_i are the vertices of T and $\sum x_i = 1$; these are called the barycentric coordinates of P with respect to T (see [92]). Any affine transformation of $\mathbf{R}^3$ (for instance, any isometry) may be represented by a 4×4 matrix acting on barycentric coordinates, where the columns of the matrix are the barycentric coordinates of the image of the V_i under the transformation. Therefore composition of transformations corresponds to matrix multiplication.

Since the reflection ϕ_1 in the face $x_1 = 0$ leaves V_2, V_3, and V_4 fixed and sends V_1 to $2C - V_1 = 2(\frac{1}{3}(V_2 + V_3 + V_4)) - V_1$, where C is the centroid of $\triangle V_2 V_3 V_4$ (see Figure 5.7), the matrix of ϕ_1 is

$$\begin{bmatrix} -1 & 0 & 0 & 0 \\ 2/3 & 1 & 0 & 0 \\ 2/3 & 0 & 1 & 0 \\ 2/3 & 0 & 0 & 1 \end{bmatrix}.$$

The other three reflections have matrices that are permutations of this one.

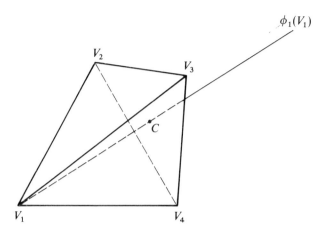

Figure 5.7.

We claim that the matrix of $\phi_{i_1} \cdots \phi_{i_s}$, where $s \geqslant 1$ and adjacent terms are distinct, has entries that are polynomials in $2/3$ with integral coefficients. Moreover, the polynomials in the i_sth column are not identically zero and have leading coefficients ± 1, while the other polynomials have lower degree than the corresponding entry in the i_sth column. This claim is easily proved by induction on s. It is obviously true for $s = 1$. Consider what happens when the matrix of a word of length s, which has the claimed form, is multiplied on the right by the matrix of ϕ_i, assuming the word does not end in ϕ_i. All columns except the ith are unchanged, while a_{ki} becomes $-a_{ki} + (2/3)\sum_{\ell \neq i} a_{k\ell}$. Exactly one of the $a_{k\ell}$, $\ell \neq i$, is a polynomial in $2/3$ whose degree dominates that of the rest of the kth row; multiplication by $2/3$ preserves this domination property and leaves the leading coefficient unchanged. It follows that the product has the desired form.

Now, a polynomial in $2/3$ with leading coefficient ± 1 cannot vanish, since taking a common denominator yields that the numerator is congruent to $\pm 2^m$ modulo 3. Since every column of the identity matrix has a zero, it follows that no nontrivial word with distinct adjacent terms can equal the identity. The proof for higher dimensions is identical, with $2/3$ replaced by $2/n$. $\square$

Corollary 5.11

(a) *The group generated by the ϕ_i of Theorem 5.10 does not contain a translation other than the identity.*

(b) *The terminal tetrahedron (or n-simplex, $n \geqslant 3$) of a Steinhaus snake is not a translate of the initial one.*

Proof

(a) Suppose a word w in the ϕ_i is a translation. Since four reflections are available, there is an $i \leqslant 4$ such that ϕ_i does not cancel with either end of w.

Now, because ϕ_i has order 2, the linear part of ϕ_i has order 2 also, and therefore the linear part of $\phi_i w \phi_i$ is the identity. This means that $\phi_i w \phi_i$ and w commute, because they are both translations, and this yields the relation $\phi_i w \phi_i w \phi_i w^{-1} \phi_i w^{-1} = 1$, in contradiction to the theorem just proved.

(b) Let the sequence of tetrahedra in the Steinhaus snake be $T_1, \ldots, T_m$. Then $T_2 = \phi_i(T_1)$ for some $i = 1, 2, 3$, or 4, and it can be proved by induction that $T_m = w(T_1)$ for some nontrivial word w in the ϕ_i. For suppose $T_{m-1} = w(T_1)$ and $T_m = \phi_i^*(T_{m-1})$ where ϕ_i^* is the reflection in the w-image of the ith face of T_1. Since w is affine, it is easy to check that $\phi_i^* w$ agrees with $w \phi_i$ on the four vertices of T_1, and hence $\phi_i^* w = w \phi_i$. Therefore $T_m = \phi_i^*(w(T_1)) = w \phi_i(T_1)$, and by the hypothesis that $T_m \neq T_{m-2}$, there is no cancellation in $w \phi_i$.

Now, suppose there is a translation τ of $\mathbf{R}^3$ such that $\tau(T_1) = T_m$. Then, if $T_m = w(T_1)$, the isometry $\tau^{-1} w$ maps T_1 to itself, and so permutes the four vertices of T_1. It follows that $(\tau^{-1} w)^{24} = 1$, whence for some other translation $\tau_0, \tau_0 w^{24} = 1$. But then w^{24} is a translation, contradicting part (a). The same proof works for higher dimensional snakes. $\qquad \square$

NOTES

The first extensive investigation into the situation for higher-dimensional and non-Euclidean spaces was made by Dekker [49, 50, 51], who discovered the locally commutative subgroup of SO_4 given in Theorem 5.2, and derived Corollary 5.3. He also derived Corollary 5.5 for all cases except SO_5, and provided the application to elliptic spaces discussed following Corollary 5.8. Dekker [50] also raised the question whether free non-Abelian subgroups of $G_n, n \geqslant 3$ without nontrivial fixed points on $\mathbf{R}^n$ exist. An affirmative solution was announced by Mycielski [156], but this was premature, and the independent isometries of Theorem 5.7 were discovered shortly thereafter, independently, by Dekker [52] and by Mycielski and Świerczkowski [170]. The two rotations of $\mathbf{R}^3$ at the beginning of 5.7's proof were first considered by de Groot [46], who proved their independence.

Dekker [49] conjectured that the cases he could not settle were similar to the ones he could, and this was confirmed, albeit 25 years later, by Deligne and Sullivan [57] who constructed free non-Abelian groups without fixed points in SO_n for all even $n \geqslant 4$. And Borel [20] confirmed that SO_5 is not an exception as far as locally commutative free groups are concerned by providing such a free group of rank 2 in SO_5. More generally, Borel deals with actions of Lie groups, and he obtains results relating the size of the set of fixed points to the Euler characteristic.

Dekker considered the hyperbolic plane as points on the hyperboloid $x^2 + y^2 - z^2 = -1$, and showed [50] that the group of orientation-preserving isometries is locally commutative and has two independent elements. The two independent isometries were obtained by substituting cosh and sinh for cos and sin in the rotations ϕ and ψ given at the beginning of the proof of Theorem 5.7 (see Theorem 6.4(c)). Dekker [52] posed the question whether the hyperbolic plane (and hence $H^n, n \geqslant 2$) has two independent isometries such that the group they generate has no nontrivial fixed points. As pointed out by Borel [20] (who obtained more general results for actions of Lie groups), this problem is easily solved using the half-plane model and linear fractional transformations. In fact (see [123, p. 15]), the elliptic elements of any discrete subgroup of PSL_2 have finite order, whence no free, discrete subgroup of PSL_2 contains any elliptic elements.

Theorem 5.9 and the decompositions illustrated in Figures 5.1–5.4 are due to the author [171], following a suggestion by Mycielski that tilings of H^2 with respect to discrete groups might be useful in obtaining paradoxical decompositions.

The problem on stacks of tetrahedra is due to Steinhaus. It was originally solved by Świerczkowski, using the group of Theorem 2.1. Dekker [53] simplified Świerczkowski's approach by proving Theorem 5.10 and deriving from it the solution to the Steinhaus problem in all dimensions greater than or equal to three. These results were later rediscovered by Mason [139].

Chapter 6

Free Groups of Large Rank: Getting a Continuum of Spheres from One

The Banach-Tarski Paradox shows how to obtain two spheres, or balls, from one, and it is clear how to get any finite number of balls: just duplicate repeatedly, lifting the subsets of the new balls back to the original. After all, the joy in owning a duplicating machine is being able to use it more than once. Alternatively, one need only consider the weak system of congruences: $A_{2j} \cong \bigcup\{A_i : 1 \leqslant i \leqslant 2n, i \neq 2j - 1\}, j = 1, \ldots, n$. By the remarks following 4.12, there is a partition of the sphere into sets A_i that satisfy this system with respect to rotations, and therefore $A_{2j} \cup A_{2j-1} \sim S^2$. Need we stop at just finitely many copies? What about infinitely many, even uncountably many? Using the existence of infinitely many independent rotations ($\sigma^i \tau \sigma^{-i}, i = 0, 1, 2, \ldots$, where σ, τ are independent), it is not hard to see how the results of Chapter 4 on systems of congruences can be made to yield the solvability of any countably infinite weak system by a partition of S^2. Hence S^2 can be partitioned into countably many sets, each of which is SO_3-equidecomposable (using just two pieces) with S^2. But even stronger transfinite duplications are possible. One can get a continuum of spheres from one: the sphere can be partitioned into sets B_α, as many sets as there are points on the sphere (i.e., $2^{\aleph_0}$), such that each B_α is SO_3-equidecomposable with the sphere.

While transfinite duplications do not add to our knowledge of finitely additive measures, these results do lead to a deeper understanding of free groups of rotations. Moreover, as in the two previous chapters, the techniques necessary for transfinite duplications lead to interesting geometrical applications of a different sort.

The first step is to generalize the results of Chapter 4 on general actions of free groups. Since those results do not depend on the finiteness of the system of congruences, this generalization is quite straightforward.

Theorem 6.1 (AC). *Let κ be any cardinal, and suppose F, a group acting on an infinite set X, is free of rank κ, with free generators σ_α, $\alpha < \kappa$. Let $\bigcup\{A_\beta : \beta \in L_\alpha\} \cong \bigcup\{A_\beta : \beta \in R_\alpha\}$, $\alpha < \kappa$, be any proper system of κ congruences involving at most $|X|$ sets A_β. If F has no nontrivial fixed points on X, then X may be partitioned into sets A_β satisfying the system, with σ_α witnessing the αth congruence. If the system is weak, then a solution exists provided only that F's action on X is locally commutative.*

Proof. The technique by which Theorem 4.5 was derived from Theorem 4.2, and 4.12 from 4.11, makes no use of the finiteness of the system; thus it suffices to show that the generalization of 4.11 is valid. The proof that F can be partitioned to satisfy the system can be done exactly as in 4.11, that is, by induction on the length of words in F. To put 1 and some given word w in the same piece, assuming the system is weak, just consider the group, F_0, generated by the finitely many σ_α that appear in w. By Theorem 4.11, F_0 can be partitioned to satisfy the congruences to be witnessed by its generators (the finiteness of r in 4.11 is unimportant), and 1 and w will be in the same piece. Then words u in $F \backslash F_0$ can be assigned inductively so that all congruences are satisfied, starting the induction with the longest terminal string of u that is in F_0. □

It should be pointed out that in the proof of Theorem 6.1, as in the proofs of the corresponding results of Chapter 4, there is no guarantee that each of the sets of the constructed partition is nonempty. Indeed, the proof never really uses the fact that there are no more than $|X|$ sets A_β, and if there were more, some would have to be empty. This is not a serious problem though, because most of the particular systems of interest imply that the sets are nonempty (for example, the system asserting m-divisibility or the one asserting the existence of a paradoxical decomposition using four pieces). Moreover, if one weakens the conclusion of 6.1 slightly by not insisting that the congruences be witnessed by the specific generators σ_α, then a partition can be found that solves the system and contains only nonempty sets. This can be proved by using the local commutativity of F to find a free subgroup, F_0, of the same rank such that there are $|X|$ F_0-orbits in X, each consisting of nonfixed points of $F_0 \backslash \{1\}$ (for details see [49, Theorem 2.9]).

There is a partial converse to Theorem 6.1 in the style of Theorem 4.8, which showed that if one particular weak system is solvable, then the group must have a locally commutative free subgroup of rank 2. The proof of the following theorem is essentially identical to that of 4.8, with σ_α, $\alpha < \kappa$, defined just as σ, τ are, and replacing σ, τ in the proof.

Theorem 6.2. *If G acts on X and X splits into κ ($\kappa \geqslant 2$) sets, each of which is G-equidecomposable with X using only two pieces, then G has a free subgroup of rank κ that is locally commutative on X.*

The rotation groups, SO_n, all have size $2^{\aleph_0}$, so a free group of rotations cannot have greater rank. In fact, locally commutative free groups of this maximum possible rank exist in SO_n if $n \geqslant 3$, and combined with Theorem 6.1, this fact will yield the desired generalizations of our earlier results on systems of congruences to continuum-sized systems. In Theorem 6.4(a) we shall construct very explicitly a set of $2^{\aleph_0}$ independent rotations of S^2. Then we shall introduce a different approach, more general but also less direct, that yields large locally commutative free subgroups of SO_n for all $n \geqslant 3$. We have seen how transcendental numbers can be used to get two independent rotations of S^2 (see first part of 5.7's proof); in order to construct $2^{\aleph_0}$ independent rotations, we shall need a large set of algebraically independent numbers. Recall that a set X of reals (or complex numbers) is *algebraically independent* if $p(x_1, \ldots, x_n) \neq 0$ for any $x_1, \ldots, x_n \in X$ and polynomial p with rational coefficients that is not identically zero.

Theorem 6.3. *There is a set $\{v_t : t \in (0, \infty)\}$ of algebraically independent numbers.*

Proof. If M is an infinite set of real numbers, then the set of reals that are algebraic over M, that is, the set of all $x \in \mathbf{R}$ for which $p(x_1, \ldots, x_n, x) = 0$ for some rational polynomial p and $x_i \in M$, has the same size as M. This follows from the countability of the number of possibilities for p. Hence an algebraically independent set that is smaller than the continuum can always be extended to a larger algebraically independent set. This allows a continuum-sized set of algebraically independent reals to be constructed by transfinite induction, although this assumes the existence of a well-ordering of the reals. Alternatively, it follows that a maximal algebraically independent set of reals (which exists by Zorn's Lemma) has cardinality $2^{\aleph_0}$.

These approaches use the Axiom of Choice and, while it is true that that axiom will be used in the application of this result to paradoxical decompositions, it is noteworthy that the desired set can be constructed in a much more effective manner that does not require Choice. Letting $[x]$ denote the greatest integer less than or equal to x, define, for any real $t > 0$,

$$v_t = \sum_{n=1}^{\infty} 2^{(2^{[t_n]} - 2^{n^2})}.$$

The series is dominated by a geometric series with ratio $\frac{1}{2}$ once $n > t$, so v_t is a well-defined real; the numbers v_t are sometimes called *von Neumann numbers*, after their discoverer. To get some feeling for v_t consider its binary expansion: when n is large enough, the 2^{n^2} term dominates the exponent, and so the tail of v_t's binary expansion has its 1's spaced very far apart. This is enough to guarantee, in the same way that the transcendence of the Liouville number, $\sum_{n=1}^{\infty} 10^{-n!}$, is proved (see [176, §7.4]), that each v_t is transcendental. Moreover, if $s < t$, then as can easily be seen by choosing two rationals between s and t, $[tn] - [sn] \to \infty$ as $n \to \infty$; hence the nth 1 in v_s's expansion occurs exponentially sooner than the nth 1 in v_t's expansion. This implies that v_t is much more closely approximable by rationals than any v_s, $s < t$, and as proved by von Neumann [244], this yields the algebraic independence of the v_t. □

In [162], Mycielski proved a general theorem, without the Axiom of Choice, about relations in separable metric spaces; this yields an alternative, more abstract, proof of the previous result of von Neumann, which is simpler in that the number theory details are avoided. Mycielski's theorem will be discussed further in Theorem 6.5. One application of von Neumann numbers is to answer the question: can one prove the existence of an uncountable, proper subfield of **R** without using the Axiom of Choice? The subfield generated by $\{v_t : 0 < t < 1\}$ is such a field. We now show how algebraically independent numbers can be used to construct a continuum of independent rotations.

Theorem 6.4

(a) *For any $n \geqslant 3$, there are independent rotations ρ_t, $0 < t < 1$, in SO_n such that the free group they generate is locally commutative on S^{n-1}. If n is even, then the group has no nontrivial fixed points on S^{n-1}.*

(b) *If $n \geqslant 3$, there are independent isometries ρ_t, $0 < t < 1$, of $\mathbf{R}^n$ such that the subgroup of G_n that they generate has no nontrivial fixed points in $\mathbf{R}^n$.*

(c) *For any $n \geqslant 2$, there are independent isometries, ρ_t, $0 < t < 1$, of hyperbolic n-space, H^n, that generate a locally commutative group.*

Proof

(a) We shall use von Neumann numbers to prove (a) in the case $n = 3$. Although this approach works for larger n, a different proof for the general case will be given following Theorem 6.5. Let v_t, $0 < t \leqslant 1$ be the algebraically independent von Neumann numbers of Theorem 6.3. Since SO_3 is locally commutative, it is sufficient to construct $2^{\aleph_0}$ independent rotations. To do this, let $\theta_t = 2 \arctan v_t$; then $\sin \theta_t = 2v_t/(1 + v_t^2)$ and $\cos \theta_t = (1 - v_t^2)/(1 + v_t^2)$. For any θ, let $\sigma(\theta)$, $\tau(\theta)$ denote the two independent rotations in SO_3 as defined at the beginning of Theorem 5.7's proof. Now, the desired independent rotations ρ_t may be defined by $\rho_t = \sigma(\theta_t)\tau(\theta_1)\sigma(\theta_t)^{-1}$. These rotations have the same angle of rotation, θ_1, but have different axes, though the axes all lie in the x-y plane. To prove the independence of the ρ_t, suppose $w = \rho_{t_1}^{m_1} \cdots \rho_{t_s}^{m_s} = 1$. Each entry of w's matrix representation, a_{ij}, is a rational polynomial in $\sin \theta_1$, $\cos \theta_1$ and those $\sin \theta_t$, $\cos \theta_t$ occurring in w, that is, with t equal to one of the t_k. If each polynomial a_{ij} vanished (or equalled 1, in case $i = j$) upon substitution of any angles for the $\theta_t, t < 1$, that occur, then the matrix would remain the identity when each of these θ_t is replaced by $k\theta_1$, k being the least index such that $t_k = t$. Since this substitution transforms each $\rho_{t_k}^{m_k}$ in w to $\sigma(k'\theta_1)\tau(\theta_1)^{m_k}\sigma(k'\theta_1)^{-1}$, which equals $\sigma(\theta_1)^{k'}\tau(\theta_1)^{m_k}\sigma(\theta_1)^{-k'}$, w is transformed to a nontrivial word in $\sigma(\theta_1)^{\pm 1}$, $\tau(\theta_1)^{\pm 1}$ that, by the choice of σ, τ, cannot equal the identity. Hence at least one entry a_{ij} of w is a rational polynomial in $2v_1/(1 + v_1^2)$, $(1 - v_1^2)/(1 + v_1^2)$ and $2v_t/(1 + v_t^2)$, $(1 - v_t^2)/(1 + v_t^2)$, where $t \in \{t_1,\ldots,t_s\}$, that does not vanish (or equal 1, if $i = j$) for arbitrary values of the v_t. It follows that when the equation $a_{ij} = 0$ (or 1) is turned into $p = 0$ by taking a common denominator, p is a nontrivial polynomial in finitely many of the v_t, contradicting the algebraic independence of the von Neumann numbers.

(b) The proof of this case can be derived from Theorem 5.7 in a way similar to part (a), that is, using von Neumann numbers and conjugation. We omit the details since this will be proved following Theorem 6.5 in a more abstract (but overall simpler) way.

(c) It suffices to consider H^2, for the natural embedding of H^2 into H^n (see Theorem 5.9) induces an embedding of isometry groups that never adds fixed points, and hence preserves local commutativity. Moreover, since the action of PSL_2 on the upper half-plane is locally commutative (see p. 68), it suffices to produce a continuum of independent (orientation-preserving) isometries. The construction is simpler if we change the model of the hyperbolic plane from the upper half-plane in $\mathbf{C}$ to points on the upper sheet $(z > 0)$ of the hyperboloid $x^2 + y^2 - z^2 = -1$ in $\mathbf{R}^3$. This model has as its metric $d(P,Q) = \text{arccosh}(-g(P,Q))$ where g denotes the inner product $p_1q_1 + p_2q_2 - p_3q_3$, and isometries may be represented by nonsingular 3×3 matrices A such that

$$A^T \begin{bmatrix} 1 & 0 & 0 \\ 0 & 1 & 0 \\ 0 & 0 & -1 \end{bmatrix} A = \begin{bmatrix} 1 & 0 & 0 \\ 0 & 1 & 0 \\ 0 & 0 & -1 \end{bmatrix},$$

and A sends the upper sheet of the hyperboloid to itself, rather than to the lower sheet (see [14, §3.7]). Now it may be proved exactly as in the first part of Theorem 5.7's proof that ϕ and ψ are independent (orientation-preserving) isometries of H^2, where

$$\phi = \begin{bmatrix} 1 & 0 & 0 \\ 0 & \cosh\theta & \sinh\theta \\ 0 & \sinh\theta & \cosh\theta \end{bmatrix}, \qquad \psi = \begin{bmatrix} \cosh\theta & 0 & \sinh\theta \\ 0 & 1 & 0 \\ \sinh\theta & 0 & \cosh\theta \end{bmatrix}$$

and $\cosh\theta$ is transcendental. We may now use exactly the same technique as in part (a) (i.e., von Neumann numbers and conjugation), using the hyperbolic trigonometric functions and their identities to define a continuum of matrices that correspond to isometries and are independent. $\qquad\square$

We now outline an entirely different approach to the construction of large free groups that is more modern and less computational. This approach can be applied to all the cases of Theorem 6.4. The key is the following theorem of Mycielski.

Theorem 6.5. *Let X be a complete, separable metric space with no isolated points and let $\mathscr{R} = \{R_i : i < \infty\}$ be a set of relations on X, that is, each $R_i \subseteq X^{m_i}$ for some positive integer m_i. Suppose further that each R_i is a nowhere dense subset of X^{m_i}. Then there is a subset F of X such that $|F| = 2^{\aleph_0}$ (in fact, F is perfect, i.e., closed and without isolated points), and for each sequence of m_i distinct elements $x_1, \ldots, x_{m_i}$ of F, $(x_1, \ldots, x_{m_i}) \notin R_i$.*

Proof. We shall construct a tree of nonempty open subsets of X having the following form and satisfying items (1)–(4):

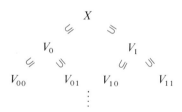

(1) $\bar{V}_{si} \subseteq V_s$ for each sequence s and $i = 0, 1$.
(2) $V_{s0} \cap V_{s1} = \varnothing$ for each s.
(3) If $\ell(s)$ denotes the length of the sequence s, then the diameter of V_s is at most $1/\ell(s)$.
(4) If $s_1, \ldots, s_{m_j}$ are distinct sequences of the same length r, and $j \leqslant r$, then $V_{s_1} \times V_{s_2} \times \cdots \times V_{s_{m_j}} \cap R_j = \varnothing$.

The existence of such a tree yields the desired set F as follows. For each infinite binary sequence b, completeness of X implies that $\bigcap \{\bar{V}_s : s \subseteq b\}$ is nonempty and condition (3) yields that this intersection must contain but a single point, x_b. Let $F = \{x_b : b$ is an infinite binary sequence$\}$; F is perfect and has size $2^{\aleph_0}$. Any m_j-tuple of distinct elements of F comes from distinct nodes at a level past the jth, whence condition (4) implies that the m_j-tuple is not in R_j.

To construct the tree, use induction on levels. Suppose V_s is defined for all s of length n. Since X has no isolated points, V_s contains at least two points, which can be separated by open balls V'_{s0} and V'_{s1}, chosen small enough so that conditions (1)–(3) are satisfied. We shall thin these two balls repeatedly to satisfy (4). Consider each $j \leqslant n + 1$ in turn. If $m_j > 2^{n+1}$, do nothing; (4) is vacuously satisfied with respect to R_j. Otherwise, consider each of the m_j-sized subsets of the 2^{n+1} sequences of length $n + 1$ in turn. If A is such a subset, then $\prod \{V'_{si} : si \in A\}$ is open, and so must contain an m_j-tuple not in $\bar{R}_j$. There must be an open set, U, containing this m_j-tuple that remains disjoint from $\bar{R}_j$, and so we may replace each V'_{si} by a smaller nonempty open set so that $\prod \{V'_{si} : si \in A\} \subseteq U$. This guarantees that (4) is satisfied with respect to R_j and A, the set of distinct sequences. Since we may repeat the thinning process to take care of all sets A and all $j \leqslant n + 1$, this yields the next level of the tree as desired.   □

The preceding theorem is valid for all complete metric spaces, but this assumes the Axiom of Choice. For separable spaces Choice is avoided because there is a countable basis of open sets that can be well-ordered; hence the choices in the proof of Theorem 6.5 can be made by using this well-ordering and choosing the first basic open set that works. This theorem can be strengthened to show that most (i.e., a comeager set in an appropriate topology) perfect subsets of X satisfy the conclusion (see [164]). Note that Theorem 6.5 yields an alternative proof of Theorem 6.3. For each nonzero polynomial $p(x_1, \ldots, x_n)$ with integer coefficients, let R_p be the subset of $\mathbf{R}^n$ consisting of all zeros of p. Then R_p is closed and contains no nonempty open set (otherwise p is identically zero), so R_p is nowhere dense. Since there are only countably many polynomials with integer coefficients, we may apply Theorem 6.5 to the collection of all the sets R_p to get an algebraically independent set of size $2^{\aleph_0}$.

Now, SO_n is a complete, separable metric space, but in order to apply Theorem 6.5, we must view SO_n as a real analytic manifold (of dimension $\frac{1}{2}n(n - 1)$). In fact, SO_n is an analytic submanifold of $\mathbf{R}^{n^2}$ under the embedding induced by considering the entries of the matrix corresponding to an element of SO_n. Furthermore, the product SO_n^m is an analytic submanifold of $\mathbf{R}^{mn^2}$. Hence the mn^2 real-valued functions on SO_n^m, each of which gives the (i, j)th entry in the kth rotation, are analytic. Moreover, SO_n is connected (and path-connected).

Now, let n be even and consider the problem of getting large free groups in SO_n without fixed points. Let w denote a reduced group word ($w \neq 1$) in m variables $x_1, \ldots, x_m$. Define $R_w \subseteq SO_n^m$ to consist of those m-tuples $(\sigma_1, \ldots, \sigma_m)$ such that $w(\sigma_1, \ldots, \sigma_m)$ has 1 as an eigenvalue (i.e., has a fixed point on S^{n-1}), and let $\mathscr{R}$ be the (countable) collection of all such sets R_w. It remains only to show that each R_w is nowhere dense; for then Theorem 6.5 may be applied to obtain a set F, which clearly will be a set of free generators for a subgroup of SO_n without nontrivial fixed points. We need the following claim.

Claim. For any word w as in the preceding paragraph, there is an analytic $f : SO_n^m \to \mathbf{R}$ such that $R_w = f^{-1}(\{0\})$.

Proof. First we observe that $(\sigma_1, \ldots, \sigma_m) \in R_w$ if and only if $\det(w(\sigma_1, \ldots, \sigma_m) - I) = 0$. Because the inverse of a matrix in SO_n equals its transpose, $w(\sigma_1, \ldots, \sigma_m)$ may be expressed as n^2 polynomials (with integer coefficients) in the mn^2 entries of the σ_i. Since the determinant is also a polynomial function in the entries of a matrix, the vanishing of the determinant is equivalent to the vanishing of a single polynomial in the mn^2 entries of the σ_i. Finally, since the functions giving the entries are analytic, it follows that R_w is the zero-set of a single analytic function on SO_n^m. $\square$

This claim immediately yields that R_w is closed, so all that remains is to prove that R_w has empty interior. But it is an easy consequence of the connectedness of SO_n^m that if an analytic function vanishes on a nonempty open set, then it vanishes on all of SO_n^m. Hence if R_w contains a nonempty open set, R_w must equal SO_n^m. But this is a contradiction to Theorem 5.4 which yields independent σ, ρ such that no word in σ, ρ has a fixed point on S^{n-1}. It follows that $(\rho, \sigma\rho\sigma^{-1}, \sigma^2\rho\sigma^{-2}, \ldots, \sigma^{m-1}\rho\sigma^{-(m-1)})$ lies in $SO_n^m \setminus R_w$. (Alternatively, one can avoid Theorem 5.4 and obtain a contradiction by using Theorem 1 of [20]; see [171].)

The case of odd n, where large locally commutative free groups are sought, can be handled in two ways. First observe that the preceding technique easily solves the SO_3 case. Because SO_3 is locally commutative on S^2, all that is needed is a perfect set of free generators, and the technique yields such a set if R_w is defined to be the set of m-tuples $(\sigma_1, \ldots, \sigma_m) \in SO_3^m$ such that $w(\sigma_1, \ldots, \sigma_m)$ is the identity. Now, one can obtain the higher-dimensional result by appealing to [20, p. 162], where it is shown that SO_3 may be represented as a subgroup H of SO_{n+1} $(n \geqslant 2)$ where H's action on S^n is locally commutative. The result then follows from the existence of the perfect set in SO_3.

Alternatively, one can imitate the proof for even n. Let R_w consist of those m-tuples $(\sigma_1, \ldots, \sigma_m) \in SO_n^m$ such that $w(\sigma_1, \ldots, \sigma_m)$ is the identity. Now, let u, v be any two reduced group words in variables $x_1, \ldots, x_r$ that do not commute as abstract words. Then let $R_{u,v}$ consist of all r-tuples $(\sigma_1, \ldots, \sigma_r)$ such that $u(\sigma_1, \ldots, \sigma_r)$ and $v(\sigma_1, \ldots, \sigma_r)$ have a common fixed point on S^{n-1}. Let $\mathscr{R}$ consist of all these sets R_w and $R_{u,v}$. Then $\mathscr{R}$ is countable, and so as in the previous case, it remains to show that each R_w and $R_{u,v}$ is nowhere dense. For then Theorem 6.5 yields a subset F of SO_n that avoids R_w and $R_{u,v}$. The avoidance of R_w means that F is a set of free generators. Hence noncommuting words in F correspond to abstract non-commuting words u, v used to define $R_{u,v}$. Since F avoids $R_{u,v}$, this means that two noncommuting rotations in the group generated by F cannot share a fixed point; that is, this group is locally commutative. By Theorem 5.4, locally commutative free groups of finite rank exist, and it follows that each R_w and $R_{u,v}$ is not equal to all of SO_n^m. Therefore, as in the previous case, the fact that these sets are nowhere dense is an immediate consequence of the following claim and the fact that an analytic function that vanishes on an open set vanishes everywhere.

Claim. The sets R_w and $R_{u,v}$ previously defined are of the form $f^{-1}(\{0\})$ for some analytic f.

Proof. The condition $w(\sigma_1, \ldots, \sigma_m) = 1$ is equivalent to the simultaneous vanishing of n^2 polynomials in the mn^2 entries of the σ_i. This condition is equivalent to the vanishing of a single polynomial by using the trick of summing squares: $\sum p_i^2 = 0$ if and only if each $p_i = 0$. The claim for R_w then follows as in the previous case. Finally, membership of $(\sigma_1, \ldots, \sigma_m)$ in $R_{u,v}$ is equivalent to the existence of a nonzero solution to a homogeneous system of $2n$ equations in n unknowns; the system is the one expressing the fact that u and v have a common eigenvector for the eigenvalue 1. Now, such a system has a nontrivial solution if and only if the determinant of the coefficients in each subsystem of size n vanishes (easy linear algebra exercise). Hence membership in $R_{u,v}$ is equivalent to the simultaneous vanishing of $\binom{2n}{n}$ polynomials in the entries. As before, this is equivalent to the vanishing of a single polynomial, which proves the claim. $\square$

The general technique just discussed can also be used to prove Theorem 6.4(b). We need to work in SG_3, the connected subgroup of G_3 consisting of those isometries whose linear part has determinant $+1$. An element of SG_3 may be viewed as an element of SL_4 as follows. If (a_{ij}) represents the linear part of $\sigma \in SG_3$ and (v_1, v_2, v_3) is the vector corresponding to the translational part, then σ may be identified with

$$\begin{bmatrix} a_{11} & a_{12} & a_{13} & v_1 \\ a_{21} & a_{22} & a_{23} & v_2 \\ a_{31} & a_{32} & a_{33} & v_3 \\ 0 & 0 & 0 & 1 \end{bmatrix}.$$

Hence SG_3 may be viewed as an analytic submanifold of $\mathbf{R}^{12}$ and a proof along the same lines as for spheres can be carried out. The key point is that an element σ of SG_3 has a fixed point if and only if a certain polynomial in the entries of the matrix vanishes. To see that such a polynomial exists, recall that σ has a fixed point if and only if the translation vector is perpendicular to the axis of the rotation given by (a_{ij}). But it is shown in Appendix A (Theorem A.6) that the rotation axis of (a_{ij}) is parallel to $(a_{32} - a_{23}, a_{13} - a_{31}, a_{21} - a_{12})$. Hence σ has a fixed point if and only if

$$v_1(a_{32} - a_{23}) + v_2(a_{13} - a_{31}) + v_3(a_{21} - a_{12}) = 0.$$

All the groups of Theorem 6.4 have rank $2^{\aleph_0}$, and using the Axiom of Choice, they all have free subgroups of any smaller rank. Hence, by applying Theorem 6.1, we get the following corollary. The elliptic spaces are handled by viewing the groups of 6.4(a) as groups of isometries of L^n, as in Chapter 5. Part (a) of Corollary 6.6 for the pair (S^n, O_{n+1}) is proved by combining the locally commutative group of 6.4(a) with the availability of the antipodal map as in 4.15 and 5.5 That technique requires that no element of the group sends a point P to $-P$; that is, no element has -1 as an eigenvalue. Since this condition is satisfiable for groups of finite rank (see the proof of Corollary 5.5), it may be built into the proof of 6.4(a) that uses Theorem 6.5. Simply add to $\mathcal{R}$ the relations R'_w, consisting of m-tuples such that $w(\sigma_1, \ldots, \sigma_m)$ has -1 as an eigenvalue. Since membership in R'_w may be expressed as a polynomial, the crucial claim remains valid.

Corollary 6.6 (AC)

(a) *Let (X, G) be any of the following pairs: (S^n, SO_{n+1}), where $n \geq 2$ and n is odd; (S^n, O_{n+1}), where $n \geq 2$; $(\mathbf{R}^n, G_n)$, where $n \geq 3$; L^n and its isometry group, where $n \geq 2$ and n is odd. Then any proper system of at most $2^{\aleph_0}$ congruences involving at most $2^{\aleph_0}$ sets is solvable, using G, by a partition of X.*

(b) *The same is true when restricted to weak systems in the following cases: (S^n, SO_{n+1}), L^n and its isometry group, and H^n and its isometry group, where $n \geq 2$.*

(c) *In all the cases of (a) and (b), X may be partitioned into $\{A_\alpha : \alpha < 2^{\aleph_0}\}$ such that $A_\alpha \sim_G X$ using two pieces.*

For spheres and Euclidean, hyperbolic, and elliptic spaces, this corollary shows that in all cases where the set or space is paradoxical, one can get a continuum of copies of the set from one.

A consequence of the analysis of the sets R_w in the preceding approach to large free groups is finer information about free groups of finite rank. Consider the case of SO_n, $n \geq 4$ and n even. Theorem 5.4 shows that free non-Abelian subgroups without nontrivial fixed points exist, but in fact the set of pairs that generate such subgroups is dense in the space of all possible pairs. This is because each R_w (see page 80), where w is a word in two variables, is nowhere dense. Hence the union of these R_w is meager in $SO_n \times SO_n$ and the complement of this union is comeager and, by the Baire Category Theorem, dense. Similarly, the fact that R_w and $R_{u,v}$ are nowhere dense yields that for any $n \geqslant 3$, the set of pairs (σ, τ) in $SO_n \times SO_n$ such that σ and τ freely generate a locally commutative subgroup of SO_n is comeager. So, while the specific examples of independent pairs of isometries constructed by Hausdorff et al. are necessary to establish these results, we see that such pairs are quite abundant.

While the remarks of the previous paragraph are valid in $\mathbf{R}^3$ as well, the situation in $\mathbf{R}^4$ and beyond is less clear. The proof of Theorem 6.4(b) uses the fact that the existence of a fixed point of an element of SG_3 is a polynomial condition; it is not clear that this is true for SG_4. Thus Theorem 6.4(b) is valid for all $n \geqslant 3$, but it is not known that the set of pairs from G_4 that freely generate a group without nontrivial fixed points is comeager. Moreover, the technique used by Borel [20] does not apply, since his results are for semisimple groups.

The hyperbolic case provides an interesting counterpoint as far as large free groups without nontrivial fixed points are concerned, as well as for the related question on the density of independent pairs. Returning to the upper half-plane model of hyperbolic 2-space, H^2, recall that the orientation-preserving isometries correspond to linear fractional transformations that we identify with PSL_2 and that may be classified as parabolic, hyperbolic, or elliptic; the latter are the only ones with fixed points in H^2. Another useful classification is the discrete/nondiscrete classification of subgroups of PSL_2. A subgroup G of PSL_2 is *discrete* if the collection of matrices corresponding to elements of G contains no convergent sequence of distinct matrices. An easy fact (see [123, p. 12]) is that a cyclic subgroup of PSL_2 is discrete if and only if it contains no elliptic element of infinite order. Also, note that a discrete subgroup must be countable.

If a free subgroup of PSL_2 is discrete, it can have no elliptic elements, for such an element would be of infinite order and so would generate a nondiscrete subgroup. It is essentially this fact that was used in Chapter 5 to show that any free subgroup of $PSL_2(\mathbf{Z})$ has no elliptic elements. It follows from work of Siegel [209] that the converse of this fact holds too.

Theorem 6.7.   *A free subgroup of PSL_2 is discrete if and only if it has no elliptic elements. Hence any group of isometries of H^2 generated by an uncountable set of independent elements contains an isometry fixing a point in H^2.*

Proof.   The forward direction was already proved, and the reverse direction for cyclic free groups is a consequence of the fact about cyclic subgroups of PSL_2 previously stated. We sketch the proof of the remaining case, noncyclic free groups, making use of several elementary facts about PSL_2

such as: (1) two transformations commute if and only if they have the same set of fixed points in $\mathbf{C} \cup \{\infty\}$, and (2) if u, w are hyperbolic with exactly one common fixed point x, then $uwu^{-1}w^{-1}$ is parabolic and fixes x (see [123, §I.1]).

Now, suppose G has no elliptic elements, is nondiscrete, and has rank at least two (with a free basis containing, say, σ and τ). First, we claim that G has two elements u, w such that each is near the identity, u and w do not share a fixed point in $\mathbf{C} \cup \{\infty\}$, and u is hyperbolic. To prove this, choose u to be near the identity, and conjugating u by some σ^n if necessary, assume that $u = \sigma^{\pm 1} \cdots \sigma^{\pm 1}$. Let $w = \tau u \tau^{-1}$. Then u and w have different leftmost terms and so are not powers of a common word; hence (see the proof of Theorem 4.5) they do not commute. If u is hyperbolic, so is w and they cannot share a fixed point. If they did, $uwu^{-1}w^{-1}$ and $wuw^{-1}u^{-1}$ would each be parabolic fixing the same point, and hence would commute, contradicting the fact that they have different leftmost terms (there is no cancellation in the formation of these commutators). If u is parabolic, so is w and we may replace u by uw or uw^2, one of which will be hyperbolic, and start again; that is, conjugate u by σ^n if necessary, let $w = \tau u \tau^{-1}$, and use the previous argument. Thus, in either case, we get a pair u, w as claimed.

Finally, using the fact that u, w are near the identity and assuming, conjugating if necessary, that u is represented by $\begin{bmatrix} a & 0 \\ 0 & a^{-1} \end{bmatrix}$, it is easy to check by a computation that either $v = uwu^{-1}w^{-1}$ or $v_1 = uvu^{-1}v^{-1}$ is elliptic.

If M is a set of independent isometries of H^2, then so is $\{\sigma^2 : \sigma \in M\}$, and each σ^2 is orientation-preserving. This reduces the second assertion to a statement about PSL_2, which follows from the first part of the theorem since an uncountable subgroup of PSL_2 is necessarily nondiscrete.   □

With more work one can improve Theorem 6.7 to: A nonsolvable, nondiscrete subgroup of SL_2 is dense in SL_2. This result yields (the reverse direction of) Theorem 6.7 as a corollary since the elliptics ($-2 < $ trace < 2) form a nonempty open subset of SL_2 and since a free group of rank 2 is not solvable. The stronger theorem, which was pointed out to the author by A. Borel and D. Sullivan, is proved as follows. Let F be the subgroup in question, let $\bar{F}$ denote the closure of F in SL_2 and let H be the component of the identity in $\bar{F}$. Assume, to get a contradiction, that $\bar{F} \neq SL_2$. Since F is nondiscrete, $H \neq \{1\}$, whence H is a one- or two-dimensional Lie subgroup of SL_2. Using Lie algebras it can be shown that H is conjugate to either SO_2 or a subgroup of the group of upper triangular matrices. In either case, it is easy to check that the normalizer of H in SL_2 is solvable, contradicting the fact that the nonsolvable group F normalizes H.

Theorem 6.7 is in direct contrast to the cases of Theorem 6.4(a) and (b) where uncountable free groups without fixed points are constructed for other spaces. The general technique, using nowhere dense sets and analytic functions, used to deduce Theorem 6.4(a) and (b) from the corresponding results about free groups of rank 2 might lead one to expect that whenever there is a pair of independent elements generating a group without fixed points, there is in fact continuum many such elements. But this is false in the case of H^2. Where does the general proof break down? The elliptic transformations in PSL_2 form an open set ($|\text{trace}| < 2$) and are therefore not nowhere dense. Unlike the spherical and Euclidean cases, the (orientation-preserving) isometries of H^2 with a fixed point are not the zero set of an analytic function. The statement that $\begin{bmatrix} a & b \\ c & d \end{bmatrix}$ corresponds to an isometry with a fixed point is expressed by a polynomial inequality ($(a + d)^2 < 4$) not by a polynomial equality, as happens in SO_n and in the group of orientation-preserving isometries of $\mathbf{R}^3$. As pointed out following Figure 5.4, this difference is also reflected in the fact that the set of pairs from PSL_2 that generate a free group without nontrivial fixed points in H^2 is not dense.

Concerning higher dimensional hyperbolic spaces, it is easy to see that in the isometry group of H^n, $n \geq 4$, there do exist free groups of continuum rank without nontrivial fixed points in H^n. This is because the action of the Euclidean isometry group G_{n-1} on $\mathbf{R}^{n-1}$ can be mimicked in H^n. If $\sigma \in G_{n-1}$, then send the point $(x_1, \ldots, x_{n-1}, t)$ of H^n (using the upper half-space model) to $(\sigma(x_1, \ldots, x_{n-1}), t)$. This yields an isometry of H^n that, assuming σ fixes no point of $\mathbf{R}^{n-1}$, has no fixed points. Hence the assertion about H^n, $n \geq 4$, follows from Theorem 6.4(b), the corresponding result about $\mathbf{R}^n$, $n \geq 3$.

The case of H^3, however, requires a more subtle approach. The action of $PSL_2(\mathbf{C})$ on $\mathbf{C} \cup \{\infty\}$ as linear fractional transformations can be extended to $\mathbf{R}^3 \cup \{\infty\}$ by identifying $\mathbf{C} \cup \{\infty\}$ with $\mathbf{R}^2 \cup \{\infty\}$ and using some geometry related to the fact that each linear fractional transformation is a Möbius transformation of $\mathbf{R}^2 \cup \{\infty\}$. Such an extension preserves H^3 and corresponds to the group of orientation-preserving isometries of H^3. Alternatively, one can use an approach based on quaternions; see [14] for both approaches. It turns out that the orientation-preserving isometries of H^3 without a fixed point correspond (except for the identity) to matrices whose trace is real and lies in the interval $(-2, 2)$. Such elements of $PSL_2(\mathbf{C})$ are called *elliptic*.

Theorem 6.8. *There is a free subgroup of $PSL_2(\mathbf{C})$ of rank $2^{\aleph_0}$ that contains no elliptic element. Hence there is a rank $2^{\aleph_0}$ free group of isometries of H^3 whose action on H^3 is without nontrivial fixed points.*

Proof. There is no loss in working in $SL_2(\mathbf{C})$ rather than $PSL_2(\mathbf{C})$. For a nonidentity reduced word w, let

$$R_w = \{(\sigma_1, \ldots, \sigma_m) \in SL_2(\mathbf{C})^m : w(\sigma_1, \ldots, \sigma_m) \text{ is elliptic}\}.$$

As in the previous proofs using 6.5, it is sufficient to prove that R_w is nowhere dense. The complication is that R_w is not the zero-set of a polynomial function of the $8m$ real numbers giving the entries of the σ_i. Rather, if $a_1, \ldots, a_{8m}$ are these reals, then $(\sigma_1, \ldots, \sigma_m) \in R_w$ if and only if $p(a_1, \ldots, a_{8m}) = 0$ and

$q(a_1, \ldots, a_{8m}) < 4$, where the polynomial p is an expression for the imaginary part of the trace while q equals the square of the real part of the trace.

Now, suppose R_w fails to be nowhere dense. Then the closed set $\{(\sigma_1, \ldots, \sigma_m) : p(a_1, \ldots, a_{8m}) = 0\}$, which contains R_w, must contain a non-empty open set. Since $SL_2(\mathbf{C})^m$ is a connected, real analytic submanifold of $\mathbf{R}^{8m}$, this means that p must be identically zero on $SL_2(\mathbf{C})^m$. But this means that the trace of $w(\sigma_1, \ldots, \sigma_m)$ is real for all choices of $\sigma_1, \ldots, \sigma_m$, which leads to a contradiction as follows.

We shall need the fact, due to Magnus and Neumann [132, 174] (see remarks following Proposition 7.1), that the two matrices $\rho = \begin{bmatrix} 1 & 1 \\ 1 & 2 \end{bmatrix}$ and $\tau = \begin{bmatrix} 5 & 2 \\ 2 & 1 \end{bmatrix}$ are free generators of a subgroup of SL_2 that, except for the identity, consists only of hyperbolic elements; in other words, the absolute value of the trace of any nontrivial word in ρ, τ is greater than 2. Returning to the problem at hand, define for each $z \in \mathbf{C}$,

$$\rho_z = \frac{1}{1 + z - z^2} \begin{bmatrix} 1 & z \\ z & 1 + z \end{bmatrix} \quad \text{and} \quad \tau_z = \frac{1}{1 + 4z - 4z^2} \begin{bmatrix} 1 + 4z & 2z \\ 2z & 1 \end{bmatrix}.$$

Then define the complex function

$$f(z) = \text{trace}\big(w(\rho_z, \tau_z \rho_z \tau_z^{-1}, \ldots, \tau_z^{m-1} \rho_z \tau_z^{-(m-1)})\big).$$

Then $f(z)$ is a rational function, whose denominator has four zeros. Let Ω be a region containing 0 and 1 but excluding these four zeros. Then f is analytic in Ω, so by the Open Mapping Theorem, $f(\Omega)$ either is just a single point or contains a nonempty open set. If the trace of $w(\sigma_1, \ldots, \sigma_m)$ is always real, only the first possibility can hold. But this is the desired contradiction because $f(0) = \text{trace}(I) = 2$ while $f(1)$ is the trace of a reduced nonidentity word in ρ and τ, whence $f(1) \neq 2$. $\qquad \square$

Figure 6.1 summarizes the results about large free groups. Note that the lack of an uncountable nonelliptic free group for H^2 leaves unanswered the question whether uncountable systems of congruences are solvable. A specific case of this problem is discussed on page 88.

Corollary 6.6(b) yields the κ-divisibility of S^2 with respect to SO_3 whenever $3 \leqslant \kappa \leqslant 2^{\aleph_0}$, and this can be combined with splittings of S^1 to obtain the κ-divisibility of higher dimensional spheres, without having to use free locally commutative groups in the higher-dimensional rotation groups.

Corollary 6.9 (AC).   *For any cardinal κ satisfying $3 \leqslant \kappa \leqslant 2^{\aleph_0}$ and any $n \geqslant 1$, S^n is κ-divisible with respect to SO_{n+1}.*

Proof.   For S^2, this follows by applying 6.6(b) to the weak system $A_0 \cong A_\alpha, \alpha < \kappa$. Once the case of S^1 is handled, the divisibility of $S^n, n > 2$, follows by combining the partitions of S^1 and S^2, as was done in Theorem 5.6 (or, apply Corollary 6.6(b) for larger n). For S^1, the result is obvious if $\kappa < \aleph_0$,

X	Spheres (S^n)		Elliptic spaces (L^n)	Hyperbolic spaces (H^n)	Euclidean spaces ($\mathbf{R}^n$)
G	Orientation-preserving isometries (SO_{n+1})	Isometries (O_{n+1})	Isometries	Isometries	Isometries (G_n)
X is G-equidecomposable with $2^{\aleph_0}$ copies of X using four pieces for each copy $\Leftrightarrow$ G has a free locally commutative subgroup of rank $2^{\aleph_0}$ $\Leftrightarrow$ all weak $2^{\aleph_0}$-sized systems of congruences are solvable.	2, 3, 4, …	2, 3, 4, …	2, 3, 4, …	2, 3, 4, …	3, 4, 5, …
G has a free subgroup of rank $2^{\aleph_0}$ without nontrivial fixed points on X.	3, 5, 7, …	3, 5, 7, …	3, 5, 7, …	3, 4, 5, …	3, 4, 5, …
All proper $2^{\aleph_0}$-sized systems of congruences are solvable.	3, 5, 7, …	2, 3, 4, …	3, 5, 7, …	2:? 3, 4, 5, …	3, 4, 5, …

Figure 6.1. Except for the one open problem, the table is complete in that the nonappearance of a certain dimension means that the phenomenon is known not to exist.

and the $\aleph_0$-divisibility of S^1 is implicit in the classical construction of a non-Lebesgue measurable subset of the circle: the sets M_i in the proof of Theorem 1.5 are pairwise rotationally-congruent. For $\aleph_0 < \kappa < 2^{\aleph_0}$, the same proof as for $\kappa = \aleph_0$ works, except that the countable subgroup of rotations through rational multiples of π must be replaced by a κ-sized subgroup of SO_2. Such a subgroup is easily obtainable by taking any collection of κ rotations of the circle and considering the group they generate.   □

This result seems to indicate that all spheres have the same divisibility properties (into three or more pieces), but in fact each S^n, $n \geqslant 2$, has a divisibility property that S^1 does not. It is a consequence of Corollary 6.6(a) that if $n \geqslant 2$, then S^n has a subset E such that, for each cardinal κ with $2 \leqslant \kappa \leqslant 2^{\aleph_0}$, S^n is divisible into κ pieces, each of which is O_{n+1}-congruent to E. In other words, E is a half of S^n and a third of S^n and $\cdots$ and a $2^{\aleph_0}$th part of S^n! To obtain E simply choose a partition of S^n into sets A_α, $1 \leqslant \alpha \leqslant 2^{\aleph_0}$, satisfying the following proper system of congruences:

$$A_1 \cong A_\alpha, \alpha < 2^{\aleph_0}$$

$$A_1 \cong \bigcup \{A_\beta : \kappa \leqslant \beta \leqslant 2^{\aleph_0}\}, \quad \kappa \text{ a cardinal and } 2 \leqslant \kappa \leqslant 2^{\aleph_0}.$$

Then let $E = A_1$. For any κ between 2 and $2^{\aleph_0}$ inclusive, S^n splits into the sets A_α, where $1 \leqslant \alpha < \kappa$, and the set $\bigcup \{A_\beta : \kappa \leqslant \beta \leqslant 2^{\aleph_0}\}$, all of which are congruent to E. If one deletes the single congruence $A_1 \cong A_2 \cup A_3 \cup \cdots$, the resulting system is weak (see [155]) and is therefore solvable in all the cases of Theorem 6.4. Hence there is a subset of S^2, for example, that is simultaneously a third, a quarter, ..., a $2^{\aleph_0}$th part of S^2 with respect to the group of rotations. Since a finitely additive, O_2-invariant measure on S^1 of total measure one exists (Corollary 10.9), such a subset of the circle cannot exist: for $n < \infty$ an nth part of S^1 would have to have measure $1/n$, and for $\kappa \geqslant \aleph_0$ a κth part would have measure zero. Hence S^1 does not have this simultaneous divisibility property that the higher-dimensional spheres do.

By the preceding remarks, H^2 has a subset that is simultaneously a κth part of H^2 for all κ satisfying $3 \leqslant \kappa \leqslant 2^{\aleph_0}$. And, because PSL_2 has a non-elliptic free subgroup of rank $\aleph_0$, H^2 has a subset that is a κth part of H^2 for all κ satisfying $2 \leqslant \kappa \leqslant \aleph_0$. But it is not known if there is a set that is simultaneously a half and a $2^{\aleph_0}$th part of H^2; because of Theorem 6.7, this cannot be proved by appealing to a nonelliptic free group of rank $2^{\aleph_0}$, since such do not exist in PSL_2.

LARGE FREE SEMIGROUPS OF ISOMETRIES

The isometry group of the Euclidean plane, being solvable, does not contain any free non-Abelian subgroup, but we have seen that free subsemigroups of rank 2 exist. In the Sierpiński-Mazurkiewicz Paradox (1.7–1.9) such a subsemigroup was used to construct a nonempty paradoxical subset of the plane. The von Neumann numbers (or any set of algebraically independent numbers) allow us to define a free semigroup of isometries of much larger rank, obtaining a refinement of the Sierpiński-Mazurkiewicz Paradox that still avoids the Axiom of Choice. First, we give the straightforward generalization of Proposition 1.9.

Proposition 6.10. *Suppose a group G acts on X, and G contains a (necessarily free) semigroup generated by σ_α, $\alpha \in I$, with the following property: for some $x \in X$, any two nontrivial words in the σ_α with different leftmost terms yield different elements of X when applied to x. Then some nonempty subset, E, of X contains pairwise disjoint sets A_α, $\alpha \in I$, such that $\sigma_\alpha^{-1}(A_\alpha) = E$.*

Proof. Let S be the subsemigroup generated by the σ_α, and let E be the S-orbit of x. Then the hypothesis guarantees that the sets $\sigma_\alpha(E)$ are pairwise disjoint subsets of E, and $\sigma_\alpha^{-1}(\sigma_\alpha(E)) = E$. The fact that the hypothesis on the σ_α implies freeness is proved in the same way as in 1.8.   □

Theorem 6.11

(a) *There are isometries of $\mathbf{R}^2$, $\{\sigma_t : 0 < t \leqslant 1\}$, such that the subsemigroup of G_2 that they generate satisfies the hypothesis of Proposition 6.10, with $x = (0,0)$.*

(b) *There is a nonempty subset E of $\mathbf{R}^2$ that may be partitioned into a continuum of sets, each of which is congruent to E.*

Proof

(a)   As usual, let $\theta_t = 2 \arctan v_t$, $0 < t \leqslant 1$, where the v_t are the von Neumann numbers. Let u_t be the complex number $e^{i\theta_t}$; we identify $\mathbf{R}^2$ with $\mathbf{C}$ to simplify the proof. Note that the u_t are algebraically independent. For if $P = 0$, where P is a rational polynomial in finitely many of the u_t, then since $u_t = (1 - v_t^2 + 2v_t i)/(1 + v_t^2)$, we may take a common denominator to get $Q = 0$ where Q is a polynomial over $\mathbf{Q}(i)$ in the corresponding v_t. Q cannot be identically 0, for otherwise P would vanish when each of its variables is set equal to u_1, contradicting the transcendence of u_1 (which follows from the transcendence of its real part). Hence the v_t satisfy an algebraic relation over $\mathbf{Q}(i)$ and, since i is algebraic, over $\mathbf{Q}$ as well, a contradiction.

For $0 < t < 1$, let $\sigma_t(z) = u_t z + u_t$, and let $\sigma_1(z) = u_1 z$; these complex functions correspond to isometries of $\mathbf{R}^2$. Suppose $w_1 = \sigma_{t_1} \cdots \sigma_{t_r}$ and $w_2 = \sigma_{s_1} \cdots \sigma_{s_l}$ are two nontrivial words in σ_t, $0 < t \leqslant 1$, with $t_1 \neq s_1$; hence we may assume $t_1 \neq 1$. Then $w_1(0)$ and $w_2(0)$ are polynomials in the u_{t_i}, u_{s_i}, respectively, and $w_1(0)$ has but a single term of degree 1, u_{t_1}, while $w_2(0)$ has no such term, if $s_1 = 1$, or also a single such term, u_{s_1}. Therefore, if $w_1(0) = w_2(0)$, subtraction would yield a polynomial relation among finitely many of the algebraically independent numbers u_t, a contradiction.

(b) Apply Proposition 6.10 to the semigroup of part (a). The sets $\sigma_\alpha(E)$ form, in general, a partition of $E \setminus \{x\}$. But in this particular case, $\sigma_1(0,0) = (0,0)$, so $(0,0) \in \sigma_1(E)$ and we really do have a partition of E. $\square$

Another question raised regarding the Sierpiński-Mazurkiewicz Paradox was whether there could be an uncountable subset of the plane that is paradoxical using two pieces; the example in Theorem 1.7 is countable. Here, too, von Neumann numbers allow the desired extension to be obtained without the Axiom of Choice.

Theorem 6.12.   *There is a subset of the plane, with the cardinality of the continuum, that is paradoxical using two pieces. Assuming the Axiom of Choice, such sets exist having any infinite cardinality* $\leqslant 2^{\aleph_0}$.

Proof.   Let u_t, $0 < t \leqslant 1$, be the algebraically independent complex numbers of modulus 1 defined from the von Neumann numbers in Theorem 6.11 (a); let u denote u_1. Let $\sigma(z) = uz$ and $\tau(z) = z + 1$, and let S be the subsemigroup of G_2 (again, $\mathbf{R}^2$ is identified with $\mathbf{C}$) generated by σ and τ. Then the desired set E may be defined as

$$E = \{w(z) : w \in S \text{ and } z = \sigma^{-k}(u_t), 0 < t < 1, k = 0, 1, 2, \ldots\}.$$

Clearly, $E \supseteq \sigma(E) \cup \tau(E)$ and $\sigma^{-1}(\sigma(E)) = E = \tau^{-1}(\tau(E))$. Moreover, since each $\sigma^{-k}(u_t) = \sigma(\sigma^{-(k+1)}(u_t)) \in \sigma(E)$, $E = \sigma(E) \cup \tau(E)$. It remains to prove that $\sigma(E)$ and $\tau(E)$ are disjoint, but it is easy to see, since words w correspond to polynomials with coefficients of the form mu^n, m, $n \in \mathbf{N}$, that an equality $\sigma w_1 \sigma^{-k}(u_t) = \tau w_2 \sigma^{-\ell}(u_s)$ implies a nontrivial polynomial relation among u, u_t, and u_s (or just u, u_t, if $s = t$), in violation of their algebraic independence. Assuming Choice, $(0, 1)$ may be well ordered in type $2^{\aleph_0}$ and so we need only replace $\{u_t : 0 < t < 1\}$ by any κ-sized subset (κ infinite) to get an example of size κ. $\square$

SETS CONGRUENT TO PROPER SUBSETS

To conclude this chapter, we discuss some geometric problems whose solutions are closely connected with free groups of isometries. These problems arise from the possibility of a set being congruent to a proper subset. To summarize some background about this notion, recall that we have already seen, and made much use of, the fact that a subset of the circle can be congruent to a proper subset: $D = \{\rho^n(1,0) : n \in \mathbf{N}\}$ is congruent to $D \setminus \{(1,0)\}$ if ρ is a rotation about the origin of infinite order. In general, call a set in any metric space *monomorphic* if it is not congruent to a proper subset. Recall that congruence may be witnessed by partial isometries: A is congruent to B if there is a distance-preserving bijection from A to B. In any compact metric space, an isometry from A to B can always be extended uniquely to one from $\bar{A}$ to $\bar{B}$.

Now, it is easy to see that all bounded subsets of the real line are monomorphic although, of course, $\mathbf{N}$ is congruent to $\mathbf{N}\backslash\{0\}$. As for the plane, we have the bounded example, D, of a monomorphic subset of $\mathbf{R}^2$. An interesting result of Lindenbaum shows that topologically, the example D is as simple as possible. Since D is countable, it is an F_σ set, although it is not a G_δ: D is dense in the circle, so any G_δ set that contains D must contain the whole circle. In fact, no such (bounded) example exists that is simultaneously an F_σ and a G_δ. Following the notation of descriptive set theory for the Borel hierarchy, let $\underset{\sim}{\Delta}{}^0_1$, denote sets that are both F_σ and G_δ sets.

Theorem 6.13. *Any $\underset{\sim}{\Delta}{}^0_1$ subset of a compact metric space is monomorphic; hence any bounded $\underset{\sim}{\Delta}{}^0_1$ subset of $\mathbf{R}^n$ is monomorphic.*

Proof. First we prove that closed sets in a compact space, X, are monomorphic. Suppose $\sigma(A) \subseteq A$, where σ is a distance-preserving function on A. For $a \in A$, let $\bar{\sigma}(a) = \{\sigma^n(a) : n \in \mathbf{Z}$ and $\sigma^n(a)$ is defined$\}$. If $\bar{\sigma}(a)$ is finite, then $a = \sigma^n(a)$ for some $n > 0$, whence $a \in \sigma(A)$. If $\bar{\sigma}(a)$ is infinite, then by compactness of X, $\bar{\sigma}(a)$ has a limit point. It follows that for any $\varepsilon > 0$, there are distinct $m, n \in \mathbf{Z}$ with $d(\sigma^m(a), \sigma^n(a)) < \varepsilon$, whence, assuming $m < n$, $d(a, \sigma^{n-m}(a)) < \varepsilon$. This yields that a is a limit point of $\{\sigma^n(z) : n > 0\}$ and since $\sigma(A)$, which is congruent to a closed set, must itself be closed, $a \in \sigma(A)$. We have proved that $A \subseteq \sigma(A)$, and therefore that A is monomorphic.

For any subset A of X define the *residue* of A, A_R, to be $A \cap \overline{\bar{A}\backslash A}$. If $\sigma(A) = B \subseteq A$ where σ is a distance-preserving function on A (which, by compactness, is assumed to be defined on $\bar{A}$), then we claim that $\sigma(A_R) \subseteq A_R$ and $A\backslash B \subseteq A_R\backslash\sigma(A_R)$. By the result on closed sets just proved, $\sigma(\bar{A}) = \bar{B}$ cannot be a proper subset of $\bar{A}$; therefore $\bar{A} = \bar{B}$ and $\bar{B}\backslash B = \bar{A}\backslash B \supseteq \bar{A}\backslash A$, whence $\overline{\bar{B}\backslash B} \supseteq \overline{\bar{A}\backslash A}$. But σ takes $\bar{A}\backslash A$ to $\bar{B}\backslash B$, and hence also takes $\overline{\bar{A}\backslash A}$ to $\overline{\bar{B}\backslash B}$; since these sets are closed, this means $\overline{\bar{B}\backslash B} = \overline{\bar{A}\backslash A}$. The fact that σ is a homeomorphism from $\bar{A}$ to $\bar{B}$ yields that $\sigma(A_R) = (\sigma(A))_R$, whence $\sigma(A_R) = B_R = B \cap \overline{\bar{B}\backslash B} \subseteq A \cap \overline{\bar{A}\backslash A} = A_R$. For the second part of the claim, $A\backslash B \subseteq \bar{A}\backslash B = \bar{B}\backslash B \subseteq \overline{\bar{B}\backslash B} = \overline{\bar{A}\backslash A}$, so $A\backslash B \subseteq A_R$. This proves the claim since $A\backslash B$ is disjoint from $\sigma(A)$, and hence from $\sigma(A_R)$. (In fact, it can be shown that $A\backslash B = A_R\backslash\sigma(A_R)$.)

Now, for any set A we may define a sequence of residues: $A_0 = A$, $A_{\alpha+1} = (A_\alpha)_R$, and $A_\gamma = \bigcap_{\alpha<\gamma} A_\alpha$ for limit ordinals γ. There must be some ordinal β such that $A_\beta = A_{\beta+1} = \dots$. The Baire Category Theorem, which is valid in complete, and hence in compact, metric spaces, can be used to show that if A is $\underset{\sim}{\Delta}{}^0_1$, then $A_\beta = \emptyset$ (see [91, §30] or [117, §§12, 34]). (The converse is true in all metric spaces. If A_β vanishes, then $A \in \underset{\sim}{\Delta}{}^0_1$; see

[91, §30].) The claim shows that a counterexample to the fact that A is monomorphic can be carried through the successor stages of the sequence of residues, and it is easy to cross the limit stages as well. Since the empty set is obviously monomorphic, A must be too.   □

The two examples, $\mathbf{N}$ and D, of sets congruent to a proper subset led Sierpiński to wonder if a set could contain two or more points, each of whose deletion, separately, leaves a set congruent to the original. He proved the following result for the line; the case of $\mathbf{R}^2$, which is somewhat more complicated and whose proof we omit, was proved by Straus.

Theorem 6.14.   *There is no subset E of either $\mathbf{R}^1$ or $\mathbf{R}^2$ such that for two distinct points P and Q in E, E is congruent to $E \backslash \{P\}$ and to $E \backslash \{Q\}$.*

Proof.   We give only the proof for $\mathbf{R}^1$, and refer the reader to [225] for $\mathbf{R}^2$. Suppose $\sigma(E) = E \backslash \{P\}$ and $\tau(E) = E \backslash \{Q\}$, where $\sigma, \tau \in G_1$ and $P \neq Q$. Since reflections have order 2, and hence their squares map E to E, σ and τ must be translations, say by a, b, respectively. Since $Q \in E \backslash \{P\} = E + a$ and $a \neq 0$, $Q - a \in E \backslash \{Q\} = E + b$. Then $Q - a - b \in E$; it follows that $Q - b \in E + a \subseteq E$ and $Q \in E + b = E \backslash \{Q\}$, a contradiction.   □

This proof uses the commutativity of translations or, more precisely, the fact that the equation $x^2 y^2 x^{-2} y^{-2} = 1$ is universally satisfied in G_1. The proof for $\mathbf{R}^2$ likewise uses a universal equation, namely $[[x^2, y^2], [x^2, y^2]] = 1$ where $[,]$ denotes the commutator (see Appendix A), but also uses the geometric form of elements of G_2. That the existence of a universal equation does not suffice is shown by the following example. Let G be the group generated by $\begin{bmatrix} 2 & 0 \\ 0 & 1 \end{bmatrix}$ and $\begin{bmatrix} 1 & 0 \\ 0 & 2 \end{bmatrix}$ and let $E = \{(0, 2^n), (2^n, 0) : n \in \mathbf{N}\}$. Then $E \backslash \{(0, 1)\}$ and $E \backslash \{(1, 0)\}$ are G-congruent to E despite the fact that G is Abelian. Nonetheless, in free groups of rank 2 or more, where of course no nontrivial equation is universally satisfied, sets congruent to any maximal proper subset do exist. This fact was used by Mycielski to construct a subset of $\mathbf{R}^3$ that is congruent to the remainder after the deletion of any finite set of points. This construction is similar to that of paradoxical decompositions, although the Axiom of Choice is not needed; first show that the set exists in a free group, and then transfer it to a set on which the group acts without fixed points.

Definition 6.15.   *If G acts on X, then a nonempty subset E of X is a Sierpiński set if for any finite D there is $\sigma \in G$ such that $\sigma(E) = E \backslash D$.*

Theorem 6.16

(a) *A free group F of rank 2 has a Sierpiński set (with respect to its action on itself by left multiplication), as does any group with a subgroup isomorphic to F.*

(b) *If F, as in (a), acts on X without nontrivial fixed points, then X has a Sierpiński set.*

(c) *Any sphere in $\mathbf{R}^3$ contains a Sierpiński set with respect to its group of rotations.*

Proof

(a) If F is freely generated by τ and ρ, let $\sigma_i = \rho^i \tau \rho^{-i}$ for $i \in \mathbf{N}$; the σ_i are independent. Let F_ω be the subgroup of F generated by the σ_i, and let $\{D_n : n < \omega\}$ enumerate the finite subsets of F_ω. We need a corresponding enumeration of the σ_i so that no word in D_n begins on the left with the nth term of this enumeration. Such may easily be obtained by choosing for the nth term the σ_i of least index i that has not yet been chosen and that has the desired property as regards D_n. Let $\{\sigma_n : n < \omega\}$ now be used to denote this new enumeration of the generators of F_ω. Now, let

$$V_n = \{\phi\sigma_n^{-m}\delta : \delta \in D_n; m = 1, 2, \ldots; \phi \in F_\omega \text{ and } \phi \text{ does not end in } \sigma_n\}.$$

Then $\sigma_n V_n = V_n \cup D_n$ while $\sigma_k V_n = V_n$ if $k \neq n$. It follows that $E = F_\omega \backslash \bigcup_n V_n$ is the desired Sierpiński set: for any finite $D = D_n \subseteq F_\omega$, $\sigma_n(E) = E \backslash D_n$.

(b) Let E be a Sierpiński set in F and let x be an arbitrary element of X. Then $\{w(x) : w \in E\}$ is a Sierpiński set in X.

(c) Let τ, ρ be any two independent rotations of S^2 (2.1), and let F be the group they generate. If D is the set of fixed points on S^2 of elements of $F \backslash \{1\}$, then D is countable. By (b) then, there is a Sierpiński set in $S^2 \backslash D$ with respect to F. $\qquad\square$

The proof above extends easily to free groups of larger rank: if F is a free group of rank κ, and λ any cardinal satisfying $\kappa^\lambda = \kappa$, then F has a subset E such that for any $\leqslant \lambda$-sized subset D of E, there is some $\sigma \in F$ with $\sigma(E) = E \backslash D$. But Mycielski [161] showed that much more is possible. First of all, the action of the free group need only be assumed to be locally commutative. And, more strikingly, one can arrange things so that small sets can be added as well as deleted. We omit the proof (see [161]), which involves an inductive construction more intricate than in 6.16.

Theorem 6.17 (AC). *Suppose F, a free group of rank κ, is locally commutative in its action on X, and λ is a cardinal satisfying $\kappa^\lambda = \kappa$. Then there is a subset E of X such that for any two sets, $D_1, D_2 \subseteq X$ with $|D_i| \leqslant \lambda$, there is some $\sigma \in F$ with $\sigma(E) = (E \backslash D_1) \cup D_2$.*

Corollary 6.18 (AC). *Each of S^n, L^n, H^n ($n \geqslant 2$), or $\mathbf{R}^n$ ($n \geqslant 3$) has a subset that, from the point of view of isometries, is unchanged by adding or deleting countably many points.*

The corollary is an immediate consequence of 6.17 and 6.4.

The techniques of the preceding results are limited to groups with free non-Abelian subgroups, since such groups are the only ones containing Sierpiński sets (compare this with 1.12 and 4.9).

Theorem 6.19.   *If G has a Sierpiński set E with respect to its left action on itself, then G has a free subgroup of rank 2.*

Proof.   Since $E\sigma$, for any $\sigma \in G$, is also a Sierpiński set, we may assume that $1 \in E$. Now, if $\sigma \in G$ takes E to $E\backslash\{1\}$, then $\sigma = \sigma 1 \in E$, so there is some $\tau \in G$ with $\tau E = E\backslash\{\sigma\}$. We claim that σ, τ are independent. Since σ and τ cannot have finite order, any nontrivial vanishing word in $\sigma^{\pm 1}$, $\tau^{\pm 1}$ can be transformed by inversion and conjugation into one of the form $w = \sigma^{m_1}\tau^{n_1}\sigma^{m_2}\cdots\tau^{n_k}$ where $k \geqslant 1$, $m_1 > 0$, and all exponents are nonzero integers. Choose such a word w such that k is as small as possible and, given k, $\sum_{i=1}^{k} m_i + n_i$ is minimal as well.

We shall prove by induction that each right-hand end-segment of w belongs to E. The first such segment is simply τ or τ^{-1}, both of which belong to E by $\tau E = E\backslash\{\sigma\}$. Assume that the proper end-segment u of w belongs to E. Then clearly σu and τu lie in E. Moreover, by the minimality of w, $u \neq 1$; hence $\sigma^{-1}u \in E$. Finally, if $\tau^{-1}u$ is the next subword and does not lie in E, then $u = \sigma$, and it may easily be checked that σu^{-1}, which equals the identity, is shorter than w with respect to one of the two minimality conditions on w. This covers all four cases, completing the induction. But this means $\sigma^{-1} = \sigma^{m_1-1}\tau^{n_1}\cdots\tau^{n_k} \in E$, violating $\sigma E = E\backslash\{1\}$.   □

The preceding proof assumed only that E is a left translate of two of its point-deleted subsets; in fact, it is sufficient to assume that E is a left or right translate of two such subsets (see [226]).

NOTES

Infinite sets of congruences were first investigated for partitions of S^2, independently by Dekker and de Groot [55, 56] and by Mycielski [155]. Theorem 6.1, in its generality, was formulated and proved by Dekker [48]. The extension of Theorem 6.1 ensuring that the sets A_α are all nonempty is due to Dekker [49], as is the converse to 6.1, Theorem 6.2 [48].

Lebesgue and Steinitz had realized that $2^{\aleph_0}$ algebraically independent reals exist, but Mazurkiewicz [141] asked if the existence of a proper uncountable subfield of the reals could be proved without using the Axiom of Choice. Von Neumann [244] answered this question affirmatively by giving a constructive proof of Theorem 6.3. The alternative, less computational approach using Theorem 6.5 is due to Mycielski [162].

Sierpiński [212] was the first to construct an uncountable free group of isometries; he used algebraically independent numbers to prove Theorem 6.4

for SO_3. Earlier, Nisnewitsch [175] had proved that any free group is isomorphic to a subgroup of $GL_2(K)$ for some field K. Mycielski [155] used Sierpinski's large free group to solve large sets of congruences on S^2 (Corollary 6.6 for S^2). Simultaneously, Dekker and de Groot obtained these results for S^2 ([46, 47, 55, 56]) and Dekker [49, 50] carried out the extension to the other spaces of 6.6, in those cases where he had solved the problem for finite systems of congruences (see Chapter 5 Notes). The idea of using Theorem 6.5 to obtain large free groups (thus generalizing the results of Deligne and Sullivan for SO_n, n even, and Borel for SO_5 given in Chapter 5) is due to Mycielski [171]. The fact, stated after Corollary 6.6, that the set of pairs generating certain free subgroups of SO_n is comeager follows from Borel's work [20]; he used a different technique to show that the sets R_w are nowhere dense.

Theorem 6.7 on uncountable free subgroups of PSL_2 is due to Siegel [209]; the proof presented here owes much to correspondence with R. Riley. For extensions of Siegel's result see [100]. Theorem 6.8 is due to Mycielski and Wagon [171]. The observation that free groups in G_3 can be used to prove Theorem 6.8 in H^4 and beyond is due to A. Borel.

Steinhaus first raised the question of whether the set E of the Sierpiński-Mazurkiewicz Paradox (1.7) could be taken to be uncountable. This was answered affirmatively by Ruziewicz [197], but he needed the Axiom of Choice. Still using Choice, Lindenbaum [125, p. 327] proved that there is a nonempty subset of the plane that is equidecomposable with κ many copies of itself, for any $\kappa \leqslant 2^{\aleph_0}$. This was all prior to the discovery of von Neumann numbers. Von Neumann [244] deduced from these numbers a result that eliminated the Axiom of Choice from Ruziewicz's work, and Sierpiński [214] refined Ruziewicz's result even more, giving the proof of Theorem 6.12. Moreover, Sierpiński [214] saw how von Neumann numbers could be used to eliminate choice from Lindenbaum's theorem (Theorem 6.11).

The definition of monomorphic sets is due to Lindenbaum [124], who proved Theorem 6.13. The example of a subset of the circle that is congruent to a proper subset of itself is due to Tarski [230]; see [124, p. 217]. *The Scottish Book* [140, p. 67] contains a weaker version of 6.13 due to Banach and Ulam.

In [218] and [219, pp. 7–10] (see also [217]), Sierpiński proved Theorem 6.14 for $\mathbf{R}^1$ and gave an erroneous construction of a set E in the plane congruent to $E \backslash \{P\}$ and $E \backslash \{Q\}$ for distinct P, $Q \in E$. The error was discovered by Mycielski (see [220; 219, p. 116]), who went on to show [153, 154] that Sierpinski sets exists in $\mathbf{R}^3$ (Theorem 6.16). The proof of 6.16 given here is from [225], where Straus settled the question for $\mathbf{R}^2$ (Theorem 6.14). The extensions of these results contained in Theorem 6.17 and Corollary 6.18 are due to Mycielski [161], although the existence of a Sierpiński set in the hyperbolic plane was proved earlier by Viola [242]. Mycielski [157, 160] also investigated these types of questions in more general analytic manifolds. Theorem 6.19, which shows that the collection of groups that do not contain Sierpinski sets coincides with the class NF, is due to Straus [226].

Chapter 7

Paradoxes in Low Dimensions

Because the isometry groups of the line and plane do not contain any free non-Abelian subgroups, the construction of the Banach-Tarski Paradox does not work in these low dimensions. Moreover, the solvability of G_1 and G_2 will allow us to construct isometry-invariant, finitely additive measures in $\mathbf{R}^1$ and $\mathbf{R}^2$ (Corollary 10.9) that show that no bounded set with interior is paradoxical. (Recall that in Theorem 1.7 an unbounded paradoxical subset of the plane was constructed.) But if we enlarge the group of allowable transformations, then a paradoxical decomposition of the unit square, for example, can be constructed. Of course this is obvious if we enlarge the group to all bijections from $\mathbf{R}^2$ to $\mathbf{R}^2$, so we seek an enlarged group that preserves the paradoxical nature of the construction. Since the construction doubles the size of a set, it is natural to restrict ourselves to area-preserving transformations. In this chapter, we show how this can be done in $\mathbf{R}^1$ and $\mathbf{R}^2$, and present a paradoxical construction of a different sort for the line.

Recall that the group, A_n, of affine transformations of $\mathbf{R}^n$ consists of transformations of the form $\sigma = \tau\ell$ where τ is a translation and ℓ is a linear transformation. Moreover, if the determinant of σ is defined to be $\det(\ell)$, then σ magnifies area by the factor $|\det(\sigma)|$. Thus SA_n, the group of affine transformations of determinant $+1$, consists of the affine transformations that preserve area and orientation. The group SA_2 is quite a bit larger than the subgroup of orientation-preserving isometries, as we shall see in a moment,

but in dimension one no new transformations appear. Any affine transformation, $ax + b$, of determinant ± 1 is, in fact, an isometry.

The following proposition gives a simple example of a free non-Abelian
subgroup of SA_2 (in fact, of $SL_2(\mathbf{Z})$; hence SA_2 is quite different from the
solvable isometry group, G_2. This proposition can also be proved geometrically (see remarks following Figure 5.2(b)) by considering the two matrices as
linear fractional transformations of the upper half of the complex plane.

Proposition 7.1. *The two transformations in* $SL_2(\mathbf{Z})$ *defined by the*
matrices $A = \left[\begin{smallmatrix} 1 & 2 \\ 0 & 1 \end{smallmatrix}\right]$ *and* $B = \left[\begin{smallmatrix} 1 & 0 \\ 2 & 1 \end{smallmatrix}\right]$ *are independent.*

Proof. Since B has infinite order, conjugation by B shows that if any
nontrivial word in $A^{\pm 1}, B^{\pm 1}$ equals the identity, there must be one of the form
$w = B^{m_1} A^{m_2} B^{m_3} \cdots B^{m_s}$, where the m_i are nonzero integers and $s > 1$ is odd. Let
(a_2, b_2) be the first row of A^{m_2}, (a_3, b_3) the first row of $A^{m_2} B^{m_3}$, and so on to
(a_s, b_s). Then $(a_2, b_2) = (1, \pm 2^{|m_2|})$, $b_i = b_{i-1}$ for i odd, and $a_i = a_{i-1}$ for i even.
Now, by induction one sees easily that for odd $i \geqslant 3$, $|a_i| > |b_i|$, while for even
$i \geqslant 2$, $|b_i| > |a_i|$. Therefore, for any odd $i \geqslant 5$, $|a_i| > |b_i| = |b_{i-1}| > |a_{i-1}| =$
$|a_{i-2}|$, whence $|a_i| \geqslant |a_{i-2}| + 2$. Since $|a_3| > |b_2| \geqslant 2$, and since s is odd, it
follows that $|a_s| \geqslant s$. But then the first row of w is $(1, 0) \left[\begin{smallmatrix} a_s & b_s \\ * & * \end{smallmatrix}\right] = (a_s, b_s)$ which,
since $|a_s| \geqslant s > 1$, is not $(1, 0)$ and so w cannot equal the identity. $\qquad\square$

The proposition and its proof remain valid if 2 is replaced by any larger
number, integer or not. The method of 7.1 can be generalized to a wider class
of matrices as follows. Say that a matrix entry is *dominant* if it is strictly greater
in absolute value than all other entries. Then $A = (a_{ij})$ and $B = (b_{ij})$, two
members of $SL_2(\mathbf{Z})$, are independent if a_{12} dominates A and b_{21} dominates B.
This result, due to K. Goldberg and M. Newman [77], is proved by a
generalization of the proof of 7.1.

Proposition 7.1 can be improved in a way relevant to paradoxical
decompositions. Since any element of SL_2 fixes the origin, the whole plane is
not SL_2-paradoxical. But is $\mathbf{R}^2 \backslash \{\mathbf{0}\}$ SL_2-paradoxical? Yes, because the action
of SL_2 on $\mathbf{R}^2 \backslash \{\mathbf{0}\}$ is locally commutative; this was proved following
Corollary 4.6. If F is the group generated by the independent pair of 7.1, then
Theorem 4.5 implies that $\mathbf{R}^2 \backslash \{\mathbf{0}\}$ is F-paradoxical using four pieces; indeed,
by 4.12, any weak system of congruences is solvable by a partition of $\mathbf{R}^2 \backslash \{\mathbf{0}\}$
and transformations in F. Since $\mathbf{R}^2 \backslash \{\mathbf{0}\}$ is equidecomposable with $\mathbf{R}^2$ using a
translation of one unit to the right, it follows that $\mathbf{R}^2$ is SA_2-paradoxical. But
in order to obtain partitions of $\mathbf{R}^2 \backslash \{\mathbf{0}\}$ to solve all proper systems of
congruences, we seek, because of 4.12, a free group of rank 2 that has no
nontrivial fixed points; that is, no nonidentity element of the group can have 1
as an eigenvalue. For elements of SL_2 this is equivalent to saying that the trace
of any nonidentity element is not 2. Magnus [132] showed that the free
generators $\left[\begin{smallmatrix} 1 & 1 \\ 1 & 2 \end{smallmatrix}\right]$, $\left[\begin{smallmatrix} 5 & 2 \\ 2 & 1 \end{smallmatrix}\right]$, considered by Neumann [174] in 1933, generate a

subgroup of $SL_2(\mathbf{Z})$ with this property. In fact (see page 86), any nontrivial word w in these two generators is such that $|tr(w)| > 2$. Hence all (countable) proper systems of congruences are solvable by a partition of $\mathbf{R}^2 \backslash \{0\}$ using $SL_2(\mathbf{Z})$. Although, as pointed out, $\mathbf{R}^2$ is SA_2-paradoxical, it is not known whether $\mathbf{R}^2$ is SA_2-paradoxical using just four pieces. This is equivalent to asking whether SA_2 has a locally commutative free subgroup of rank 2. It is possible that the two free generators $\begin{bmatrix} x \\ y \end{bmatrix} \mapsto \begin{bmatrix} 1 & 2 \\ 0 & 1 \end{bmatrix} \begin{bmatrix} x \\ y \end{bmatrix} + \begin{bmatrix} 0 \\ 1 \end{bmatrix}$ and $\begin{bmatrix} x \\ y \end{bmatrix} \mapsto \begin{bmatrix} 1 & 0 \\ 2 & 1 \end{bmatrix} \begin{bmatrix} x \\ y \end{bmatrix} + \begin{bmatrix} 1 \\ 0 \end{bmatrix}$ generate such a subgroup, but it is not clear that the group generated by these two transformations is locally commutative. However, SA_2 does not have a free subgroup of rank 2 whose action on $\mathbf{R}^2$ is fixed-point free: any fixed-point free subgroup of SA_2 is solvable (see [171, §9]).

Our immediate goal, however, is to obtain a paradoxical decomposition of a set of positive finite area, such as the unit square. We shall show that this can be done using any independent pair in $SL_2(\mathbf{Z})$ and translations, thus staying within SA_2.

Let J be the half-open unit square $[0, 1) \times [0, 1)$. The general results of Chapter 4 apply to a group acting on a set, but $SL_2(\mathbf{Z})$ does not act on the square. This difficulty can be handled by using a planar version of arithmetic modulo 1; since all translations are affine, this will not take us out of SA_2. More precisely, let $\approx$ denote the equivalence relation on $\mathbf{R}^2$ determined by the lattice $\mathbf{Z}^2 : P \approx Q$ if and only if $Q = P + (m, n)$ for integers (m, n). For any $P \in \mathbf{R}^2$ let $\hat{P}$ be the unique point in J such that $\hat{P} \approx P$. Let $SA_2(\mathbf{Z})$ denote the group of affine transformation $\tau\ell$ where τ is any translation and $\ell \in SL_2(\mathbf{Z})$; $SA_2(\mathbf{Z}) = \pi^{-1}(SL_2(\mathbf{Z}))$ where $\pi : A_2 \to GL_2$ is the canonical homomorphism. For any $\sigma \in SA_2(\mathbf{Z})$ let $\hat{\sigma}$ be the function with domain J defined by $\hat{\sigma}(P) = \widehat{\sigma(P)}$.

Proposition 7.2. *The mapping $\sigma \mapsto \hat{\sigma}$ is a homomorphism from $SA_2(\mathbf{Z})$ to the group of area-preserving, piecewise affine bijections from J to J. When restricted to $SL_2(\mathbf{Z})$, this homomorphism is an isomorphism onto its image.*

Proof. Clearly, $\hat{\sigma} : J \to J$. The following property of $\approx$, which follows easily from the fact that $\ell(\mathbf{Z}^2) \subseteq \mathbf{Z}^2$ for $\ell \in SL_2(\mathbf{Z})$, is the crux of the proof: if $P \approx Q$ and $\sigma \in SA_2(\mathbf{Z})$ then $\sigma(P) \approx \sigma(Q)$. It follows that $\hat{\sigma}$ is one-to-one on J. Since $\sigma(J)$ is a parallelogram, $\hat{\sigma}$ decomposes into finitely many functions $\tau_i\sigma$ where τ_i is a translation by a point in $\mathbf{Z}^2$; hence $\hat{\sigma}$ is piecewise in $SA_2(\mathbf{Z})$, and so preserves area. Moreover, any $P \in J$ equals $\hat{\sigma}(Q)$ where $Q = \sigma^{-1}(P)$, so $\hat{\sigma}(J) = J$.

Now, if $P \in J$, then again using the crucial property of $\approx$,

$$\hat{\sigma}_1\hat{\sigma}_2(P) = \hat{\sigma}_1\widehat{\sigma_2(P)} = \widehat{\sigma_1(\widehat{\sigma_2(P)})} = \widehat{\sigma_1(\sigma_2(P))} = \widehat{\sigma_1\sigma_2}(P).$$

This shows that $\sigma \mapsto \hat{\sigma}$ is a homomorphism. To see that the transformation $\ell \mapsto \hat{\ell}$, $\ell \in SL_2(\mathbf{Z})$, is an isomorphism it is sufficient to check that the only transformation of $SL_2(\mathbf{Z})$ that gets taken to 1_J, the identity on J, is 1, the identity in $SL_2(\mathbf{Z})$. Suppose $\hat{\ell} = 1_J$. There is some translation τ such that on a polygon, $\ell = \tau\ell$. Hence $\tau\ell$ is the identity on three noncollinear points; since $\tau\ell$ is affine, this means that $\tau\ell = 1$. But then $\ell = \tau^{-1}$, which is impossible unless ℓ is the identity, as required. $\qquad\square$

If F is any free subgroup of $SL_2(\mathbf{Z})$, then by this proposition, $\hat{F} = \{\hat{\ell} : \ell \in F\}$ is a free group of the same rank, acting on J. Each element of F fixes the origin, and may fix other points as well. But since the fixed point sets are not too large, they may be absorbed analogously to the way a similar problem was handled in the Banach-Tarski Paradox (3.9 and 3.10). To do this we need to bring in translations.

Theorem 7.3 (*The von Neumann Paradox for the Plane*) (AC). *If σ_1, σ_2 are any two independent elements of $SL_2(\mathbf{Z})$ and G is the subgroup of $SA_2(\mathbf{Z})$ generated by σ_1, σ_2 and T, the group of all translations, then J is G-paradoxical. Moreover, any two bounded subsets of $\mathbf{R}^2$ with nonempty interior are G-equidecomposable.*

Proof. Let F be the group generated by σ_1 and σ_2, and let $\hat{F}$ and $\hat{T}$ be the images of F and T under the reduction modulo $\mathbf{Z}^2$. If D is the set of points in J that are fixed by some nonidentity element of $\hat{F}$, then $\hat{F}$ acts without nontrivial fixed points on $J\backslash D$. Since $\hat{F}$ is freely generated by $\hat{\sigma}_1, \hat{\sigma}_2$ (Proposition 7.2), it follows from 1.10 that $J\backslash D$ is $\hat{F}$-paradoxical. Each element of $\hat{F}$ is piecewise in G (since $T \subseteq G$), which implies that $J\backslash D$ is G-paradoxical. By Proposition 3.4 then, the proof will be complete once we show that J and $J\backslash D$ are T-equidecomposable.

Since every point in D is a fixed point of an affine map, and hence is contained in an affine subspace, and since F is countable, D splits into $D_0 \cup D_1$ where D_0 is a countable set of points in J and D_1 is a countable set of line segments. Now, the technique of Theorem 3.9 shows that for any countable subset C of J, $J \sim_T J\backslash C$ using two pieces. For it suffices to find a translation τ such that $C \cap \hat{\tau}^n(C) = \varnothing$ for any $n \geq 1$, and this may easily be done once C is enumerated as $\{P_0, P_1, P_2, \ldots\}$. For each i, discard the countably many translations τ such that $\tau(P_i) \approx P_j$ for some $j < i$; a nondiscarded translation is as required. We shall reduce the general case, $J\backslash D$, to the case of a deleted countable set by finding a countable set C such that $J\backslash D \sim_T J\backslash C$. It will then follow that $J\backslash D \sim_T J$, whence $J\backslash D \sim_T J$, completing the proof that J is G-paradoxical.

We claim that there is some translation τ such that $D \cap \hat{\tau}^n(D)$ is countable for any integer $n \neq 0$. Enumerate the segments in D_1 as $S_0, S_1, S_2, \ldots$. Suppose n, as well as i, j with $0 \leq i, j$ and $i \neq j$ are given integers. If S_i and S_j are not parallel, we do nothing. For then, no matter what τ is, $\hat{\tau}^n(S_j)$ intersects S_i in at most a single point. But if they are parallel, choose a point P on S_j and discard all translations τ such that for some $(m_1, m_2) \in \mathbf{Z}^2$, $\tau^n(P)$ lies on the doubly infinite line containing the segment $S_i + (m_1, m_2)$. This guarantees that if τ is not discarded, then $\hat{\tau}^n(S_j) \cap S_i = \varnothing$. For each of the countably many triples n, i, j, at most countably many lines of translations (where T is identified with $\mathbf{R}^2$) are discarded, and hence altogether only countably many lines of translations are discarded. Therefore on any nondiscarded line only countably many translations are discarded, and so there must be one, call it τ, left over. Now, τ is as desired; for if $D \cap \hat{\tau}^n(D)$ is uncountable, then some segment in D_1 must overlap with one in $\hat{\tau}^n(D_1)$, and any translation for which this could happen was discarded.

Let $C = \bigcup \{D \cap \hat{\tau}^n(D) : n = \pm 1, \pm 2, \ldots \}$. C is a countable subset of D and it follows from the claim about τ that the sets $\hat{\tau}^n(D \backslash C)$, $n \geqslant 0$, are pairwise disjoint, and disjoint from C. If A is the complement, in $J \backslash C$, of the union of this sequence of sets, then

$$J \backslash C = \left(\bigcup_{n=0}^{\infty} \hat{\tau}^n(D \backslash C) \right) \cup A \sim_{\hat{T}} \left(\bigcup_{n=1}^{\infty} \hat{\tau}^n(D \backslash C) \right) \cup A = J \backslash D,$$

as required.

The G-equidecomposability of any two bounded sets with nonempty interior follows from the Banach-Schröder-Bernstein Theorem (3.5) exactly as in the proof of 3.11, using squares instead of balls. Note that for any transformation s of the form $s(x, y) = (\alpha x, \alpha y)$, where $\alpha > 0$, $sGs^{-1} \in G$. It follows that a square of any size is G-paradoxical if the unit square is, and this allows the method of 3.11 to be used.   □

The proof of Theorem 7.3 is much simpler if the free generators σ_1 and σ_2 are such that F has no nontrivial fixed points in $\mathbf{R}^2 \backslash \{0\}$ (e.g., the Magnus-Neumann matrices discussed after Proposition 7.1). For then D, the set of fixed points of $\hat{F} \backslash \{1\}$, consists only of a countable set of points, and the last half of the proof is unnecessary.

The von Neumann Paradox shows how paradoxical decompositions in the plane completely analogous to the Banach-Tarski Paradox can be obtained if we are willing to step outside the isometry group. In fact, and this will be important for later applications (see remarks following Corollary 11.18), adjoining a certain single linear transformation to the isometry group G_2 is sufficient. More precisely, let σ be the shear of the plane defined by $\sigma(x, y) = (x + y, y)$. Then $\sigma = \begin{bmatrix} 1 & 1 \\ 0 & 1 \end{bmatrix} \in SL_2(\mathbf{Z})$. Let G_2^* be the group generated by G_2 and σ; then Theorem 7.3 holds with respect to G_2^*. Since $\begin{bmatrix} 1 & 2 \\ 0 & 1 \end{bmatrix}$ and $\begin{bmatrix} 1 & 0 \\ 2 & 1 \end{bmatrix}$ are both in G_2^* (the former is σ^2, the latter is $\rho \sigma^{-2} \rho^{-1}$, where ρ is the rotation $\begin{bmatrix} 0 & -1 \\ 1 & 0 \end{bmatrix}$), and since $G_2^* \supseteq T$, the hypothesis, and hence the conclusion of Theorem 7.3 are valid.

The translations apparently are crucial to the proof of 7.3, for there seems to be no obvious way to eliminate the set of fixed points. Moreover, without translations it is not clear how to reduce the action of a free subgroup of $SL_2(\mathbf{Z})$ to a bounded set.

Question 7.4. Is the interior of the unit square SL_2-paradoxical, or perhaps even F-paradoxical, where F is the free group generated by $\left[\begin{smallmatrix} 1 & 2 \\ 0 & 1 \end{smallmatrix}\right]$ and its transpose?

CIRCLE-SQUARING

The study of equidecomposability in the plane leads to one of the most perplexing and intractable problems of this whole area. As we have mentioned several times, there is an isometry-invariant, finitely additive measure $\mu : \mathscr{P}(\mathbf{R}^2) \to [0, \infty)$ with $\mu(J) = 1$. Such a measure must agree with area on all squares and hence, by approximations in the style of Riemann sums, on all nice regions (precisely, the Jordan measurable sets; see 9.6). It follows that two such regions can only be equidecomposable if their areas are equal, and for polygons Theorems 3.2 and 3.8 show that this condition is sufficient as well as necessary. But what about curved regions?

Question 7.5 (Tarski's Circle-Squaring Problem). Is a circle (with interior) equidecomposable with a square (necessarily of the same area)*?

There is very little known about this question, which was posed in 1925. It is not clear what sort of invariant for equidecomposability could be devised that would distinguish the circle from a square of the same area, and no construction known seems to be relevant to a proof of equidecomposability. Indeed, the situation seems not so different from that of the Greek geometers who considered the classical, straight-edge and compass form of the circle-squaring problem. Polygons present no great difficulty, but the circle seems totally unmanageable.

There are some partial results of interest, though none give any clue to whether one answer to 7.5 is more plausible than the other. The most noteworthy partial result is that the circle cannot be squared using a mathematical scissors. To make this precise, let us use the term *disc* to denote a topological disc in the plane, that is, a Jordan curve (simple, closed path) together with its interior. Call two discs *scissors-congruent* if one can be decomposed into finitely many pairwise interior-disjoint discs that, ignoring boundaries, can be rearranged by isometries to form the other. This generalizes the definition of congruent by dissection (3.1) in which the discs are assumed to be polygons; that is, the scissors may cut only on straight lines. Thus the Bolyai-Gerwien Theorem (3.2) implies that a square is scissors-congruent with any polygon of the same area. Dubins, Hirsch, and Karush [62] investigated whether the ability to cut along arbitrary Jordan curves allows a square to be transformed into a circle. They proved that such a decomposition is impossible.

Theorem 7.6. *A square is not scissors-congruent to a circle.*

* R. J. Gardner (Dhahran) has answered this question (and Question 3.14) negatively if the group is assumed to be discrete.

In fact, Dubins, Hirsch, and Karush established that a circle (more generally, an ellipse) is scissors-congruent to no other convex region. Theorem 7.6 has the following corollary for the set-theoretic context of Tarski's Problem.

Corollary 7.7.   *A circle is not equidecomposable to a square if the pieces of the decomposition are restricted to interiors of Jordan curves or arcs of such curves.*

The restrictions of this result are quite strong; more relevant to Tarski's problem would be a proof of nonequidecomposability using Borel sets. For an example of two discs that are not scissors-congruent but are equidecomposable (using Borel sets) see [A7].

There is a similarity between Question 7.5 and Question 3.14, since they both deal with the equidecomposability of elementary figures in a context that precludes paradoxical decompositions. A related question appears at the end of the next chapter (8.15), and it yields an unsolved problem of this type for the line. Returning briefly to scissors-congruence, we give a proof of a special case of 7.6. The full proof of 7.6 is an intricate argument in plane topology, but there is a much simpler proof if the curves in the theorem (and the corollary) are assumed to be rectifiable.

Proof of Theorem 7.6, Rectifiable Case.   Let C be a fixed circle. If E is a union of finitely many pairwise interior-disjoint discs, each of which is the interior and boundary of a rectifiable Jordan curve, then the boundary of E, ∂E, is rectifiable as well. Hence we may define a number, $\mu(E)$, as follows. If A is an arc of ∂E, then A is called convex relative to E if the convex hull of A is contained in E; A is called concave relative to E if the convex hull of A is disjoint from the interior of E. For any point P on ∂E let $f_E(P) = +1(-1)$ if P is contained in the interior of an arc, A, of ∂P such that A is convex (concave) relative to E, and A is congruent to an arc of C. Then let $\mu(E) = \int_{\partial E} f_E \, ds$, the integral of f_E on ∂E with respect to an arc length parametrization of ∂E. Now, μ is clearly isometry-invariant, and because of the cancellation that ensues when a concave piece of arc is matched with a convex piece of arc, $\mu(E_1 \cup E_2) = \mu(E_1) + \mu(E_2)$ if E_1 and E_2 are interior-disjoint. It follows that $\mu(D_1) = \mu(D_2)$ whenever D_1 and D_2 are scissors-congruent discs. Since $\mu(C)$ is the circumference of C while μ assigns any square the value 0, this proves the rectifiable case of the theorem.   □

A positive circle-squaring result can be derived from Theorem 7.3: the circle can be squared if, in addition to isometries, the shear $\sigma = \begin{bmatrix} 1 & 1 \\ 0 & 1 \end{bmatrix}$ can be used. This is a consequence of 7.3 and the remarks following its proof. But the addition of σ changes the problem so dramatically that this result really has little bearing on Tarski's problem. After all, if the shear is available, then any two discs are equidecomposable, regardless of their size.

Another extension of the isometry group allows one to obtain a sort of approximation to the desired squaring of the circle. Let G be the group of similarities of the plane, that is, the group generated by all isometries and all magnifications from a point. Since isometries are affine, G consists of all transformations $d\sigma$ where $\sigma \in G_2$ and $d = \begin{bmatrix} \alpha & 0 \\ 0 & \alpha \end{bmatrix}$, $\alpha > 0$, is a magnification from the origin. A similarity $\begin{bmatrix} \alpha & 0 \\ 0 & \alpha \end{bmatrix}\sigma$, where $\sigma \in G_2$, is called ε-*magnifying* if $1 - \varepsilon \leqslant \alpha \leqslant 1 + \varepsilon$.

Theorem 7.8. *For any $\varepsilon > 0$ the circle is G-equidecomposable with a square of the same area, and the similarities used to transform the pieces are ε-magnifying. Moreover, the pieces in the decomposition are Borel sets.*

Proof. Let C be the unit circle (with interior) and S the square of side-length $\sqrt{\pi}$. By the Banach-Schröder-Bernstein Theorem (3.5) and its constructive proof, it is sufficient to find a piecewise ε-magnifying similarity from C to a subset of S, and a piecewise ε-magnifying similarity from C to a set containing S. For then $C \leqslant S$ and $S \leqslant C$, and since the proof of 3.5 introduces no new similarities, this implies that $C \sim_G S$ using ε-magnifications.

Choose n so big that the circle, when shrunk radially by $1 - \varepsilon$, fits inside a regular n-gon inscribed in C. Since by Tarski's version of the Bolyai-Gerwien Theorem (3.2 and 3.8), the polygon is equidecomposable (using isometries) with a square smaller than S, C can be packed into S, too, provided it is preshrunk by $1 - \varepsilon$. To get a superset of S, choose n so large that the circle, when expanded radially by $1 + \varepsilon$, contains the regular n-gon circumscribed about C, and proceed in the same way. It is easy to see that the results used in this proof—Theorems 3.2, 3.5, and 3.8—do not introduce any non-Borel sets.   □

Of course, Tarski's Problem is equivalent to demanding that the preceding theorem be true with $\varepsilon = 0$. Among the many measures constructed in Part II will be a finitely additive extension, μ, of two-dimensional Lebesgue measure with the property that $\mu(s(A)) = \alpha^2 \mu(A)$ for each similarity s, where α is the amount by which s magnifies distances (see Corollary 11.5). It follows that Theorem 7.8 is false if the square's area does not equal that of the circle.

PARADOXES IN $\mathbf{R}^1$

The idea of allowing affine maps, so fruitful in the plane, does not help with the line since the measure-preserving affine transformations are just the isometries. How much does the isometry group have to be expanded before a paradoxical decomposition of an interval arises? We are not interested in transformations that do not preserve Lebesgue measure: there is nothing surprising about the fact that the function $2x$ can be used to duplicate an interval. Let $G(\lambda)$ be the group of all bijections f from $\mathbf{R}$ to $\mathbf{R}$ such that both f and f^{-1} are Lebesgue measurable and preserve Lebesgue measure: $\lambda(f(A)) = \lambda(f^{-1}(A)) = \lambda(A)$ if A is measurable. This group, a rather large extension of the isometry group, is rich enough to produce paradoxical decompositions on the line.

Theorem 7.9 (AC).   *Any interval is $G(\lambda)$-paradoxical. Any two bounded subsets of $\mathbf{R}$ with nonempty interior are $G(\lambda)$-equidecomposable.*

Proof.   It is a general fact of measure theory that there is a bijection $f : [0, 1) \to S^2$ such that both f and f^{-1} take measurable sets to measurable sets, and both f and f^{-1} preserve measure (where the measure on S^2 is $\lambda/4\pi$, normalized surface Lebesgue measure). This follows from [196, p. 327, Theorem 9], for example. These properties of f imply that if $\sigma \in SO_3$, then $f^{-1}\sigma f$ is a measure-preserving bijection of $[0, 1)$ to itself (which, by periodic extension, may be considered as an element of $G(\lambda)$). Now, the Banach-Tarski Paradox states that S^2 is SO_3-paradoxical, via σ_i and A_i, say. It follows that $[0, 1)$ is $G(\lambda)$-paradoxical, using pieces $f^{-1}(A_i)$ and transformations $f^{-1}\sigma_i f$. This technique applies as well to any half-open interval, and since any open or closed interval is $G(\lambda)$-equidecomposable with a half-open interval, this yields the first part of the theorem. The second part follows by the usual technique using the Banach-Schröder-Berstein Theorem (see 3.11).   □

Von Neumann found an entirely different sort of paradox on the line, based on using linear fractional transformations as a way of bringing free groups of 2×2 matrices to bear. Let L be the group of linear fractional transformations of $\mathbf{R} \cup \{\infty\}$ of the form $x \mapsto (ax + b)/(cx + d)$, $ad - bc = 1$, where the usual arithmetic of ∞ is used to handle $x = \infty$ or $x = -d/c$. This is a group action of L on $\mathbf{R} \cup \{\infty\}$, and any $\sigma \in L$ is a strictly increasing continuous function when restricted to an interval of $\mathbf{R}$ not containing $-d/c$. In fact, as with the upper half-plane of $\mathbf{C}$, L is isomorphic to PSL_2, but there will be no harm in what follows if we identify each $\sigma \in L$ with the coefficient matrix $\left[\begin{smallmatrix} a & b \\ c & d \end{smallmatrix}\right] \in SL_2$.

Definition 7.10.   *A transformation $\sigma \in L$ will be called an ε-contraction with respect to an interval of $\mathbf{R}$, if $|\sigma(x) - \sigma(y)| \leqslant \varepsilon|x - y|$ for all x, y in the interval.*

The linear fractional transformations do not contain any measure-preserving maps except the isometries, but Lebesgue measure, λ, behaves nicely with respect to contractions ($\varepsilon < 1$) in the sense that it shrinks.

Proposition 7.11.   *If $\sigma \in L$ is an ε-contraction on $[a, b]$ and A is a measurable subset of $[a, b]$, then $\lambda(\sigma(A)) \leqslant \varepsilon\lambda(A)$.*

Proof. This is easily proved by considering the outer measure, λ^*. Assume $A \subseteq (a, b)$. If A is covered by open subintervals (a_i, b_i) of (a, b), then the intervals $(\sigma(a_i), \sigma(b_i))$ cover $\sigma(A)$, and each $\sigma(b_i) - \sigma(a_i) < \varepsilon(b_i - a_i)$. Hence $\lambda^*(\sigma(A)) \leqslant \varepsilon\lambda(A)$, and since σ^{-1} is a measurable function, $\sigma(A)$ is a measurable set and $\lambda(\sigma(A)) = \lambda^*(\sigma(A))$. $\qquad\qquad\qquad\Box$

Now, if all sets were Lebesgue measurable, by the preceding result an interval could not be paradoxical using contractions. Indeed, the image of $[0, 1]$ under a piecewise ε-contraction with respect to $[0, 1]$, using measurable pieces, has measure at most ε. But the next theorem gives such a construction for arbitrarily small ε; of course, nonmeasurable sets appear in the decomposition.

Theorem 7.12 (*The von Neumann Paradox for the Line*) (*AC*). *The unit interval is paradoxical using contractions; in fact, for any $\varepsilon > 0$, there is a paradoxical decomposition of $[0, 1]$ that uses ε-contractions with respect to $[0, 1]$ to transform the sets that partition the interval. Moreover, for any $\varepsilon > 0$ and any two sets $A, B \subseteq \mathbf{R}$ that are bounded and have nonempty interior, there is a partition of A into $A_1 \cup \cdots \cup A_n$ such that $\{\sigma_i(A_i)\}$ is a partition of B for some $\sigma_i \in L$, which are all ε-contractions with respect to an interval containing A.*

The proof requires the following lemma about free groups of piecewise linear fractional transformations. Note that the existence of two algebraically independent numbers in any interval is an easy consequence of the countability of the set of rational polynomials; von Neumann numbers are not needed.

Lemma 7.13. *If α, β are any two algebraically independent numbers with $0 < \alpha < \beta < 1$, then $\hat{\sigma}$ and $\hat{\tau}$ are independent bijections of $[0, \gamma)$ where*

$$\gamma = \frac{\beta(2 + \beta)}{(1 + \beta)(2 - \beta)}, \quad \hat{\sigma}(x) = \frac{(1 + \beta)x}{(2 - \beta)x + \dfrac{1}{1 + \beta}},$$

and

$$\hat{\tau}(x) = \begin{cases} x + \alpha & \text{if } x + \alpha < \gamma \\ x + \alpha - \gamma & \text{otherwise.} \end{cases}$$

Proof. Since $\gamma > \beta > \alpha$, τ is just the reduction modulo γ of the translation $\tau(x) = x + \alpha$. And $\hat{\sigma}$ is the restriction to $[0, \gamma)$ of $\sigma \in L$ where σ fixes 0 and γ. Hence $\hat{\sigma}, \hat{\tau}$ generate a group, F, of bijections of $[0, \gamma)$ that are piecewise in L. Suppose w is a nontrivial word in $\hat{\sigma}^{\pm 1}$, $\hat{\tau}^{\pm 1}$ that equals the identity on $[0, \gamma)$. Let $\tau_0(x) = x + \alpha - \gamma$. Then on some subinterval of $[0, \gamma)$, $w = \sigma^{m_1}\rho^{n_1}\sigma^{m_2}\rho^{n_2} \cdots$ where the m_i, n_i are nonzero integers and each ρ is either τ

or τ_0. Since $w = 1$, and since γ is a rational function of β, there must be four rational functions induced by the multiplications of powers of $\begin{bmatrix} 1 & \alpha \\ 0 & 1 \end{bmatrix}$, $\begin{bmatrix} 1 & \alpha - \gamma \\ 0 & 1 \end{bmatrix}$, and $\begin{bmatrix} 1+\beta & 0 \\ 2-\beta & 1/(1+\beta) \end{bmatrix}$ in w, $R_{ij}(\alpha, \beta)$, $i, j = 1, 2$, which equal 0 if $i \neq j$, or 1 if $i = j$. Both α and β are algebraically independent so these rational functions must be 0 or 1, and hence w yields the identity matrix no matter what values are chosen for α and β. Letting $\alpha = 2$ and $\beta = 0$ yields that $\gamma = 0$, so that w is a nontrivial word in $\begin{bmatrix} 1 & 2 \\ 0 & 1 \end{bmatrix}$ and $\begin{bmatrix} 1 & 0 \\ 2 & 1 \end{bmatrix}$, which equals the identity, and this contradicts Proposition 7.1. $\qquad\square$

Proof of 7.12. Suppose $\varepsilon > 0$ is given. We shall define a piecewise ε-contraction from $[0, 1)$ to a set containing $[0, 2]$, and this, together with an extra translation, induces the desired decomposition of $[0, 1]$. The transformation will be defined by choosing β sufficiently small, shrinking $[0, 1)$ to $[0, \gamma)$, and iterating a paradoxical decomposition of $[0, \gamma)$, using the transformations of the lemma, often enough to cover $[0, 2]$.

First we claim that if $\alpha, \beta, \gamma, \hat{\sigma}, \hat{\tau}$, and F are as in the lemma, then $[0, \gamma)$ is L-paradoxical using $(1 + \beta)^2$-contractions. This follows from the fact that D, the set of nontrivial fixed points of F, is countable (transformations in L fix at most two points), and hence the usual absorption technique can be used to get a translation τ_1 such that $[0, \gamma) \sim [0, \gamma) \backslash D$ using $\hat{\tau}_1$. By Propositions 1.10 and 3.4 then, $[0, \gamma)$ is paradoxical using σ and translations. Since $d\sigma/dx = ((2 - \beta)x + 1/(1 + \beta))^{-2}$, which is at most $(1 + \beta)^2$ on $[0, \gamma)$, the Mean Value Theorem implies that this is a paradoxical decomposition of the sort claimed.

Now, repeating this paradoxical decomposition sufficiently often yields enough copies of $[0, \gamma)$ to cover $[0, 2]$. More precisely, each iteration doubles the number of copies of $[0, \gamma)$, so this will require $m = [-\log_2\gamma]$ iterations. The amount by which all this magnifies distances is $((1 + \beta)^2)^m$. Therefore the final piecewise linear fractional transformation from $[0, 1)$ to a superset of $[0, 2]$, obtained by preceding the paradoxical decompositions by $x \mapsto \gamma x$, will shrink distances by $\gamma(1 + \beta)^{2m}$. Since $\gamma/\beta \to 1$ as $\beta \to 0$, this quantity is no greater, in the limit as $\beta \to 0$, than $\beta/(1 + \beta)^{\log_2\beta}$, which is easily seen to approach 0, since the denominator approaches 1. Hence if β is chosen so small that $\gamma(1 + \beta)^{2m} \leq \varepsilon$, the transformation from $[0, 1)$ to a superset of $[0, 2]$ will be a piecewise ε-contraction as required.

Note that if $\varepsilon < 1$ then, because $\varepsilon^n < \varepsilon$, the paradoxical decomposition has the curious property that if it is iterated to obtain more copies of the unit interval (or a larger interval), then the final transformations contract even more than the original.

Finally, suppose A and B are as in the last assertion of the theorem, with A contained in an interval I, and containing an interval $[a, b]$. Let $\varepsilon > 0$ be given. By the proof of the Banach-Schröder-Bernstein Theorem, it is sufficient to find a piecewise ε-contraction from A to a subset of B, and another piecewise ε-contraction from A (or a subset of A) to a superset of B. The former is trivial: use $\delta x + n$ where $\delta \leqslant \varepsilon$ is small enough that this transformation takes I to an interval contained in B. For the latter, let $0 < \delta < \min\{1, (b - a)\varepsilon\}$ and iterate the previous construction to get a piecewise δ-contraction from $[0, 1)$ to an interval large enough to cover B. If we precede this transformation by the affine $1/(b - a)$-contraction taking $[a, b)$ to $[0, 1)$, we get a piecewise ε-contraction from $[a, b)$ to a set containing B, as required. $\square$

Because the von Neumann Paradox for the line does not refer to a group action in the same way as the earlier paradoxical decompositions, we cannot use it to derive a result about the negligibility of intervals with respect to finitely additive measures. But it does imply that no finitely additive measure defined on all sets of reals and normalizing the unit interval can have the property of shrinking under contractions. In fact, because $[0, 1]$ is equidecomposable using ε-contractions with arbitrarily many copies of itself, we get the following corollary. The additional assertion about measure-preserving transformations follows from the other linear paradox, Theorem 7.9.

Corollary 7.14 (AC). *Suppose $\mu \colon \mathscr{P}(\mathbf{R}) \to [0, \infty]$ is a finitely additive measure with $\mu([0, 1]) = 1$. Then for any $\varepsilon, K > 0$, there is a set $A \subseteq [0, 1]$ with $\mu(A) > 0$ and an ε-contraction, σ, with respect to $[0, 1]$ such that $\mu(\sigma(A)) \geqslant K\mu(A)$. Furthermore, there is a set $B \subseteq [0, 1]$ and a Lebesgue measure-preserving bijection τ of $[0, 1]$ to itself such that $\mu(\tau(B)) \neq \mu(B)$.*

Because of the strong version of the Banach-Tarski Paradox, it is clear how to get a version of Theorem 7.12 in $\mathbf{R}^3$: shrink a ball radially as much as desired and then use isometries to get as large a ball as desired. Sierpiński [213, 216] has shown how a contraction-type paradox in the plane can be derived directly from the Banach-Tarski Paradox of the sphere. He proved that for any $r > 0$ there is a bijection f from the unit disc to the disc of radius r that, piecewise, contracts distances.

NOTES

The idea of expanding the isometry groups in a way that produces a generalization of the Banach-Tarski Paradox in the plane and on the line is due to von Neumann [246]. He showed that the unit square was SA_2-paradoxical; although he made use of the fact that $SL_2(\mathbf{Z})$ has pairs of independent elements, his final proof used pairs in $SL_2 \backslash SL_2(\mathbf{Z})$, defined from small algebraically independent numbers. The version of Theorem 7.3 presented here and the observation that the addition of a single shear to the planar isometry group is sufficient to produce paradoxes are due to Wagon [219]. This latter result was motivated by work of Rosenblatt [195] on the uniqueness of Lebesgue measure as a shear-invariant measure, which will be discussed in Chapter 11.

Proposition 7.1 is due to Sanov [203], and the extension to larger off-diagonal numbers is due to Brenner [25] (see also [135, p. 100]). Brenner also characterized the matrices that appear in the group generated by $\left[\begin{smallmatrix} 1 & m \\ 0 & 1 \end{smallmatrix}\right]$ and its transpose ($m \geqslant 2$). Further results on independent pairs of 2×2 matrices may be found in [27, 77, 130, 131, 132, 134, 190]. Independent pairs in $PSL_2(\mathbf{Z})$ that generate a group consisting of only hyperbolic elements were considered by Magnus [132], who used work of Neumann [174]; see also [167].

The circle-squaring problem of Tarski appears in [232], and has been mentioned often in the literature: [107], [217], [248]. Theorem 7.6 is due to Dubins, Hirsch, and Karush [62] (see also [202]). For various generalizations of the notion of scissors-congruence see Sah [201]. The idea of using the Banach-Schröder-Bernstein Theorem to attack an approximate form of the circle-squaring problem using ε-magnifications is due to Klee; Theorem 7.8 as stated here is due to Henle and Wagon.

The existence of paradoxes on the line, using linear fractional transformations, was proved by von Neumann [246]. Similar results for the plane were obtained by Sierpiński [216]. Another result related to pathology involving contractions of nonmeasurable sets can be found in [101].

Chapter 8

The Semigroup of Equidecomposability Types

Certain proofs and theorems involving equidecomposability would be much simplified if we could add sets. For instance, if X could literally be added to X to form $2X$, then the fact that X is paradoxical could be stated simply as $X = 2X$. In fact, this can be done if we expand the group action appropriately so that multiple copies of X can be formed. This new context for discussing equidecomposability will allow us to state and prove theorems that otherwise would be very cumbersome. One of these is a cancellation law for equidecomposability that has several uses, the most important of which is its use in Tarski's theorem (9.2) relating paradoxical decompositions and invariant measures. Another application will be a proof that any two subsets of S^2 with nonempty interior are equidecomposable using rotations. This expanded context for equidecomposability will also yield a simpler proof of Theorem 4.5 that a locally commutative action of a free non-Abelian group is paradoxical.

Definition 8.1. *Suppose the group G acts on X. Define an enlarged action as follows. Let $X^* = X \times \mathbf{N}$ and let $G^* = \{(g, \pi) : g \in G \text{ and } \pi \text{ is a permutation of } \mathbf{N}\}$, and let the group G^* act on X^* by $(g, \pi)(x, n) = (g(x), \pi(n))$. If $A \subseteq X^*$, then those $n \in \mathbf{N}$ such that A has at least one element with second coordinate n are called the* levels *of A.*

The action of G^* extends that of G, and treats all the levels in the same way; for example, if $E \subseteq X$, then $E \times \{n\}$ is G^*-congruent to $E \times \{m\}$. Moreover, G-equidecomposability and G^*-equidecomposability are closely related: if $E_1, E_2 \subseteq X$, then $E_1 \sim_G E_2$ if and only if $E_1 \times \{m\} \sim_{G^*} E_2 \times \{n\}$ for all $m, n \in \mathbf{N}$. Now, the equidecomposability class of some $E \subseteq X$ in $\mathscr{P}(X^*)$ with respect to G^* is much more valuable than the $\sim_G$-equivalence class of E in $\mathscr{P}(X)$. Since copies of E at different levels can be identified with E, the set $E \times \{0, 1\}$ serves as a representative of what we intuitively would like to call $2E$, and if E is G-paradoxical, then $E \times \{0\} \sim_{G^*} E \times \{0, 1\}$, that is, $E = 2E$. We make this more precise by defining an addition operation for those subsets of X^* having only finitely many levels.

Definition 8.2.   *Let* G, X, G^*, X^* *be as above.*

(a) *A subset A of X^* is called* bounded *if it has only finitely many levels. The equivalence class with respect to G^*-equidecomposability of a bounded $A \subseteq X^*$ is called the* type *of A, and denoted $[A]$. The collection of types of bounded sets will be denoted by $\mathscr{S}$.*

(b) *For $[A], [B] \in \mathscr{S}$, define $[A] + [B]$ to be $[A \cup B']$, where B' is an upward shift of B so that the levels of B' are disjoint from those of A; that is, $B' = \{(b, m + k) : (b, m) \in B\}$, where k is sufficiently large.*

It is a simple matter to check that $+$ is well defined: $[A] + [B]$ is independent of the choice of representatives A, B, and of the permutation used to shift B upward. Moreover, $+$ is commutative and associative, whence $(\mathscr{S}, +)$ is a commutative semigroup, called the *type semigroup*. Note that $[\varnothing] = \mathbf{0}$ serves as an identity for $+$. If $E \subseteq X$, then $[E]$ is used to denote $[E \times \{0\}]$.

For any commutative semigroup with identity, there is a natural way of multiplying elements by natural numbers: $n\alpha = \alpha + \alpha + \cdots + \alpha$ with n summands. Also, there is a natural ordering given by $\alpha \leqslant \beta$ if and only if $\alpha + \gamma = \beta$ for some γ in the semigroup. Note that $[A] \leqslant [B]$ if and only if $A \preccurlyeq B$, that is, A is G^*-equidecomposable with a subset of B. These semigroup operations satisfy many familiar axioms: $n(m\alpha) = (nm)\alpha$, $(n + m)\alpha = n\alpha + m\alpha$, $n(\alpha + \beta) = n\alpha + n\beta$, $n\alpha \leqslant n\beta$ if $\alpha \leqslant \beta$, $n\alpha \leqslant m\alpha$ if $n \leqslant m$, and $\alpha + \gamma \leqslant \beta + \gamma$ if $\alpha \leqslant \beta$. The fact that only bounded subsets of X^* were considered when types were formed means that the type semigroup satisfies an Archimedean condition with respect to $[X]$: for each $\alpha \in \mathscr{S}$ there is some $n \in \mathbf{N}$ such that $\alpha \leqslant n[X]$. The following proposition shows that the addition of types behaves as one would expect with respect to set-theoretic unions.

Proposition 8.3.   *Suppose A and B are bounded subsets of X^*. Then $[A] + [B] \geqslant [A \cup B]$, with equality if $A \cap B = \varnothing$.*

Proof. Consider the last assertion first. If B' is any upward shift of B, then $B \sim B'$, so $A \cup B \sim A \cup B'$, using two pieces. Hence $A \cup B$ is in $[A] + [B]$, as required. It follows from this that arbitrary bounded sets satisfy $[A] + [B] = [A \cup B] + [A \cap B]$, whence $[A] + [B] \geqslant [A \cup B]$. □

In the context of the type semigroup, the Banach-Schröder-Bernstein Theorem takes on the simple form: if $\alpha, \beta \in \mathscr{S}$, then $\alpha \leqslant \beta$ and $\beta \leqslant \alpha$ imply $\alpha = \beta$. And a type can be called paradoxical if $2\alpha = \alpha$; hence $E \subseteq X$ is G-paradoxical if and only if $[E] = 2[E]$ in $\mathscr{S}$. A subtle point arises here since from a measure-theoretic point of view, it might make sense to say $\alpha \in \mathscr{S}$ is paradoxical if for some $k, (k + 1)\alpha \leqslant k\alpha$ (which is equivalent to $(k + 1)\alpha = k\alpha$). This latter condition on α would, like $2\alpha = \alpha$, imply the nonexistence of certain measures normalizing α, but in a general semigroup the two conditions are not equivalent. One would need a certain sort of cancellation law in order to deduce $2\alpha = \alpha$ from $(k + 1)\alpha = k\alpha$. One of the pleasant aspects of equidecomposability theory is that an appropriate cancellation law is valid for the type semigroup, and these two conditions on α are equivalent. The proof, however, is quite intricate, unlike that of the other major algebraic law, the Banach-Schröder-Bernstein Theorem (see Theorem 8.7).

Our first application of this formalism is to give a different proof that a locally commutative action of a free group of rank 2 on a set X yields a paradoxical decomposition. This approach will be much simpler than the proof given in Chapter 4 (Theorem 4.5), although it does not yield the stronger result that four pieces suffice. First, we generalize Proposition 1.10 slightly, and it is here that the type semigroup is useful.

Definition 8.4. *Suppose the groups $H_1, \ldots, H_n$ all act on X. These actions are called* jointly free *if for each $x \in X$ there is at least one $i = 1, \ldots, n$ such that x is not a fixed point of any element of $H_i \backslash \{1\}$.*

Proposition 8.5 (AC). *Suppose G acts on X in such a way that the actions of $H_1, \ldots, H_n$ on X are jointly free, where each H_i is a free subgroup of G of rank 2. Then X is G-paradoxical.*

Proof. Let $D_i = \{x \in X : h(x) = x \text{ for some } h \in H_i \backslash \{1\}\}$, and let $\delta_i = [X \backslash D_i]$ with respect to G^* and X^*. By Proposition 1.10, each $X \backslash D_i$ is H_i-paradoxical, so $\delta_i = 2\delta_i$. Because the actions are jointly free, $X = \bigcup (X \backslash D_i)$, and so by 8.3, $[X] \leqslant \sum \delta_i$. Now for each i,

$$[X] = [(X \backslash D_i) \cup D_i] = \delta_i + [D_i] = 2\delta_i + [D_i] = [X] + \delta_i.$$

Therefore

$$[X] = [X] + \delta_1 = [X] + \delta_2 + \delta_1$$
$$= \ldots = [X] + \delta_n + \ldots + \delta_1 \geqq 2[X] \geqq [X],$$

whence $[X] = 2[X]$ and X is G-paradoxical. □

Corollary 8.6 (AC).   *If F's action on X is locally commutative and F is a free group of rank 2 freely generated by σ and τ, then X is F-paradoxical.*

Proof.   We shall show that if G, a free group of rank 4 freely generated by $\rho_0, \rho_1, \rho_2, \rho_3$, acts on X and is locally commutative, then X is G-paradoxical. Since F contains such a free subgroup ($\rho_i = \sigma^i \tau \sigma^{-i}$), this yields the corollary. Let H_1 be the subgroup of G generated by ρ_0 and ρ_1, and let H_2 be the subgroup generated by ρ_2 and ρ_3. By Proposition 8.5, it suffices to show that the actions of H_1, H_2 are jointly free. If not, then some $x \in X$ is fixed by w_1 and w_2 in H_1, H_2, respectively, with $w_i \neq 1$. But G's action is locally commutative, so $w_1 w_2 w_1^{-1} w_2^{-1} = 1$ in G, contradicting the independence of $\rho_0, \rho_1, \rho_2, \rho_3$. $\square$

A CANCELLATION LAW

We now wish to discuss a cancellation law for $\mathscr{S}$ that, like the Banach-Schröder-Bernstein Theorem, is a powerful tool for proving sets equidecomposable. Let $\mathscr{S}$ denote the type semigroup of an arbitrary group action.

Theorem 8.7 (Cancellation Law) (AC).   *If, for $\alpha, \beta \in \mathscr{S}$ and a positive integer n, $n\alpha = n\beta$, then $\alpha = \beta$.*

Before diving into the details of the proof, we illustrate the power of this law by deriving some corollaries. First, we show that finitely many copies of a set cannot be packed into fewer copies of the set, unless the set is paradoxical.

Corollary 8.8 (AC).   *If $\alpha \in \mathscr{S}$ and $n \in \mathbf{N}$ satisfy $(n + 1)\alpha \leqslant n\alpha$, then $2\alpha = \alpha$.*

Proof.   Substituting the hypothesized inequality into itself yields: $n\alpha \geqslant (n + 1)\alpha = n\alpha + \alpha \geqslant (n + 1)\alpha + \alpha = n\alpha + 2\alpha$. Repeating this substitution, we eventually obtain that $n\alpha \geqslant n\alpha + n\alpha = 2n\alpha$. Since $n\alpha \leqslant 2n\alpha$, we have $n\alpha = 2n\alpha = n(2\alpha)$ and hence the cancellation law yields $\alpha = 2\alpha$. $\square$

In Chapter 3 we showed how the Banach-Tarski Paradox for balls can be combined with the Banach-Schröder-Bernstein Theorem to get the equidecomposability of arbitrary bounded subsets of $\mathbf{R}^3$ with nonempty interior (3.11). But that technique does not suffice to get the SO_3-equidecomposability of arbitrary subsets of S^2 with interior. The problem is that an open subset of S^2 does not contain a sphere, and so it is not evident that it contains any paradoxical sets. The cancellation law allows us to show rather easily that any open subset of S^2 is paradoxical.

Corollary 8.9 (AC). *If G acts on X and $E \subseteq X$ is G-paradoxical, then E is G-equidecomposable with any subset A of E with the property that $g_1 A \cup \cdots \cup g_n A \supseteq E$ for some $g_1, \ldots, g_n \in G$. Hence any such A is G-paradoxical and any two such subsets of E are G-equidecomposable.*

Proof. Work in $\mathscr{S}$. The hypothesis on A yields $n[A] \geqslant [E]$, while $[E] = 2[E] = \cdots = n[E]$ since E is paradoxical. So $n[A] \geqslant n[E] \geqslant n[A]$ ($A \subseteq E$), whence $n[A] = n[E]$, and the cancellation law now implies that $[A] = [E]$, that is, $A \sim_G E$. Hence A is G-paradoxical because E is, and if B is another subset of E rich enough so that finitely many copies cover E, then $B \sim E$ too, whence $B \sim A$. $\qquad\square$

Corollary 8.10 (AC). *If $n \geqslant 2$, then any two subsets of S^n, each of which has nonempty interior, are SO_{n+1}-equidecomposable, and any such subset is SO_{n+1}-paradoxical.*

Proof. This follows from 8.9 and 5.1, since sufficiently many copies of a nonempty open subset of S^n cover S^n. $\qquad\square$

It is not completely clear which subsets E of S^2 are SO_3-equidecomposable with all of S^2. Recall (Theorem 3.9) that $S^2 \sim S^2 \backslash D$ for any countable D, and so a set with empty interior can be equidecomposable with S^2. On the other hand, if $E \sim S^2$ then E cannot be nowhere dense, nor can E have Lebesgue measure zero; this is because these two properties are preserved under equidecomposability.

The cancellation law, like the Banach-Schröder-Bernstein Theorem, is motivated by the corresponding result for cardinality, that is, arbitrary bijections, namely: $n|X| = n|Y|$ implies $|X| = |Y|$. This more basic result is due to Bernstein [16] who proved it without using the Axiom of Choice (see [98, p. 158] for the beautiful proof in the case $n = 2$). Note that both Bernstein's Theorem and the classical Schröder-Bernstein Theorem ($|X| \leqslant |Y| \leqslant |X|$ implies $|X| = |Y|$) are quite easy if the Axiom of Choice is assumed. For, in the case of the former, Choice implies that $n|X| = |X|$ for X infinite, and for the latter, $|X| \leqslant |Y| \leqslant |X|$ implies that $\aleph_\alpha \leqslant \aleph_\beta \leqslant \aleph_\alpha$, where $\aleph_\alpha, \aleph_\beta$ are the cardinals corresponding to X, Y respectively, whence $\aleph_\alpha = \aleph_\beta$. What is noteworthy about the two results is that they are theorems of ZF rather than ZFC. In the case of the Schröder-Bernstein Theorem, the Choiceless treatment proved extremely valuable, as it led to Banach's useful generalization to equidecomposability; the proof using $\aleph_\alpha, \aleph_\beta$ is of no help with equidecomposability types. With the Bernstein cancellation law, too, the proof using Choice ($n|X| = |X|$) does not help us deal with arbitrary types in $\mathscr{S}$, but the more constructive proof does. There is a catch, though. Bernstein's proof, unlike that of the Schröder-Bernstein Theorem, does not carry over verbatim to the context of $\mathscr{S}$. The necessity of *finite* decompositions when working in $\mathcal{S}$ requires the Axiom of Choice to be used at one point in the proof. The interested reader should compare Kuratowski's proof of 8.7 for $n = 2$ [116] with Bernstein's Theorem as presented by Jech [98, p. 158].

We follow a rather different route, pioneered by D. König. He saw [109] that results on matchings in infinite graphs (which use Choice) could be applied to cancellation laws, and after Kuratowski published a proof of 8.7 for $n = 2$, König showed [110] how results of Valkó and himself could be used to prove 8.7 for all $n \geq 1$.

A *graph* is a set V of vertices with a collection E of unordered pairs of distinct elements of V, called edges. We wish to allow graphs to have multiple edges; thus a single pair may appear repeatedly in E. A graph is *bipartite* if the vertex set splits into two pieces so that each edge has one vertex in each piece. The *degree* of a vertex v is the number of edges containing v, where multiple edges count multiply, and a graph is *k-regular* if all vertices have degree k. A *perfect matching* in a graph is a collection of edges spanning all vertices such that no two of these edges have a vertex in common. The cancellation law 8.7 turns out to be a simple consequence of the following basic result of graph theory.

Theorem 8.11 (König's Theorem) (AC).   *A k-regular bipartite graph* $(k < \infty)$ *has a perfect matching.*

Proof.   For finite graphs, this result appears in most graph theory texts and is a consequence of a result commonly known as The Marriage Theorem (the reader may consult [19, p. 73] or [128, Problems 7.4, 7.10, 7.14]). In fact, the latter reference contains three proofs of this result, the simplest of which proceeds by induction on the number of vertices in one of the two parts.

Because only k-regular graphs are considered, we may reduce the general case to the case of countably many vertices (and edges). This is because for each finite n, at most k^n vertices can be reached from a given vertex by a path of length n, and hence the connected component of any vertex is at most countable. Now, we may split any graph into its connected components, and any union of a set of matchings, one for each component, will be a matching of the entire graph. Thus it remains to prove the result under the assumption that V and E are countably infinite. Note that the countability of the edges in a component depends on first selecting a vertex from which to enumerate paths; doing this simultaneously for all components requires the Axiom of Choice.

Suppose then that V and E are countable, say $E = \{e_n : n \in \mathbf{N}\}$. Consider the collection of all finite sequences of 0's and 1's ordered by: $s_1 \leq s_2$ if s_1 is an initial segment of s_2. If s is such a sequence of length n, s will be called *good* if there is some finite, k-regular graph (V', E') where V' contains all vertices appearing in $\{e_0, \ldots, e_{n-1}\}$, E' contains $\{e_0, \ldots, e_{n-1}\}$, and (V', E') has a perfect matching M such that, for $i < n$, $e_i \in M$ if and only if $s(i) = 1$.

Now, if $s_1 \leqslant s_2$ and s_2 is good, so is s_1. Moreover, there are infinitely many good sequences, at least one of each length. This is because any finite bipartite graph of maximum degree k can be extended to a finite, k-regular bipartite graph. Add some vertices to equalize the two parts and then add edges wherever necessary to push each degree up to k. Such a k-regular extension has a perfect matching by the finite case, and such a matching induces a good sequence of length n, the sequence being defined according to which of the e_i are included in the matching. We may now define an infinite sequence $s : \mathbb{N} \to \{0, 1\}$ by induction so that each initial segment of s is good; let $s(n)$ be whichever of 0 or 1 satisfies the condition; $s(0)s(1)\ldots s(n)$ is good and has infinitely many good extensions. If they both work, let $s(n) = 0$. This infinite sequence induces a set of edges $\{e_n \in E : s(n) = 1\}$ that is easily seen to be a perfect matching. To see that a given vertex v is covered, use the fact that the initial segment s of length n is good, where n is so large that all of the k neighbors of v lie in $\{e_0, \ldots, e_{n-1}\}$.   □

The cancellation law is a consequence of König's Theorem. In fact, just like the Schröder-Bernstein Theorem, the proof works for the situation where equidecomposability is replaced by any relation on sets satisfying properties (a) and (b) of 3.5's proof.

Proof of 8.7.   If $n\alpha = n\beta$, then there are two disjoint, bounded, G^*-equidecomposable sets $E, E' \subseteq X^*$ with partitions $E = A_1 \cup \cdots \cup A_n$, $E' = B_1 \cup \cdots \cup B_n$ such that each $[A_i] = \alpha$ and each $[B_i] = \beta$. Let $\chi : E \to E'$ be the piecewise G^*-congruence witnessing $E \sim E'$, and let $\phi_i : A_1 \to A_i$, $\psi_i : B_1 \to B_i$ likewise withness G^*-equidecomposability (take ϕ_1 and ψ_1 to be the identity). For each $a \in A_1$ let $\bar{a} = \{a, \phi_2(a), \ldots, \phi_n(a)\}$ and for $b \in B_1$ let $\bar{b} = \{b, \psi_2(b), \ldots, \psi_n(b)\}$. Note that $\{\bar{a} : a \in A_1\}$, $\{\bar{b} : b \in B_1\}$ form partitions of E, E', respectively.

Form a bipartite graph by letting one part of the vertex set be $\{\bar{a} : a \in A_1\}$, while $\{\bar{b} : b \in B_1\}$ is the other part. There will be n edges emanating from each $\bar{a}$: for each $i = 1, \ldots, n$ form an edge from $\bar{a}$ to that $\bar{b}$ such that $\chi\phi_i(a) \in \bar{b}$. This graph is n-regular—each $\bar{b}$ is in n edges, one corresponding to each $\chi^{-1}\psi_i(b)$—and hence by König's Theorem has a perfect matching, M. For each vertex $\bar{a}$, there is a unique edge $\{\bar{a}, \bar{b}\}$ in M, and $\bar{a}$ is connected to $\bar{b}$ by virtue of $\chi\phi_i(a) = \psi_j(b)$ for some i, j. Let $C_{ij} = \{a \in A_1 : \{\bar{a}, \bar{b}\} \in M$ and $\chi\phi_i(a) = \psi_j(b)\}$; similarly, let $D_{ij} = \{b \in B_1 : \{\bar{a}, \bar{b}\} \in M$ and $\chi\phi_i(a) = \psi_j(b)\}$. Then $\psi_j^{-1}\chi\phi_i$ maps C_{ij} bijectively onto D_{ij}, and is a piecewise G^*-congruence since each of $\phi_i, \chi, \psi_j^{-1}$ are. Since $\{C_{ij}\}$, $\{D_{ij}\}$ partition A_1, B_1, respectively, this shows $A_1 \sim B_1$, or $\alpha = \beta$, as desired.   □

For reasons to be discussed shortly, it is a bit unfortunate that the proof of König's Theorem, and hence of the cancellation law, uses the Axiom of Choice. It is easy to see that König's Theorem for 2-regular graphs is equivalent to the Axiom of Choice for collections of sets, each of which has size 2. For if X is a collection of pairs, form a bipartite graph with parts A, B as follows: $A = \{(a, x) : a \in x \in X\}$, $B = \{(0, x), (1, x) : x \in X\}$ with each (a, x) adjacent to each (i, x). This bipartite graph is 2-regular, and a matching induces a choice function f for X by letting $f(x)$ be the $a \in x$ such that (a, x) is matched to $(0, x)$. The other direction is proved by observing that each connected component in a 2-regular bipartite graph has exactly two matchings. Choosing one for each component yields a matching for the entire graph. Now, the Axiom of Choice for collections of pairs is known to be independent of ZF set theory (see [98, §5.4]); hence König's Theorem cannot be proved in ZF alone.

EQUIDECOMPOSABILITY WITH RESTRICTIONS
ON THE PIECES

In earlier chapters, we have seen that it may be natural or useful when dealing with equidecomposability to restrict the pieces in some way; see, for example, Questions 3.12 and 3.13 or Corollary 7.7. In such situations a semigroup of types can still be formed, but its algebraic structure may not be as nice as the case of arbitrary pieces. Suppose a group G acts on a set X in such a way that G leaves a certain subalgebra $\mathscr{A}$ of $\mathscr{P}(X)$ invariant, that is, $\sigma(A) \in \mathscr{A}$ whenever $A \in \mathscr{A}$ and $\sigma \in G$. (By a subalgebra we mean a collection of subsets of X containing ϕ, X and closed under finite union and intersection, and complementation in X.) Question 3.12 deals with the case where $\mathscr{A}$ consists of those subsets of the sphere having the Property of Baire and G is the group of rotations. Further examples arise by letting $\mathscr{A}$ consist of all Borel subsets of a metric space (where G is a group of isometries), or all Borel subsets of a topological group acting on itself by left multiplication. In these examples $\mathscr{A}$ is a σ-algebra; that is, $\mathscr{A}$ is closed under countable unions and intersections. For such $\mathscr{A}$ it is easy to see that the proof of the Banach-Schröder-Bernstein Theorem (3.5) applies to G-equidecomposability in $\mathscr{A}$; the proof uses a countable union at one point, and so the sets it constructs remain in $\mathscr{A}$ provided $\mathscr{A}$ is a σ-algebra.

To form a semigroup of equidecomposability types for $\mathscr{A}$, let X^* and G^* be defined as in 8.1, but consider only those bounded subsets A of X^* such that each level of A corresponds to a set in $\mathscr{A}$. Precisely, let $\mathscr{A}^* = \{A \subseteq X^* : \text{for some } n \in \mathbb{N}, A = \bigcup_{m < n} A_m \times \{m\} \text{ where each } A_m \in \mathscr{A}\}$ and define the type of $A \in \mathscr{A}^*$ to consist of all sets $B \in \mathscr{A}^*$ such that $B \sim_{G^*} A$ using pieces in $\mathscr{A}^*$. Let the collection of such types be denoted by $\mathscr{S}(\mathscr{A})$. If we define $+$ in $\mathscr{S}(\mathscr{A})$ as we did in $\mathscr{S}$(Definition 8.2), then $\mathscr{S}(\mathscr{A})$ is a commutative semigroup. The proof of the cancellation law, however, does not carry over to $\mathscr{S}(\mathscr{A})$, even if $\mathscr{A}$ is a σ-algebra. This is because the proof used the Axiom of Choice, which in general produces sets outside of $\mathscr{A}$: the sets C_{ij}, D_{ij}, which witnessed $A_1 \sim B_1$, were defined from the matching M, which need not lie in $\mathscr{A}$ since it was nonconstructively provided by König's Theorem.

In $\mathbf{R}^n$ the restriction to Borel pieces eliminates paradoxical decompositions of sets such as balls or cubes, because of the existence of Lebesgue measure. But there is another way of looking at equidecomposability in $\mathbf{R}^n$ that leads to an intriguing problem if $n \geqslant 2$. An open subset of $\mathbf{R}^n$ is called *regular-open* if it equals the interior of its closure; loosely speaking, the open set can have no cracks. The union of two regular-open sets is not necessarily regular-open, but there is another natural binary operation that is very similar to union. If A and B are regular-open, let $A \vee B$ be $\text{int}(\overline{A \cup B})$, where $\text{int}(E)$ denotes the interior of E; it is easy to check that $A \vee B$ is regular-open and $\vee$ is associative. Moreover, $A_1 \vee A_2 \vee \cdots \vee A_n = \text{int}(\overline{A_1 \cup \cdots \cup A_n})$. The importance of this notion is that the regular-open sets in any topological space X form a complete Boolean algebra, with $A \wedge B = A \cap B$, $A' = \text{int}(X \setminus A)$ and $\sum \{A_i : i \in I\} = \text{int} \, \overline{\bigcup \{A_i : i \in I\}}$ (see [37, p. 53]). We are interested in the collection of bounded regular-open subsets of $\mathbf{R}^n$, which we denote by $\mathscr{R}$. Now, we can define equidecomposability in $\mathscr{R}$ with respect to G_n, the group of isometries of $\mathbf{R}^n$, in the usual way, using $\vee$ instead of $\bigcup$ and with the restriction that all pieces come from $\mathscr{R}$. Then let $[A]$, for $A \in \mathscr{R}$, be the collection of sets that are equidecomposable in $\mathscr{R}$ with A, and let $\mathscr{S}(\mathscr{R}) = \{[A] : A \in \mathscr{R}\}$. Unlike the general case, where it was necessary to expand X to $X \times \mathbf{N}$, this definition of $\mathscr{S}(\mathscr{R})$ is adequate for adding types. Translations can be used to obtain as many disjoint copies of a bounded set as required, whence addition of types may be defined by $[A] + [B] = [A \cup \tau(B)]$, where τ is a translation taking B to a set disjoint from A.

Now, for the same reason as given at the beginning of this section, it is unclear whether the cancellation law holds in $\mathscr{S}(\mathscr{R})$, even for the line.

Question 8.12. Does $2\alpha = 2\beta$ imply $\alpha = \beta$ in $\mathscr{S}(\mathscr{R})$, where $\mathscr{R}$ is the collection of bounded regular-open subsets of $\mathbf{R}^1$?

There is a much more basic question about $\mathscr{S}(\mathscr{R})$, however, that is unsolved in dimensions three and beyond and that might make the cancellation law quite trivially true. Namely, does $\mathscr{S}(\mathscr{R})$ consist of more than two elements? Surprisingly, it is conceivable that the only types in $\mathscr{S}(\mathscr{R})$ are $[\varnothing] = \{\varnothing\}$ and $[J]$, where J is the open unit cube. If this were so, then $[J]$ would equal $\mathscr{R} \setminus \{\varnothing\}$; that is, any bounded, nonempty regular-open set would be equidecomposable with J, in the sense of $\mathscr{R}$.

We have already seen that all bounded nonempty open sets in $\mathbf{R}^n (n \geqslant 3)$ are indeed equidecomposable—this is a consequence of the strong form of the Banach-Tarski Paradox—but that result allows the use of arbitrary sets in the decomposition. Equidecomposability in $\mathscr{R}$ uses only regular-open sets as pieces. Furthermore, the existence of a nice measure on the open sets, Lebesgue measure, does not help, because λ is not additive with respect to the join operation, $\vee$, which is used to define equidecomposability in $\mathscr{R}$. We show why additivity fails in $\mathbf{R}^1$, but the example generalizes easily to $\mathbf{R}^n$.

Proposition 8.13. *For any* $\varepsilon > 0$, *the open unit interval* $(0, 1)$ *contains disjoint regular-open sets* A_1 *and* A_2 *such that* $A_1 \vee A_2 = (0, 1)$, *but* $\lambda(A_1) + \lambda(A_2) < \varepsilon$.

Proof. Let $\{q_n : n \in \mathbf{N}\}$ enumerate the rationals in $(0, 1)$. Choose open intervals I_n such that I_n is centered at q_n, has an irrational radius less than $\varepsilon 2^{-n+1}$, and is disjoint from the previously defined intervals. But if q_n already lies in I_m, $m < n$, let $I_n = \varnothing$. Let $A = \bigcup I_n$, an open dense set with $\lambda(A) < \varepsilon$. Let I_n' be the open interval forming the left half of I_n, while I_n'' is the (open) right half. Finally, let $A_1 = \bigcup I_n'$, $A_2 = \bigcup I_n''$.

It is clear that $\overline{A_1 \cup A_2}$ contains all rationals in $(0, 1)$; hence $\overline{A_1 \cup A_2} = [0, 1]$ and $A_1 \vee A_2 = (0, 1)$. Moreover, $\lambda(A_1) = \lambda(A_2) = \frac{1}{2}\lambda(A)$, so $\lambda(A_1) + \lambda(A_2) < \varepsilon$. Since A_1 and A_2 are open, it remains to show that they are regular-open. For this, it suffices to show that any point x in the interior of $\overline{A}_1$ lies in A_1. Let I be an open interval containing x, and contained in $\overline{A}_1$. Then I is disjoint from A_2. Let $q < x$ be a rational in I; since $I \cap A_2 = \varnothing$, the interval I_n containing q must be centered at a rational to the right of I, which implies that $x \in A_1$. Similarly, $A_2 \in \mathscr{R}$. □

The previous example is based on the existence of an open set, namely A, whose boundary has positive measure. Since the boundary of an open set is nowhere dense, a finitely additive measure will be additive with respect to $\vee$ as well if, unlike λ, it vanishes on all nowhere dense sets; for the difference between $A \cup B$ and $A \vee B$ is nowhere dense. The techniques of Part II (Corollary 11.2) will show how, in $\mathbf{R}^1$ and $\mathbf{R}^2$, the solvability of the isometry group allows one to construct a finitely additive, isometry-invariant measure, μ, on all bounded subsets of $\mathbf{R}^n (n \leqslant 2)$ such that $\mu(J) = 1$ and $\mu(E) = 0$ for all meager sets E. Such a measure μ, unlike λ, *is* additive with respect to $\vee$, and hence yields that neither an open interval in $\mathbf{R}$ nor a square in the plane is paradoxical in the sense of $\mathscr{R}$. But this technique does not apply to $\mathbf{R}^3$, leaving the question whether the open unit cube in $\mathbf{R}^3$ is paradoxical in the sense of $\mathscr{R}$.

Question 8.14. If J is the open unit cube in $\mathbf{R}^3$, does $2[J] = [J]$ in $\mathscr{S}(\mathscr{R})$? Equivalently, are all bounded, nonempty, regular-open subsets of $\mathbf{R}^3$ equidecomposable in the sense of $\mathscr{R}$?

To see that the two versions of this question are equivalent, note that if J is paradoxical in $\mathscr{R}$, then so is any open cube (use similarities as at the end of 7.3's proof). Hence the technique of Theorem 3.11 can be used, that is, apply the Banach-Schröder-Bernstein Theorem. (The validity of the Banach-Schröder-Bernstein Theorem in $\mathscr{S}(\mathscr{R})$ follows from the same proof as that given in 3.5, where one replaces $A \cup B$ by $A \vee B$, $\bigcup_{n=0}^{\infty} C_n$ by $\sum C_n$ and $A \backslash B$ by $A \cap B'$.) Question 8.15 can also be formulated in a way that avoids explicit mention of regular-open sets or the join operation in $\mathcal{R}$; see Theorem 9.5 where it is shown that Question 8.14 is equivalent to Marczewski's Problem.

Paradoxical decompositions involving the join operation on regular-open sets can certainly be eliminated by a further restriction on the pieces. Let $\mathcal{R}_v$ be the collection of bounded, regular-open, Jordan measurable sets, and let v, for volume, denote Jordan measure. Recall that Jordan measure is the precursor of Lebesgue measure, using coverings by finite collections of intervals rather than countable collections. A bounded set is Jordan measurable if and only if its boundary has Jordan measure zero if and only if its characteristic function is Riemann integrable (see Appendix B). Define equidecomposability in $\mathcal{R}_v$ using the join operation ($A \vee B \in \mathcal{R}_v$ if $A, B \in \mathcal{R}_v$) and pieces in $\mathcal{R}_v$ (and the full isometry group, G_n). Since the sets in $\mathcal{R}_v$ are a mathematical analog of physical bodies, we shall say that A and B are *equidecomposable as geometric bodies* when they are equidecomposable in $\mathcal{R}_v$ in the sense just defined. By the characterization of Jordan measurable sets in terms of boundaries, $v(A \vee B) = v(A \cup B)$, whence the finite additivity of v yields that v is an invariant for equidecomposability: if $A, B \in \mathcal{R}_v$ are equidecomposable as geometric bodies, then $v(A) = v(B)$. Is the converse true; that is, is volume a complete invariant for equidecomposability?

Question 8.15.   Is it true that in each $\mathbf{R}^n$, any two sets in $\mathcal{R}_v$ having the same volume are equidecomposable as geometric bodies?

An affirmative answer would provide a generalization of the Bolyai-Gerwien Theorem that is valid in all dimensions, since the elementary notion of congruence by dissection (3.1) is a special case of equidecomposability for geometric bodies. Moreover, it would provide some evidence for a possible solution to Questions 3.13 and Tarski's Circle-Squaring Problem (7.5). For if a circle, C, and a square, J, are equidecomposable as geometric bodies, then there are nowhere dense sets P_1, P_2 such that $C \backslash P_1$ and $J \backslash P_2$ are equidecomposable using regular-open pieces; if true, this would be an interesting approximation to a solution of 7.5. Note that even for $\mathbf{R}^1$, the question of whether Jordan measure is a complete invariant for equidecomposability in $\mathcal{R}_v$ is open. For example, let A be the regular-open set $(1/3, 2/3) \cup (7/9, 8/9) \cup (25/27, 26/27) \cup \dots$. Is A equidecomposable in $\mathcal{R}_v$ to $(0, 1/2)$? Moreover, as C.A. Rogers has observed, it is not even known whether A is equidecomposable to $(0, 1/2)$ using arbitrary sets as pieces. This question may be viewed as a one-dimensional analogue of Tarski's Circle-Squaring Problem (7.5).

We close this chapter with a result that answers Question 8.15 in a very simple case. By an *interval* of $\mathbf{R}^n$ we mean a set of the form $I_1 \times I_2 \times \cdots \times I_n$, where each I_i is an interval on the line.

Proposition 8.16.   *Any two open intervals of* $\mathbf{R}^n$ *having the same volume are equidecomposable as geometric bodies. Moreover, any two intervals of the same volume are equidecomposable in the usual sense, using pieces that are Borel sets.*

Proof.   By transitivity, it suffices to show that any open interval is equidecomposable in $\mathscr{R}_v$ to a cube (of the same volume). For $\mathbf{R}^1$ this is trivial, and for $\mathbf{R}^2$ it follows by the part of the Bolyai-Gerwien Theorem that deals with rectangles. For higher dimensions, assume inductively that the result is valid for intervals of dimension $n - 1$, and let $I = I_1 \times \cdots \times I_n$ be an interval in $\mathbf{R}^n$. Denote the length of an interval by ℓ. Then, by the two-dimensional case, $I \sim K_1 \times K_2 \times I_3 \times \cdots \times I_n$ in $\mathscr{R}_v$, where $\ell(K_1) = v(I)^{1/n}$ and $\ell(K_2) = \ell(I_1) \cdot \ell(I_2)/\ell(K_1)$. And the induction hypothesis yields that $K_2 \times I_3 \times \cdots \times I_n$ is equidecomposable in $\mathscr{R}_v$ to a cube of volume $v(I)^{(n-1)/n}$. It follows that $K_1 \times K_2 \times I_3 \times \cdots \times I_n$, and hence I, is equidecomposable in $\mathscr{R}_v$ to a cube of volume $v(I)$. The assertion about Borel sets follows by replacing the phrase "equidecomposable in $\mathscr{R}_v$" in the proof with "equidecomposable using Borel sets," and invoking Theorem 3.8 instead of the Bolyai-Gerwien Theorem. Note that the proof of 3.8 uses only Borel sets. $\square$

NOTES

The idea of expanding the action of G on X to add equidecomposability types and form a semigroup is due to Tarski [235, p. 601]. He discovered a fundamental result on measures in semigroups that he applied to this particular semigroup (see 9.1 and 9.2).

Jointly free actions stem from work of von Neumann [246] who essentially proved 8.5 with the conclusion: X is G-negligible. The idea of decomposing a locally commutative action into two jointly free actions (Corollary 8.6) is due to Wagon [247].

The Cancellation Law for equidecomposability, Theorem 8.7, was proved for $n = 2$ by Kuratowski [116], who modified a proof of Bernstein's analogous law for cardinality, due to Sierpiński [210]. Banach and Tarski [11] deduced the case $n = 2^m$ from the case $n = 2$, and this was sufficient for some applications, for example, 8.9 and 8.10. See [236, p. 33] for further references to Bernstein's Theorem and for an abstract treatment of many algebraic laws derived from cardinality considerations. Dénes König [110] realized that Kuratowski's result followed from his matching theorem (8.11), which he had proved much earlier for all finite graphs and all infinite 2-regular graphs [109]. König also knew that the countable case of 8.11 for n-regular graphs yielded the general case. Finally, König and Valkó [112] proved the countable case, yielding the complete matching theorem and cancellation law. The technique used by König and Valkó is based on a result that has come to be known as the König Tree Lemma: every infinite, finitely branching tree has an infinite branch (see [173, p. 287]). In a later paper, König [111] isolated this lemma and gave several other applications, including a proof that a countable graph is n-colorable ($n < \infty$) if and only if all of its finite subgraphs are. The proof of 8.11 presented here avoided explicit mention of the Tree Lemma, since in its general form it requires the Axiom of Choice (see [98, p. 115]). The connections between the König Matching Theorem and Choice for collections of pairs were pointed out to the author by F. Galvin and H. Läuchli.

Corollary 8.10, the analog of the strong form of the Banach-Tarski Paradox for S^2 rather than $\mathbf{R}^3$, is due to Banach and Tarski [11]. Their proof was quite different than the one presented here. It used the analog of the Bolyai-Gerwien Theorem for spherical polygons [76].

Equidecomposability using regular-open sets, $\mathscr{R}$, and regular-open Jordan measurable sets, $\mathscr{R}_v$, was introduced by Mycielski [166], who posed Questions 8.12, 8.14 (which he realized is equivalent to Marczewski's Problem), and 8.15.

Proposition 8.16 is a special case of a result of Hadwiger [86, p. 25], who gave necessary and sufficient conditions for the geometric equidecomposability of two n-dimensional parallelotopes using translations. The extension of his result to set-theoretic equidecomposability was provided by Kummer [114].

PART II

Finitely Additive Measures, or the Nonexistence of Paradoxical Decompositions

Chapter 9

Transition

This chapter discusses a result of Tarski that ties the two halves of this book tightly together. The first half was devoted to constructing paradoxical decompositions; the second half is devoted to constructing invariant measures, which provide a way of showing that paradoxical decompositions do not exist. Tarski's result is that this technique is essentially the only one: if a paradoxical decomposition does not exist in a certain context, then a finitely additive, invariant measure does. Or, recalling Definition 2.4: if G acts on X and $E \subseteq X$, then E is G-paradoxical if and only if E is G-negligible.

In this chapter we shall prove this beautiful result of Tarski and discuss some applications. Then we summarize the state of equidecomposability theory when a countably infinite number of pieces are allowed, with and without restrictions on the type of pieces that may be used.

In order to prove the nontrivial direction of Tarski's Theorem—the construction of a measure in the absence of a paradoxical decomposition—we shall work in $\mathscr{S}$, the type semigroup of G's action on X. Since any finitely additive, G-invariant measure on $\mathscr{P}(X)$ is necessarily invariant under G-equidecomposability, such a measure, μ, induces a measure v from $\mathscr{S}$ into $[0, \infty]$: if $[A] \in \mathscr{S}$, let $v([A]) = \sum \mu(A_n)$, where $A = \bigcup A_n \times \{n\}$. Note that $v(\alpha + \beta) = v(\alpha) + v(\beta)$ for any $\alpha, \beta \in \mathscr{S}$. Conversely, any function $v: \mathscr{S} \to [0, \infty]$ with this additivity property induces a finitely additive, G-invariant measure on $\mathscr{P}(X)$: let $\mu(A) = v([A])$. Thus the type semigroup is a natural context for the construction of a G-invariant measure. Tarski saw how to formulate and prove a very general theorem on measures in semigroups that he applied to the type semigroup of a group action to obtain the main result of this chapter.

Recall that any commutative semigroup with an identity, $(\mathcal{T}, +, 0)$, admits an ordering defined by $\alpha \leqslant \beta$ if $\alpha + \delta = \beta$ for some $\delta \in \mathcal{T}$, and there is a natural multiplication $n\alpha$ of elements of $\mathcal{T}$ by natural numbers. If $\varepsilon \in \mathcal{T}$ is fixed, then an element, α, of $\mathcal{T}$ will be called *bounded* if for some $n \in \mathbf{N}$, $\alpha \leqslant n\varepsilon$. In the type semigroup of a group action, the ordering $\leqslant$ is a partial ordering (Banach-Schröder-Bernstein Theorem), but the following theorem applies to arbitrary commutative semigroups, even those where $\alpha \leqslant \beta$ and $\beta \leqslant \alpha$ might hold for distinct α, β.

Theorem 9.1 (AC). *Let $(\mathcal{T}, +, 0, \varepsilon)$ be a commutative semigroup with identity 0 and a specified element ε. Then the following are equivalent:*

(1) *For all $n \in \mathbf{N}$, $(n + 1)\varepsilon \nleqslant n\varepsilon$.*

(2) *There is a measure $\mu : \mathcal{T} \to [0, \infty]$ such that $\mu(\varepsilon) = 1$ and $\mu(\alpha + \beta) = \mu(\alpha) + \mu(\beta)$ for all $\alpha, \beta \in \mathcal{T}$. (Note that μ is a homomorphism of semigroups, from $\mathcal{T}$ to $([0, \infty], +)$.)*

Proof. That (2) implies (1) is clear, since any μ satisfying (2) is such that $\mu(\alpha) \leqslant \mu(\beta)$ if $\alpha \leqslant \beta$, and $\mu(n\varepsilon) = n$; hence $(n + 1)\varepsilon \leqslant n\varepsilon$ would imply $n + 1 \leqslant n$. For the other direction, we shall use a technique that will be one of the main tools for constructing measures in later chapters. The key is to exploit the compactness of the product space $[0, \infty]^{\mathcal{T}}$ (all functions from $\mathcal{T}$ into $[0, \infty]$) together with the observation that if μ fails to be a measure on $\mathcal{T}$ as in (2), then the failure is evident in some finite subset of $\mathcal{T}$. Note that the compactness of $[0, \infty]^{\mathcal{T}}$ is a consequence of Tychonoff's Theorem that products of compact spaces are compact (see [196]), which requires the Axiom of Choice. Without loss of generality, we assume that all elements of $\mathcal{T}$ are bounded (with respect to ε). For once we have a measure on the bounded elements, it may be extended by assigning the unbounded elements measure ∞. Most of the work will be in the proof of the following claim.

Claim. If $\mathcal{T}_0$ is a finite subset of $\mathcal{T}$ that contains ε, then there is a function $\mu : \mathcal{T}_0 \to [0, \infty]$ such that (i) $\mu(\varepsilon) = 1$, and (ii) if $\phi_i, \theta_j \in \mathcal{T}_0$ satisfy $\phi_1 + \cdots + \phi_m \leqslant \theta_1 + \cdots + \theta_n$, then $\sum \mu(\phi_i) \leqslant \sum \mu(\theta_j)$.

Before proving the claim, we show how it can be combined with compactness to produce the desired measure on all of $\mathcal{T}$. For any $\mathcal{T}_0$ as in the claim, let $\mathcal{M}(\mathcal{T}_0)$ consist of all $f \in [0, \infty]^{\mathcal{T}}$ satisfying (1) $f(\varepsilon) = 1$, and (2) $f(\alpha + \beta) = f(\alpha) + f(\beta)$ whenever $\alpha, \beta, \alpha + \beta \in \mathcal{T}_0$. This last additivity property is an easy consequence of property (2) of the claim, whence the claim implies that each $\mathcal{M}(\mathcal{T}_0)$ is nonempty. Now, the compactness of the product space $[0, \infty]^{\mathcal{T}}$ may be interpreted as: if a collection of closed subsets of $[0, \infty]^{\mathcal{T}}$ has the finite intersection property (any intersection of finitely many members of the collection is nonempty), then the intersection of all sets in the collection is nonempty. Each $\mathcal{M}(\mathcal{T}_0)$ is closed. This is because whether f is in $\mathcal{M}(\mathcal{T}_0)$ depends only on the finitely many coordinates mentioned in $f\upharpoonright\mathcal{T}_0$, and if f fails to satisfy (1) or (2), then either $f(\varepsilon) \neq 1$, or $f(\alpha + \beta) \neq f(\alpha) + f(\beta)$. In either case there is an open subset of $[0, \infty]^{\mathcal{T}_0}$ containing $f\upharpoonright\mathcal{T}_0$ on which the condition is violated—just vary either the ε-coordinate or the α-coordinate a little bit. Therefore $\mathcal{M}(\mathcal{T}_0)$ is closed. Note that $\mathcal{M}(\mathcal{T}_1) \cap \cdots \cap \mathcal{M}(\mathcal{T}_n) \supseteq \mathcal{M}(\mathcal{T}_1 \cup \cdots \cup \mathcal{T}_n)$, where each $\mathcal{T}_i$ is a finite subset of $\mathcal{T}$ containing ε. Since $\bigcup\mathcal{T}_i$ is also finite, the claim implies that $\mathcal{M}(\bigcup\mathcal{T}_i) \neq \varnothing$; it follows that $\{\mathcal{M}(\mathcal{T}_0) : \mathcal{T}_0 \text{ is a finite subset of } \mathcal{T} \text{ containing } \varepsilon\}$ has the finite intersection property. By compactness, there must be a μ that lies in each $\mathcal{M}(\mathcal{T}_0)$, and such a μ is a measure as desired: since $\mu \in \mathcal{M}(\{\varepsilon\})$, $\mu(\varepsilon) = 1$, and to see that $\mu(\alpha + \beta) = \mu(\alpha) + \mu(\beta)$, use the fact that $\mu \in \mathcal{M}(\{\varepsilon, \alpha, \beta, \alpha + \beta\})$.

Proof of Claim.   The proof will be done by induction on the size of $\mathcal{T}_0$. If $|\mathcal{T}_0| = 1$, then $\mathcal{T}_0 = \{\varepsilon\}$, and $\mu(\varepsilon) = 1$ is the desired function. In this case, property (ii) of the claim reduces to showing that if $m\varepsilon \leqslant n\varepsilon$, then $m \leqslant n$. But this is a consequence of (1), the theorem's hypothesis on ε; for if $m\varepsilon \leqslant n\varepsilon$ and $m \geqslant n + 1$, then $(n + 1)\varepsilon \leqslant m\varepsilon \leqslant n\varepsilon$, a contradiction. This is the only place in the proof where the hypothesis on ε is used.

Now, suppose $|\mathcal{T}_0| > 1$, and let α be any element of $\mathcal{T}_0\backslash\{\varepsilon\}$. Use the induction hypothesis to get a function v on $\mathcal{T}_0\backslash\{\alpha\}$ satisfying the claim. By our boundedness assumption, all elements are bounded by some $n\varepsilon$; hence v takes on only finite values. Define μ on $\mathcal{T}_0$ by letting μ agree with v on $\mathcal{T}_0\backslash\{\alpha\}$ and setting $\mu(\alpha) = \inf\{(\sum v(\beta_k) - \sum v(\gamma_i))/r\}$, where the inf is over all positive integers r and $\beta_1, \ldots, \beta_p, \gamma_1, \ldots, \gamma_q \in \mathcal{T}_0\backslash\{\alpha\}$ satisfying $\gamma_1 + \cdots + \gamma_q + r\alpha \leqslant \beta_1 + \cdots + \beta_p$. Since α is bounded, $\mu(\alpha)$ is the greatest lower bound of a nonempty set: if $\alpha \leqslant n\varepsilon$, then $\mu(\alpha) \leqslant n$. Note the similarity with the classical outer measure construction: $\mu(\alpha)$ is defined to be as large as possible subject to bounds it must have if the claim is to be satisfied.

Since a consequence of property (ii) of the claim is that $\mu(\alpha) \geqslant 0$ ($\varepsilon \leqslant \varepsilon + \alpha$ so $1 \leqslant 1 + \mu(\alpha)$), it remains only to prove that μ continues to satisfy property (ii). So suppose that $\phi_1 + \cdots + \phi_m + s\alpha \leqslant \theta_1 + \cdots + \theta_n + t\alpha$, where $\phi_i, \theta_j \in \mathcal{T}_0\backslash\{\alpha\}$ and $s, t \in \mathbf{N}$. If both s and t are 0, then the desired inequality follows from the fact that v satisfies the claim on $\mathcal{T}_0\backslash\{\alpha\}$. Consider first the case $s = 0$ and $t > 0$; we must show that $\sum v(\phi_i) \leqslant t\mu(\alpha) + \sum v(\theta_j)$, that is, that $\mu(\alpha) \geqslant w = (\sum v(\phi_i) - \sum v(\theta_j))/t$. Let $\gamma_1 + \cdots + \gamma_q + r\alpha \leqslant \beta_1 + \cdots + \beta_p$ be a typical inequality defining $\mu(\alpha)$; it suffices to show that $(\sum v(\beta_k) - \sum v(\gamma_\ell))/r \geqslant w$. Multiplying the given inequality $\phi_1 + \cdots + \phi_m \leqslant \theta_1 + \cdots + \theta_n + t\alpha$ by r and adding the same quantity to both sides yields

$$r\phi_1 + \cdots + r\phi_m + t\gamma_1 + \cdots + t\gamma_q \leqslant r\theta_1 + \cdots + r\theta_n + tr\alpha + t\gamma_1 + \cdots + t\gamma_q.$$

Substituting the inequality involving γ, α, β yields

$$r\phi_1 + \cdots + r\phi_m + t\gamma_1 + \cdots + t\gamma_q \leqslant r\theta_1 + \cdots + r\theta_n + t\beta_1 + \cdots + t\beta_q.$$

The induction assumption about v now yields

$$r \sum v(\phi_i) + t \sum v(\gamma_\ell) \leqslant r \sum v(\theta_j) + t \sum v(\beta_k),$$

which implies that $(\sum v(\beta_k) - \sum v(\gamma_\ell))/r \geqslant w$, as desired.

Finally, suppose $\phi_1 + \cdots + \phi_m + s\alpha \leqslant \theta_1 + \cdots + \theta_n + t\alpha$ where $s > 0$. It suffices to show that $s\mu(\alpha) + \sum v(\phi_i) \leqslant z_1 + \cdots + z_t + \sum v(\theta_j)$, where $z_1, \ldots, z_t$ are any of the numbers whose greatest lower bound defines $\mu(\alpha)$. By considering the smallest of $z_1, \ldots, z_t$ we may assume that these numbers are all the same. In other words, suppose that $\gamma_1 + \cdots + \gamma_q + r\alpha \leqslant \beta_1 + \cdots + \beta_p$, where $\gamma_\ell, \beta_k \in \mathcal{T}_0 \backslash \{\alpha\}$, and let $z = (\sum v(\beta_k) - \sum v(\gamma_\ell))/r$; then we must prove that $s\mu(\alpha) + \sum v(\phi_i) \leqslant tz + \sum v(\theta_j)$. Multiplying the ϕ, α, θ-inequality by r and adding the same quantity to both sides yields

$$r\phi_1 + \cdots + r\phi_m + rs\alpha + t\gamma_1 + \cdots + t\gamma_q$$
$$\leqslant r\theta_1 + \cdots + r\theta_n + rt\alpha + t\gamma_1 + \cdots + t\gamma_q,$$

and substituting the γ, α, β-inequality yields

$$r\phi_1 + \cdots + r\phi_m + t\gamma_1 + \cdots + t\gamma_q + rs\alpha \leqslant r\theta_1 + \cdots + r\theta_n + t\beta_1 + \cdots + t\beta_p.$$

This last inequality is a typical one used to define $\mu(\alpha)$, and it follows that $s\mu(\alpha) + \sum v(\phi_i)$ is bounded by

$$\sum v(\phi_i) + s\left(\frac{1}{rs}\right)\left(r \sum v(\theta_j) + t \sum v(\beta_k) - r \sum v(\phi_i) - t \sum v(\gamma_\ell)\right),$$

which equals $tz + \sum v(\theta_j)$, as required. $\qquad\qquad\square$

The reader familiar with transfinite induction will see that the proof could be shortened a little by using that technique rather than Tychonoff's Theorem. The claim allows one to construct μ by treating the elements of $\mathcal{T}$ one at a time, starting with ε and proceeding inductively through $|\mathcal{T}|$ steps; this was Tarski's original approach. We chose to illustrate the use of compactness, since this technique will appear several times in Part II.

Corollary 9.2 (*Tarski's Theorem*) (AC). *Suppose G acts on X and E $\subseteq X$. Then there is a finitely additive, G-invariant measure $\mu : \mathcal{P}(X) \to [0, \infty]$ with $\mu(E) = 1$ if and only if E is not G-paradoxical.*

Proof. The easy direction has already been discussed (Proposition 2.5) so assume E is not G-paradoxical and let $\mathcal{S}$ be the type semigroup of G's action on X. Then $2\varepsilon \neq \varepsilon$, where $\varepsilon = [E \times \{0\}]$, and hence Corollary 8.8 of the cancellation law for equidecomposability implies that $\mathcal{S}$, with the distinguished element ε, satisfies condition (1) of Theorem 9.1. Hence that theorem yields a measure v on $\mathcal{S}$ normalizing ε, whence $\mu(A) = v([A \times \{0\}])$ is the desired G-invariant measure on $\mathcal{P}(X)$. $\qquad\qquad\square$

Theorem 9.1 may also be applied to the case of equidecomposability where the pieces are restricted in some way. Suppose G acts on X and $\mathscr{A}$ is a G-invariant subalgebra of $\mathscr{P}(X)$. Let $\mathscr{S}(\mathscr{A})$ be the type semigroup of the induced action of G on $\mathscr{A}$, where only pieces in $\mathscr{A}$ can be used to witness the equidecomposability of sets in $\mathscr{A}$ (see p. 116). We do not have a general cancellation law in this context, and if Corollary 8.8 fails, it may be that, say, $3[E] = 2[E]$ even though $2[E] \neq [E]$ ($[E]$ stands for $[E \times \{0\}]$ in $\mathscr{S}(\mathscr{A})$). In such a case, no G-invariant measure on $\mathscr{A}$ normalizing E could exist, despite the fact that E is not G-paradoxical using pieces in $\mathscr{A}$. Nevertheless, Theorem 9.1 yields the following slightly weaker analogue of Corollary 9.2.

Corollary 9.3 (AC).   *Let G, X, $\mathscr{A}$ be as in the preceding paragraph and suppose $E \in \mathscr{A}$. Then the following are equivalent*:

(1) *For all $n \in \mathbf{N}$, $(n + 1)[E] \not\leqslant n[E]$ in $\mathscr{S}(\mathscr{A})$.*
(2) *There is a finitely additive, G-invariant measure $\mu \colon \mathscr{A} \to [0, \infty]$ with $\mu(E) = 1$.*

Proof.   If μ is as in (2), then μ induces a measure on $\mathscr{S}(\mathscr{A})$ satisfying condition (2) of Theorem 9.1 (see remarks preceding 9.1), and hence, as in that theorem, $(n + 1)[E] \not\leqslant n[E]$. For the other direction, apply Theorem 9.1 to $\mathscr{S}(\mathscr{A})$, with $\varepsilon = [E]$, and use the measure on $\mathscr{S}(\mathscr{A})$ so obtained to induce the desired measure on $\mathscr{A}$.   □

Tarski proved a more general version of these results dealing with the case of partial transformations of a set. We refer the reader to [236] for details, mentioning here one important application (see [236, p. 233]). Let $\mathscr{A}$ be the Borel subsets of a compact metric space X and let $\mathscr{F}$ consist of all partial isometries, that is, distance-preserving bijections with domain (and hence image (see remarks preceding Question 3.13)) in $\mathscr{A}$. Then Tarski's work yields the equivalence of the existence of a finitely additive, congruence-invariant (i.e., $\mathscr{F}$-invariant) Borel measure on X of total measure one with the assertion that for any n, $(n + 1)[X] \not\leqslant n[X]$ in the type semigroup formed from $\mathscr{F}$-equidecomposability. The Banach-Ulam Problem asks if such a measure exists in every compact metric space; see the discussion following Question 3.13 for some partial results.

Since Corollary 9.3 does not appeal to the cancellation law for equidecomposability, the only point that requires the Axiom of Choice is the compactness of $[0, \infty]^{\mathscr{S}(\mathscr{A})}$. If it happens that $\mathscr{A}$ is a countable subalgebra of $\mathscr{P}(X)$, however, then $\mathscr{S}(\mathscr{A})$ is countable too, and the compactness of a countable product of copies of $[0, \infty]$ can be proved without using the Axiom of Choice. Hence, in this case, Corollary 9.3 is valid without Choice. This idea will be important in some later applications of the compactness technique of Theorem 9.1, so we give the effective compactness proof here. It should be pointed out that the more general assertion that X^s is compact whenever X is compact and S is countable is equivalent to the Axiom of Choice for countable families of sets.

Proposition 9.4.   *If S is a countable set, then the product space $[0, \infty]^S$ is compact.*

Proof.   Since $[0, \infty]$ and $[0, 1]$ are homeomorphic, it suffices to show that $[0, 1]^N$ is compact. Suppose that $\{U_i\}$ is an open cover of $[0, 1]^N$ without a finite subcover; assume each U_i is a basic open set in $[0, 1]^N$. Consider the two sets $[0, 1/2] \times [0, 1] \times [0, 1] \times \ldots$ and $[1/2, 1] \times [0, 1] \times [0, 1] \times \ldots$. Choose the leftmost of these two sets that is not coverable by finitely many U_i and get four subsets of it by splitting the first coordinate in two again and splitting the second coordinate in two as well. Thus a typical set at this second stage might be $[1/4, 1/2] \times [1/2, 1] \times [0, 1] \times [0, 1] \times \ldots$. Now choose the (lexicographically) least one of these four sets that cannot be covered by finitely many U_i and get eight subsets of it by halving in three coordinates. Continue for $\aleph_0$ steps, obtaining a decreasing sequence that determines a unique element r of $[0, 1]^N$. One of the U_i contains r. But U_i is determined by open sets in finitely many coordinates, whence U_i contains one of the sets in the sequence defining r, contradiction.   □

EQUIVALENT FORMULATIONS OF MARCZEWSKI'S PROBLEM

The preceding results allow us to reformulate some of our previous questions on paradoxical decompositions in measure-theoretic terms. We shall examine Marczewski's Problem (3.12) in some detail. Let $\mathscr{B}$ denote the collection of subsets of $\mathbf{R}^n$ (or S^n) having the Property of Baire, and let the term $\mathscr{B}$-*measure* denote a finitely additive, G_n-invariant (or O_{n+1}-invariant) measure on $\mathscr{B}$ that normalizes the unit cube (or S^n). The existence of a $\mathscr{B}$-measure on $\mathbf{R}^n$ implies that the unit cube is not paradoxical using pieces with the Property of Baire. Moreover, a $\mathscr{B}$-measure on $\mathbf{R}^n$ induces a $\mathscr{B}$-measure on S^{n-1}, using the technique of adjoining radii, and hence implies that S^{n-1} is not paradoxical using pieces with the Property of Baire. Thus a $\mathscr{B}$-measure on either $\mathbf{R}^3$ or S^2 would settle Question 3.12.

Because we do not know whether the cancellation law holds for equidecomposability in the algebra of sets with the Property of Baire, it is not immediate (via 9.2) that the existence of a $\mathscr{B}$-measure can be equated with the lack of a paradoxical decomposition. Nevertheless, such an equivalence is true, at least in the case of $\mathbf{R}^n$. For there one can deduce the important Corollary 8.8 of the cancellation law. The following theorem shows how to do this, and shows as well why Marczewski's Problem for $\mathbf{R}^n$ is equivalent to a question raised earlier (8.14) about paradoxes in $\mathscr{R}$, the algebra of bounded, regular-open subsets of $\mathbf{R}^n$.

Theorem 9.5 (AC). *Let $\mathscr{S}(\mathscr{B})$, $\mathscr{S}(\mathscr{R})$ be the type semigroups with respect to G-equidecomposability in $\mathscr{B}$, $\mathscr{R}$, respectively, as defined in Chapter 8. J denotes the (open) unit cube in $\mathbf{R}^n$. The following are equivalent:*

(1) *For all $m \in \mathbf{N}$, $(m + 1)[J] \nleqslant m[J]$ in $\mathscr{S}(\mathscr{B})$.*
(2) *J is not paradoxical in $\mathscr{B}$.*
(3) *There exists a $\mathscr{B}$-measure on $\mathbf{R}^n$.*
(4) *For all $m \in \mathbf{N}$, $(m + 1)[J] \nleqslant m[J]$ in $\mathscr{S}(\mathscr{R})$.*
(5) *J is not paradoxical in $\mathscr{R}$.*
(6) *$|\mathscr{S}(\mathscr{R})| > 2$, that is, $\mathscr{S}(\mathscr{R}) \neq \{\{\varnothing\}, [J]\}$.*
(7) *There is a finitely additive (with respect to $\vee$), G_n-invariant measure $\mu : \mathscr{R} \to [0, \infty)$ with $\mu(J) = 1$.*
(8) *It is not the case that every bounded, nonempty open subset U of $\mathbf{R}^n$ contains pairwise disjoint open subsets $U_1, \ldots, U_m$ such that for suitable isometries $\rho_1, \ldots, \rho_m$, $\bigcup \rho_i U_i$ is dense in J.*
(9) *No open dense subset of J can be packed into a proper subcube of J using finitely many pieces in $\mathscr{B}$.*

Proof.   Corollary 9.3 yields (1) $\Leftrightarrow$ (3) and (2) $\Rightarrow$ (1) can be proved as follows. Suppose $(m + 1)[J] \leqslant m[J]$ in $\mathscr{S}(\mathscr{B})$; using a similarity shows that this is true for any cube. Choose a cube K small enough that m copies of K fit into J. Then choose k large enough that $(m + k)[K] \geqslant 2[J]$. Since $(m + 1)[K] \leqslant m[K]$ yields $(m + k)[K] \leqslant m[K]$, we have: $m[K] \leqslant [J] \leqslant 2[J] \leqslant (m + k)[K] \leqslant m[K]$; hence the Banach-Schröder-Benstein Theorem yields $[J] = 2[J]$. Since (1) $\Rightarrow$ (2) is obvious, this proves (1) $\Leftrightarrow$ (2) and hence the equivalence of (1), (2), and (3).

Exactly the same technique, using Theorem 9.1 in the semigroup $\mathscr{S}(\mathscr{R})$, yields the equivalence of (4), (5), and (7). Statement (5) implies that $[J] \neq 2[J]$ in $\mathscr{S}(\mathscr{R})$, yielding (6). For the converse, recall that it was proved following Question 8.14 that if J is paradoxical in $\mathscr{R}$ then $[U] = [J]$ for any nonempty $U \in \mathscr{R}$, which yields (6). Hence statements (4)–(7) are equivalent.

These two groups of statements will be tied together by considering the assertions about measures, (3) and (7). The proof that (7) $\Rightarrow$ (3) hinges on the fact that any set $E \in \mathscr{B}$ can be uniquely represented as $A \bigtriangleup P$ where A is regular-open and P is meager (see [181, p. 20]). Suppose μ is as in (7); extend μ to all regular-open sets by setting $\mu(A) = \infty$ if A is unbounded. Then define ν on $\mathscr{B}$ by $\nu(A) = \mu(E)$ where $A = E \bigtriangleup P$ is the representation just mentioned. The uniqueness of this representation and the fact that $(A_1 \vee A_2) \backslash (A_1 \cup A_2)$ is nowhere dense yield that ν is a $\mathscr{B}$-measure. The vonverse needs some auxiliary results and will be discussed after we treat (8) and (9). We shall show that a $\mathscr{B}$-measure, when restricted to $\mathscr{R}$, satisfies the conditions of (7).

To deal with (8) and (9) we prove that (5) $\Rightarrow$ (9) $\Rightarrow$ (8) $\Rightarrow$ (6). To see that (5) $\Rightarrow$ (9), suppose $A_i \in \mathscr{B}$ and $\rho_i \in G_n$ witness the packing of an open dense subset of J into a proper subcube, K. As in the proof of (7) $\Rightarrow$ (3), write

A_i as $E_i \triangle P_i$ where E_i is regular-open and P_i is meager. Then $(E_i \cap E_j) \backslash (P_i \cup P_j) \subseteq A_i \cap A_j = \varnothing$ if $i \neq j$, so $E_i \cap E_j \subseteq P_i \cup P_j$. But $P_i \cup P_j$ is meager, and so contains no nonempty open set (Baire Category Theorem; see [181, p. 2]); hence $E_i \cap E_j = \varnothing$ if $i \neq j$. Similarly, the disjointness of $\rho_i(E_i)$ and $\rho_j(E_j)$ follows from that of $\rho_i(A_i)$ and $\rho_j(A_j)$. The Baire Category Theorem also implies that $E_i \subseteq \bar{A}_i$; hence $\rho_i(E_i) \subseteq \rho_i(\bar{A}_i) = \overline{\rho_i(A_i)} \subseteq \bar{K}$, so $\rho_i(E_i) \subseteq \bar{K}$. The preceding remarks show that the packing of J into K yields that $[J] \leqslant [K]$ in $\mathscr{S}(\mathscr{R})$. Hence, if L is a cube contained in $J \backslash K$, $[K] + [L] \leqslant [K]$, and it follows that $[K] + m[L] \leqslant [K]$ for any m. Choose m so large that m copies of L cover K. Then $2[K] = [K] + [K] \leqslant [K] + m[L] \leqslant [K]$, so K is paradoxical in $\mathscr{R}$. Using a similarity yields that J is paradoxical in $\mathscr{R}$, contradicting (5). The implication (9) $\Rightarrow$ (8) is almost immediate: let U in (8) be a proper subcube of J. Finally, to prove (8) $\Rightarrow$ (6) suppose U witnesses (8). Then $V \thicksim J$ in $\mathscr{R}$ where V is the regular-open set int $(\bar{U})$. For if V contains pairwise disjoint, regular-open sets V_i with int $\bigcup \rho_i(V_i) = J$, then $\bigcup \rho_i(V_i \cap U)$ is dense in J, contradicting (8).

We now return to (3) $\Rightarrow$ (7), the proof of which uses an interesting and useful observation of Tarski. To place it in its proper context, we mention the following general problem. If μ is a finitely additive, G_n-invariant measure on the Lebesgue measurable subsets of $\mathbf{R}^n$, and $\mu(J) = 1$ where J is the unit cube, then to what extent must μ agree with Lebesgue measure, λ? This problem, which goes back to Lebesgue, was settled only recently and will be discussed in more detail later (see Theorem 11.11). Note that to have any hope of concluding that μ agrees with λ, one must restrict the discussion to bounded sets; for if μ is defined to agree with λ on the bounded Lebesgue measurable sets, but to give measure ∞ to unbounded measurable sets, then μ satisfies the hypothesis of the problem, but $\mu \neq \lambda$. First we observe that even under the weaker assumption of translation invariance, one can at least be assured that μ agrees with λ on certain elementary sets, namely those that are Jordan measurable.

Proposition 9.6. *Jordan measure, v, is the unique finitely additive, translation-invariant measure on the Jordan measurable sets in $\mathbf{R}^n$ that normalizes J.*

Proof. If E is Jordan measurable, then for any $\varepsilon > 0$ there are sets D, F such that $D \subseteq E \subseteq F$, D and E are each a union of finitely many pairwise disjoint open cubes with edges parallel to the coordinate axes, and $v(F) - v(D) < \varepsilon$ (see Appendix B). Hence, to show that μ, a given finitely additive, translation-invariant measure on the Jordan measurable sets with $\mu(J) = 1$, coincides with v, it suffices to show that $\mu(K) = v(K)$ for any properly oriented open cube K. We prove this for the plane, but the same idea works in all dimensions. Translation invariance implies that any finite set of bounded line segments gets μ-measure zero, since one could pack arbitrarily many

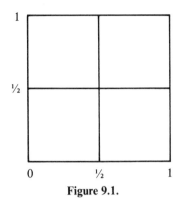

Figure 9.1.

translates of such a set into the unit square, J (see the proof of 3.8). This means we can ignore the boundaries of squares, and therefore because $\mu(J) = 1$, Figure 9.1 shows that μ is correct on any properly oriented square of side-length $1/2$. Subdividing repeatedly yields the correctness of μ on any properly oriented square of side-length $1/2^n$, and hence those of side-length $m/2^n$, m, $n \in \mathbf{N}$. Now, these dyadic rationals are dense in the positive reals; approximating an arbitrary properly oriented square by one with side-length $m/2^n$ yields the correctness of μ on all properly oriented squares, as desired. $\qquad \square$

The previous result is valid in all dimensions; Tarski observed that the Banach-Tarski Paradox could be used to provide an improvement in $\mathbf{R}^3$ and beyond. One of the main points of interest of the following lemma is that it turns the negative implications of the Banach-Tarski Paradox to positive advantage. By shifting the focus from all subsets of $\mathbf{R}^n$ to the bounded Lebesgue measurable ones, the lemma allows us to view the Banach-Tarski Paradox as giving additional information about the uniqueness of Lebesgue measure, rather than just saying that no measures of a certain sort exist on all of $\mathscr{P}(\mathbf{R}^n)$. In fact, the recent solution of the Ruziewicz Problem (discussed in Chapter 11) yields the uniqueness of Lebesgue measure as a finitely additive, invariant measure on the bounded Lebesgue measurable sets, if $n \geqslant 3$. Lemma 9.7, and hence the Banach-Tarski Paradox, is a small but crucial step in the solution. (See Theorem 11.11; Corollary 11.3 shows that Lemma 9.7, and hence the uniqueness of Lebesgue measure as a finitely additive measure, fails in $\mathbf{R}^1$ and $\mathbf{R}^2$.)

Lemma 9.7 (AC). *If $n \geqslant 3$ and μ is a finitely additive, G_n-invariant measure on the bounded, Lebesgue measurable subsets of $\mathbf{R}^n$ normalizing J (or, a $\mathscr{B}$-measure in $\mathbf{R}^n$), then μ vanishes on the bounded sets of Lebesgue measure zero (or, on the bounded meager sets).*

Proof. Suppose E is a bounded subset of $\mathbf{R}^n$ with $\lambda(E) = 0$, and K is a cube large enough to contain E. For any $\varepsilon > 0$, K is G_n-equidecomposable with a cube L of volume ε, by the strong form of the Banach-Tarski Paradox, Theorem 5.1(c). Suppose $A_1, \ldots, A_m \subseteq K$ are the pieces witnessing this decomposition. Now, by Proposition 9.6, μ agrees with volume on cubes, and hence $\mu(L) = \varepsilon$. However, this does not imply that $\mu(K) = \varepsilon$ because the pieces, A_i, will not be Lebesgue measurable, and hence not μ-measurable. But the sets $A_i \cap E$ must be μ-measurable: they are subsets of E, which has Lebesgue measure zero, and hence are Lebesgue measurable. It then follows from the packing of E into L, and the finite additivity and G_n-invariance of μ, that $\mu(E) \leqslant \mu(L)$, that is, $\mu(E) \leqslant \varepsilon$. Since ε was arbitrary, this means $\mu(E) = 0$. This proof works just as well in the case of meager and Property of Baire sets, because a subset of a meager set is meager, and hence has the Property of Baire and is μ-measurable. $\qquad\qquad\square$

The case of the preceding lemma dealing with $\mathscr{B}$-measures and meager sets allows us to prove $(3) \Rightarrow (7)$ of Theorem 9.5 under the additional assumption that $n \geqslant 3$. For if μ is a $\mathscr{B}$-measure, then Lemma 9.7 implies that μ vanishes on all meager sets. Hence all nowhere dense sets get μ-measure zero, and it follows from the remarks preceding Question 8.14 that μ itself, when restricted to $\mathscr{R}$, is additive with respect to $\vee$. In $\mathbf{R}^1$ and $\mathbf{R}^2$ $\mathscr{B}$-measures do exist, but do not necessarily vanish on the meager sets (see p. 168). But $\mathscr{B}$-measures that do vanish on the meager sets can be constructed (Corollary 11.3), and they yield the validity of (7) (and hence of $(3) \Rightarrow (7)$) in those two cases.

With the equivalences of Theorem 9.5 proved, we now reformulate Marczewski's Problem.

Definition 9.8. *A* Marczewski measure *in $\mathbf{R}^n$ (or S^n) is a $\mathscr{B}$-measure that vanishes on the bounded meager sets.*

Question 9.9 (Marczewski's Problem). Do Marczewski measures exist in $\mathbf{R}^3$ (or in any $\mathbf{R}^n$, $n \geqslant 3$)?

Lemma 9.7 shows that in $\mathbf{R}^3$ and beyond a $\mathscr{B}$-measure is necessarily a Marczewski measure, but these two notions do not coincide in $\mathbf{R}^1$ and $\mathbf{R}^2$. The introduction to Chapter 11 contains some more motivation for the definition of Marczewski measure.

See Corollary 11.3 for a construction of Marczewski measures in $\mathbf{R}^1$ and $\mathbf{R}^2$. Because of the lack of similarities on S^2, the proof of $(2) \Rightarrow (1)$ of 9.5 fails for S^2. Still, the existence of a Marczewski measure in S^2 is equivalent to the failure for all m of $(m + 1)[S^2] \leqslant m[S^2]$ using pieces with the Property of Baire.

EQUIDECOMPOSABILITY WITH COUNTABLY
MANY PIECES

Ever since Lebesgue, countably additive measures have been more prevalent in mathematics than finitely additive measures, and it is natural to ask if the theory of paradoxical decompositions can be applied to such measures. Recall from Chapter 1 that the first example of a paradox using isometries arose in Vitali's proof of the nonexistence of a countably additive, translation-invariant measure defined on all sets of reals. That decomposition used a countably infinite number of pieces, and we shall now study some further consequences of allowing infinitely many pieces in the definition of G-equidecomposability. First, we note that the question of Lebesgue measure's uniqueness among countably additive measures is much simpler than the case where λ is viewed as a finitely additive measure.

Proposition 9.10. *Lebesgue measure, λ, is the unique countably additive, translation-invariant measure on the Lebesgue measurable subsets of $\mathbf{R}^n$ that normalizes J.*

Proof. If μ is another such measure, then by Proposition 9.7, μ agrees with volume on cubes. It follows that μ agrees with λ on all open sets, since an open set is a union of countably many pairwise disjoint half-open cubes (of side-length $1/2^m$). This fact can be proved by subdividing space repeatedly, retaining the cubes that are contained in the open set; see [36, p. 28] for a detailed proof. It follows that μ agrees with λ on bounded closed sets too. Now, if A is measurable and bounded, then for any $\varepsilon > 0$ there are F, G with F closed (and bounded), G open, $F \subseteq A \subseteq G$, and $\lambda(G \backslash F) < \varepsilon$. It follows easily that $\mu(A) = \lambda(A)$. Since an unbounded measurable set is a union of countably many bounded measurable sets, μ agrees with λ on all measurable sets. $\square$

The preceding result can be generalized. Let $(X, \mathscr{A}, m)$ be a measure space with $m(X) = 1$ (i.e., m is a countably additive measure on the σ-algebra $\mathscr{A}$) and suppose m is G-invariant where G, a group acting on X, preserves $\mathscr{A}$. Suppose further that G's action is *ergodic*; that is, $A \in \mathscr{A}$ satisfies $m(A \bigtriangleup g(A)) = 0$ for all $g \in G$ only if $m(A)$ is 0 or 1. Then, using the Radon-Nikodym Theorem, it is not hard to see that any other countably additive, G-invariant measure on $\mathscr{A}$ that normalizes X and vanishes on the sets of m-measure 0 (i.e., is absolutely continuous with respect to m) must equal m.

Now let us formalize the notion of countable equidecomposability.

Definition 9.11. *Suppose a group G acts on X and A, $B \subseteq X$. Then A and B are* countably G-equidecomposable, $A \sim_\infty B$, *if there is a partition of A into countably many sets A_i, $i = 0, 1, \ldots$, and elements $\sigma_i \in G$, such that the* sets $\sigma_i(A_i)$, $i = 0, 1, \ldots$, *form a partition of B. A subset E of X is* countably *G-paradoxical if E contains disjoint sets A, B such that* $A \sim_\infty E$ *and* $B \sim_\infty E$. *If G is the group of isometries of X, then the prefix "G-" will be omitted. If all the pieces are required to lie in* $\mathscr{A}$, *a G-invariant σ-algebra of subsets of X, the sets will be called* countably *G-equidecomposable in* $\mathscr{A}$, *or* countably paradoxical *in* $\mathscr{A}$.

It is easy to see that countable equidecomposability is an equivalence relation. Moreover, since this relation satisfies properties (a) and (b) of the proof of Theorem 3.5, the proofs of the Banach-Schröder-Bernstein Theorem and the Cancellation Law carry over without modification. Indeed, as we shall discuss further, the Cancellation Law is valid even for the case where the pieces are restricted to a σ-algebra of sets. But Tarski's Theorem (Corollary 9.2) fails for countable equidecomposability, as the following example shows.

Theorem 9.12. *Let G be the group of all permutations π of ω_1, the first uncountable ordinal, with the property that $\{\alpha < \omega_1 : \pi(\alpha) \neq \alpha\}$ is finite. Then ω_1 is not countably G-paradoxical, but there is no countably additive, G-invariant measure, μ, on $\mathscr{P}(\omega_1)$ with $\mu(\omega_1) = 1$.*

Proof. Since all singletons are congruent under G's action, a measure μ as in the theorem must assign them all measure zero. But it is a well-known result of Banach and Kuratowski (see [181, p. 25]) that there exists no countably additive measure on ω_1 having total measure one and vanishing on singletons. To see that ω_1 is not countably G-paradoxical, note that if $A \subseteq \omega_1$ and $A \sim_\infty \omega_1$, then $\omega_1 \backslash A$ is countable. For if $\omega_1 \backslash A$ were uncountable, there would be some $\alpha \in \omega_1 \backslash A$ that is fixed by each σ_i, where σ_i, $i = 0, 1, \ldots$ are the permutations witnessing $A \sim_\infty \omega_1$. Such an α would be missing from each $\sigma_i(A)$, a contradiction. $\qquad\square$

Despite the previous example, there is a possibility for a generalization of Tarski's Theorem to the countably additive case. Suppose G acts on X and $\mathscr{A}$ is a G-invariant σ-algebra of subsets of X. Define an ideal I in $\mathscr{A}$ to consist of all $A \in \mathscr{A}$ such that X contains pairwise disjoint subsets $A_i \in \mathscr{A}$, $i \in \mathbf{N}$, with each $A_i \sim_\infty A$ in $\mathscr{A}$. Any countably additive, G-invariant measure on $\mathscr{A}$ that normalizes X must assign measure zero to all sets in I. Moreover, it is easy to see that X is countably G-paradoxical in $\mathscr{A}$ if and only if $X \in I$. In the example of Theorem 9.12, $\mathscr{A} = \mathscr{P}(\omega_1)$ and I consists of all countable subsets of ω_1. The proof relies on the fact that there is no countably additive measure on $\mathscr{A}$ that vanishes on I and normalizes ω_1, even without the additional requirement of G-invariance. Chuaqui has conjectured that this phenomenon is the only barrier to a generalization of Tarski's Theorem.

Question 9.13. Suppose X, $\mathscr{A}$, G, and I are as in the beginning of the preceding paragraph. Then assertion (a) below implies (b) (which implies that X is not countably G-paradoxical in $\mathscr{A}$). Does (b) imply (a)?

(a) There is a countably additive, G-invariant measure on $\mathscr{A}$ that normalizes X.
(b) There is a countably additive measure on $\mathscr{A}$ that normalizes X and vanishes on I.

Chuaqui [31] has proved a weaker version of this conjecture where, in (b), it is assumed in addition that the ideal of sets of measure zero is G-invariant.

As was done for finite equidecomposability, one can form a semigroup, $\mathscr{S}_\infty(\mathscr{A})$, of types with respect to the relation of countable equidecomposability in $\mathscr{A}$. Instead of $X \times \mathbf{N}$, one uses $X \times \omega_1$ and considers bounded sets to be those that have only countably many levels. An important aspect of countable equidecomposability is that $\mathscr{S}_\infty(\mathscr{A})$ has a richer algebraic structure than $\mathscr{S}(\mathscr{A})$: $\mathscr{S}_\infty(\mathscr{A})$ is a cardinal algebra, as defined by Tarski [236]. The structure of these algebras is worked out in detail in [236]; in particular, the Cancellation Law is valid. See [31, 33, 34] for a variety of applications of this point of view to countable equidecomposability and countably additive measures.

Turning now to the specific case of $\mathbf{R}^n$ and isometries, we shall see that countable equidecomposability is much simpler than finite equidecomposability. For instance, the existence of paradoxes is not dependent on dimension. The following result may be viewed as a generalization of Vitali's construction of a nonmeasurable set, showing even more emphatically why translation-invariant, countably additive measures defined on all subsets of $\mathbf{R}^n$ do not exist.

Theorem 9.14 (AC).   *Any two subsets of $\mathbf{R}^n$ with nonempty interior are countably equidecomposable.*

Proof. First we consider $\mathbf{R}^1$. By transitivity and the Banach-Schröder-Bernstein Theorem, it suffices to show that for any half-open interval K, $K \sim_\infty \mathbf{R}$. Vitali's construction (Theorem 1.5), restated for K, shows that K may be partitioned into countably many sets A_i, $i \in \mathbf{N}$, such that $A_i \sim_2 A_j$ for each i, j. (In fact, by a result of von Neumann (see p. 48), it can be guaranteed that the sets A_i are all congruent to each other.) It follows that for any infinite subset S of $\mathbf{N}$, $\bigcup\{A_i : i \in S\} \sim_\infty K$. Now, by splitting the family $\{A_i : i \in \mathbf{N}\}$ into infinitely many infinite subfamilies, one gets that K is countably equidecomposable with infinitely many copies of itself. It follows, upon using some additional translations, that $K \sim_\infty \mathbf{R}$. One can get a Vitali partition of $[0, 1]^n$ by simply using the Vitali partition on the first coordinate; this allows the proof for $\mathbf{R}$ to be repeated for $\mathbf{R}^n$. □

The condition on interiors in the preceding result is necessary since, for example, a meager set cannot be countably equidecomposable with a nonmeager set.

The Vitali partition shows that a countably additive measure on $\mathscr{P}([0,1])$ that has total measure one cannot be invariant with respect to translations (modulo 1). This leads to the question of whether there can be a countably additive measure on $\mathscr{P}([0,1])$ that has total measure one, but is not required to satisfy any invariance condition whatever. Of course, the principal measure determined by a point is such a measure; hence we add the natural condition (which is a consequence of translation-invariance) that points get measure zero. Since this problem does not refer at all to the geometry of the line, it is really a problem in set theory: if f is a bijection from X to Y, then X bears such a measure if and only if Y does. So let us say that the continuum is *real-valued measurable* if $2^{\aleph_0}$ bears a countably additive measure defined on all subsets, having total measure one, and vanishing on singletons.

In 1929, Banach and Kuratowski [10] showed that the Continuum Hypothesis ($2^{\aleph_0} = \aleph_1$) implies that the continuum is not real-valued measurable; their result was quoted in the proof of 9.12 (ω_1 and $\aleph_1$ are different notations for the same set). But unlike the Continuum Hypothesis, which is independent of ZFC, the usual axioms of set theory (i.e., the addition of either the Continuum Hypothesis or its negation yields a noncontradictory system), the statement that the continuum is real-valued measurable has a more complicated status. Its negation is consistent since it follows from the Continuum Hypothesis, but the statement in its positive form is connected with large cardinal axioms for set theory. It follows that the consistency of the statement cannot be derived from the consistency of ZFC alone. In fact, the existence of a real-valued measure on the continuum is equiconsistent with the existence of a (2-valued) measurable cardinal. In short, if it is noncontradictory to add a certain axiom about the existence of very very large cardinals to ZFC, then (and only then) is it noncontradictory to add the statement that the continuum is real-valued measurable. See [99] for a complete discussion of the connection between this problem in measure theory and large cardinal axioms in set theory. Other connections between large cardinals and consistency results in set theory and in measure theory are discussed in Chapter 13.

Even if there is a real-valued measure on $\mathscr{P}([0,1])$, such a measure need not extend Lebesgue measure. For example, one can use a bijection of $[0,1]$ with $[0,1/2]$ to transform the measure into one that assigns measure 0 to $[1/2,1]$. Nevertheless, if the continuum is real-valued measurable, then there is a countably additive measure on $\mathscr{P}([0,1])$ that extends Lebesgue measure; see [99, p. 302] for a proof.

Of course, Theorem 9.14 fails if the pieces in the decomposition are required to be Lebesgue measurable. On the other hand, the result remains valid if the pieces are restricted to lie in $\mathscr{B}$, the algebra of subsets of $\mathbf{R}^n$ having the Property of Baire.

Theorem 9.15 (AC). *Any two subsets of* $\mathbf{R}^n$, *each of which has the Property of Baire and nonempty interior, are countably equidecomposable in* $\mathscr{B}$.

Proof. We shall show that $K \sim_\infty \mathbf{R}^n$ in $\mathscr{B}$, where K is any cube in $\mathbf{R}^n$. By transitivity and the Banach-Schröder-Bernstein Theorem, this yields the theorem. First we show that K can be packed into any other cube L, using countably many pieces in $\mathscr{B}$. Let E be an open dense subset of K with $\lambda(E) \leqslant \lambda(L)/2$; simply get E as a union of sufficiently small pairwise disjoint intervals I_m, with I_m containing the mth point of a countable dense subset of K. Let $M = K \backslash E$, a nowhere dense set. Now, split L into two halves, L_1 and L_2, and pack E and M into L_1 and L_2, respectively, as follows (see Figure 9.2). By Proposition 8.16, each I_m is finitely equidecomposable, using Borel pieces, with an open strip of L having the same volume (see figure), and it follows that $E = \bigcup I_m$ can be packed into L_1 using countably many Borel pieces.

To pack M into L_2, use Theorem 9.14 to pack all of $\mathbf{R}^n$ into L_2 with no restriction on the pieces. But a subset of a nowhere dense set is nowhere dense, and therefore the packing of M into L_2 induced by the packing of $\mathbf{R}^n$ has its pieces in $\mathscr{B}$, as desired.

Now, we may cover $\mathbf{R}^n$ with infinitely many closed cubes K_i, each having the same volume as a given cube K, and K itself contains infinitely many pairwise disjoint open cubes L_i. The result in the previous paragraph shows how to pack each K_i into L_i, and this yields that $\mathbf{R}^n \leqslant_\infty K$. Obviously $K \leqslant_\infty \mathbf{R}^n$, so $K \sim_\infty \mathbf{R}^n$ as desired.   □

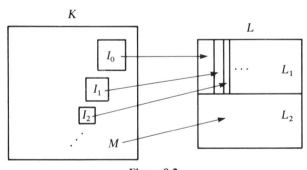

Figure 9.2.

A consequence of Theorem 9.15 is that no invariant measure on $\mathscr{B}$ that normalizes J can be countably additive. Recall that the Marczewski Problem asks whether a finitely additive, invariant measure on $\mathscr{B}$ that normalizes J exists if $n \geqslant 3$. To put it another way, while it is not known whether the unit cube in $\mathbf{R}^n$, $n \geqslant 3$, is paradoxical in $\mathscr{B}$ using finitely many pieces, the use of countably many pieces yields that it is paradoxical in $\mathscr{B}$, even in $\mathbf{R}^1$ and $\mathbf{R}^2$. Note that the result about the nonexistence of a countably additive measure on $\mathscr{B}$ could not have been obtained by the classical Vitali approach (1.5). Using the representation of sets in $\mathscr{B}$ in terms of regular-open sets (see proof of

Theorem 9.5, (7) $\Rightarrow$ (3)) it is easy to see that the unit cube cannot be partitioned into countably many sets with the Property of Baire that are pairwise congruent using translations modulo 1. Note also that is not clear that the condition on interiors in Theorem 9.15 is necessary. Perhaps any two nonmeager sets with the Property of Baire are countably equidecomposable in $\mathscr{B}$.

In $\mathbf{R}^n$, $n \geqslant 3$, one can simplify the part of 9.15's proof that packs M into L_2 by invoking the Banach-Tarski Paradox instead of Theorem 9.12. Since M is bounded, the strong form of the Banach-Tarski Paradox yields that a cube containing M, and hence M itself, may be packed into L_2 using finitely many pieces. But all subsets of M are nowhere dense, and hence in $\mathscr{B}$. If E could also be handled with finitely many pieces in $\mathscr{B}$, we would essentially have a paradoxical decomposition in $\mathscr{B}$ as in Theorem 9.5(2). But the existence of an E that could be so handled is just the negation of statement (9) of 9.5, and is equivalent to the nonexistence of a Marczewski measure.

Of course, Theorem 9.15 fails if $\mathscr{L}$, the algebra of Lebesgue measurable sets, is used instead of $\mathscr{B}$, because of the existence of the countably additive invariant measure λ. Even if two sets have the same Lebesgue measure, they need not be countably equidecomposable in $\mathscr{L}$: simply consider the empty set and any nonempty set of measure zero. But this example is essentially the only one. The following definition will allow us to show that Lebesgue measure is as complete an invariant for countable equidecomposability as is possible.

Definition 9.16. *Two measurable sets A, $B \in \mathscr{L}$ are almost countably equidecomposable in $\mathscr{L}$ if there are measure zero sets E_1, E_2 such that $A \backslash E_1 \sim_\infty B \backslash E_2$ in $\mathscr{L}$.*

The reader may prove, as an exercise, that the relation of almost countable equidecomposability in $\mathscr{L}$ is transitive. (This also follows from the discussion of quotient algebras in Chapter 11, since the relation in question is equivalent to countable equidecomposability in the quotient algebra $\mathscr{L}/N$, where N is the ideal of sets of measure zero.) This new relation clearly preserves λ and eliminates the problem of sets of measure zero. As we now show, λ is a complete invariant for this modified notion of equidecomposability.

Theorem 9.17. *If A and B are Lebesgue measurable subsets of $\mathbf{R}^n$ (or of S^n), then $\lambda(A) = \lambda(B)$ if and only if A and B are almost countably equidecomposable in $\mathscr{L}$.*

Proof. Only the forward direction requires proof. Assume $\lambda(A) < \infty$. For if $\lambda(A) = \lambda(B) = \infty$, we may partition each of A, B into countably many bounded sets such that corresponding sets have the same measure. Then the fact that corresponding summands are almost equidecomposable yields that the same is true for A and B. Now, the Lebesgue density theorem asserts that if $\lambda(E) > 0$, then for almost all $P \in E$, $\lim_{h \to 0} \lambda(E \cap C(h))/h^n = 1$, where $C(h)$ is a cube of side-length h centered at P. Choose P_1, P_2 in A, B, respectively, so that for some $h > 0$, the quotients of the preceding limit are each greater than $1/2$. Then $B \cap (A + (P_2 - P_1))$ has positive measure. This shows that whenever two sets have positive measure, some translate of one intersects the other in a set of positive measure.

Now, let ℓ be the supremum of $\lambda(\tau(A) \cap B)$ over all translations τ, and choose τ_0 so that $\lambda(\tau_0(A) \cap B) = \ell_0 \geqslant \ell/2$. Let $B_0 = \tau_0(A) \cap B$ and $A_0 = \tau_0^{-1}(B_0)$. Then replace A, B by $A \backslash A_0$, $B \backslash B_0$, respectively, and (stopping if $\lambda(A \backslash A_0) = \lambda(B \backslash B_0) = 0$) repeat to obtain A_1, B_1, and ℓ_1. Continue as long as possible, always choosing τ_m so that $\ell_m = \lambda(\tau_m(A_{m-1}) \cap B_{m-1})$ is at least half as large as is possible. If this procedure stops in finitely many steps—that is, if one of $A \backslash \bigcup_{k \leqslant m} A_k$, $B \backslash \bigcup_{k \leqslant m} B_k$ has measure zero—then both of these sets have measure zero (since $\lambda(A) = \lambda(B)$, $\lambda(A_k) = \lambda(B_k)$) and A and B are almost finitely equidecomposable in $\mathscr{L}$. Otherwise, $\bigcup A_m \sim_\infty \bigcup B_m$ in $\mathscr{L}$, and since the ℓ_m correspond to measures of pairwise disjoint subsets of B, $\sum \ell_m \leqslant \lambda(B) = \lambda(A)$. Therefore the sequence $\{\ell_m\}$ converges to 0. But this means that $\sum \ell_m = \lambda(\bigcup A_m) = \lambda(A)$, for if $\lambda(A \backslash \bigcup A_m) > 0$ then there would be a translation τ such that $\lambda(\tau(A \backslash \bigcup A_m) \cap (B \backslash \bigcup B_m)) = r > 0$. Choosing m such that $r > 2\ell_m$ yields that τ contradicts the choice of τ_m, since ℓ_m is not greater than $r/2$.

The same proof works for subsets of S^n, since the Lebesgue density theorem is valid on spheres and since any point can be taken to any other by a rotation. $\qquad\square$

It is interesting to note that in the preceding proof, it is sufficient to choose τ_m so that $\ell_m > 0$. The procedure might then continue on beyond ω steps, but one could still define $\tau_\alpha, \ell_\alpha, A_\alpha$, and B_α for infinite ordinals α. Since the ℓ_α correspond to disjoint sets, the sets A and B must be exhausted, in measure, by the A_α and B_α corresponding to $\alpha < \beta$ for some countable ordinal β. This is a consequence of the fact that a set of finite measure cannot contain uncountably many pairwise disjoint subsets of positive measure (otherwise, for some n, infinitely many of the sets would have measure greater than $1/n$).

Theorem 9.17 may be generalized to Haar measure on the Borel subsets of a locally compact topological group: if A and B are Borel sets with the same finite Haar measure, then A and B are almost countably equidecomposable using Borel pieces. The proof is the same as the one just given, once it is established that whenever $\mu(A)$ and $\mu(B)$ are nonzero (μ denotes Haar measure), then for some $g \in G$, $\mu(B \cap gA) > 0$. This, in turn, may be derived from the following two facts about Haar measure (see [36, Chap. 9]):

(1) $\mu(A^{-1}) > 0$ if $\mu(A) > 0$, and
(2) the function $F(s, t) = (s, s^{-1}t)$ is a measure-preserving (with respect to $\mu \times \mu$) homeomorphism of $G \times G$ to $G \times G$.

Since $\mu(B)\mu(A^{-1}) = (\mu \times \mu)(F^{-1}(B \times A^{-1})) = (\mu \times \mu)(\{(h, g) : h \in B \cap gA\})$, the set of g such that $\mu(B \cap gA) > 0$ must itself have positive measure.

It might even be true that two bounded subsets of **R** with the same Lebesgue measure are almost finitely equidecomposable in $\mathscr{L}$. This is not known, and is a similar question to 8.15.

One can also consider almost countable equidecomposability with no restrictions on the pieces. This yields the following variation of Theorem 9.17: any two subsets of $\mathbf{R}^n$ with positive Lebesgue inner measure are almost countably equidecomposable. For the proof of this see [11, Th. 41].

Another way to avoid the problem caused by sets of measure zero is to consider just the open sets or, more generally, sets with nonempty interior, rather than all measurable sets. We shall show that such sets are countably equidecomposable in $\mathscr{L}$ provided they have the same measure. The proof requires the following lemma, part (a) of which can also be deduced from the "sack of potatoes" theorem (see [140, p. 74]).

Lemma 9.18

    (a) If $E \subseteq \mathbf{R}^n$ has Lebesgue measure 0 and U is a given nonempty open subset of $\mathbf{R}^n$, then E is countably equidecomposable with a subset of U.

    (b) The same result is valid for subsets E of S^n having Lebesgue measure zero.

Proof

    (a) Choose a cube $K \subseteq U$. Then cover E with rectangles, R_i, whose volumes sum to less than the volume of K. Choose pairwise disjoint rectangles $R_i' \subseteq K$ such that $\lambda(R_i') = \lambda(R_i)$ and apply Proposition 8.16 to obtain the (Borel) equidecomposability of R_i with R_i'. It follows that E is countably equidecomposable to a subset of $\bigcup R_i'$.

    (b) The result is clear for S^1, and may be proved for S^2 in the same way as part (a), using the Bolyai-Gerwien Theorem for spherical polygons [76]. But the idea of Lemma 9.7 yields a proof for all S^n, $n \geqslant 2$. For the Banach-Tarski Paradox, in the strong form of Corollary 8.10, yields that $S^n \sim U$. It follows immediately that E is equidecomposable with a subset of U (using only finitely many pieces). Of course, an appeal to the Banach-Tarski Paradox requires the Axiom of Choice, and this is not necessary if one is content to use countably many, rather than finitely many, pieces. By using a stereographic projection, from the north pole, of the southern hemisphere of S^n onto a ball in $\mathbf{R}^n$, one can derive the spherical case of the lemma from the "sack of potatoes" theorem in $\mathbf{R}^n$ mentioned just prior to the lemma. The key point is that this projection sends spherical caps to balls, and vice versa. One advantage of this constructive approach that pertains to both cases is that if the set E is Borel, then the packing can be accomplished with Borel pieces. The approach via the Banach-Tarski Paradox might introduce non-Borel sets, since it uses subsets of measure zero sets.   □

Lemma 9.18 raises the question whether the packings can be effected using only finitely many pieces. As pointed out in the proof (see also 9.7), the Banach-Tarski Paradox yields that any null set in $\mathbf{R}^n$ ($n \geqslant 3$) or S^n ($n \geqslant 2$) can be packed into any nonempty open set using finitely many (necessarily measurable) pieces. In $\mathbf{R}^1$, $\mathbf{R}^2$, and S^1, however, there exist null sets that cannot be finitely packed into any open set of measure less than 1; see remarks following Corollary 11.3. A more subtle question arises if we attempt to pack a null Borel subset of $\mathbf{R}^3$, say, into arbitrarily small open sets using finitely many pieces with the Property of Baire. The Banach-Tarski Paradox does not necessarily yield pieces in $\mathscr{B}$, but if the unit cube is paradoxical in $\mathscr{B}$ (see 3.12, 9.5, and 9.9), then any bounded Borel subset of $\mathbf{R}^3$ can be packed into any nonempty open set using finitely many pieces in $\mathscr{B}$. On the other hand, if J is not paradoxical in $\mathscr{B}$, then by Theorem 9.5, a Marczewski measure exists and it follows as for $\mathbf{R}^1$ and $\mathbf{R}^2$ that some null Borel subset of J cannot be finitely packed, using pieces in $\mathscr{B}$, into any open set of measure less than one.
Lemma 9.18 can be combined with Theorem 9.17 to yield the following theorem.

Theorem 9.19. *Any two measurable subsets of $\mathbf{R}^n$ (or of S^n) with nonempty interior and with the same Lebesgue measure are countably equidecomposable in $\mathscr{L}$.*

Proof. First note that if in Theorem 9.17, A and B have nonempty interior, then we may construct the decomposition so that $A \backslash E_1$ and $B \backslash E_2$ also have nonempty interior. Simply choose the first translation τ_0 so that $\tau_0(A) \cap B$ contains an open set; the condition $\ell_0 \geq \ell/2$ may fail for this one case, but this will not affect the proof, since the sequence $\{\ell_m\}$ still converges to 0. Hence if A and B are as hypothesized, there are measure zero subsets E_1, E_2 of A, B, respectively, such that $A \backslash E_1 \sim_\infty B \backslash E_2$ in $\mathscr{L}$ and $A \backslash E_1$, $B \backslash E_2$ each contain a nonempty open set.
Now, it is sufficient to show that $A \sim_\infty A \backslash E_1$ in $\mathscr{L}$ (and the same for B and $B \backslash E_2$). But this is a consequence of the previous lemma. Choose countably many pairwise disjoint open subsets U_i of $A \backslash E_1$, and choose $F_i \subseteq U_i$ so that $F_i \sim_\infty E_1$. Then partition A into $A \backslash (E_1 \cup \bigcup F_i)$, E_1, F_0, $F_1, \ldots$. Since $E_1 \sim_\infty F_0$, $F_0 \sim_\infty F_1, \ldots$, this shows that $A \sim_\infty A \backslash E_1$ in $\mathscr{L}$ as required. Similarly, $B \sim_\infty B \backslash E_2$ in $\mathscr{L}$, completing the proof.   □

If A and B are Borel subsets of $\mathbf{R}^n$ and have nonempty interior and the same measure, then the proof of the previous theorem, together with the remark at the end of 9.18's proof, yields a decomposition using Borel sets. It follows that some of the decomposition problems posed earlier have affirmative answers if countably many pieces are allowed: the circle can be squared and a regular tetrahedron cubed using Borel sets.

As with 9.17, it seems reasonable to expect that the previous result is valid for sets of the same finite Haar measure, using Borel sets as pieces. Even for Lie groups, this problem is unsolved (although Freyhoffer [73] has obtained a solution in commutative Lie groups). Because the generalization of 9.17 to Haar measure is valid, the method of proof of 9.19 reduces the problem to generalizing 9.18.

Question 9.20. It is true that whenever E is a Borel set of Haar measure zero in a locally compact topological group G, and U is a nonempty open subset of G, then E is countably equidecomposable with a subset of U using left translations and Borel pieces?

C. Bandt [A4] has recently obtained a positive answer to this question, provided measurable, rather than just Borel, sets may be used as pieces.

NOTES

Tarski's theorem equating the existence of measures with the absence of paradoxical decompositions is presented in [235], although he obtained the result much earlier [233]. Tarski used transfinite induction. The use of the Tychonoff Theorem stems from work of Łoś and Ryll-Nardzewski [126], who used it to prove the Hahn-Banach Theorem. Tarski's Theorem is not as well known as it deserves to be, and the result has been rediscovered by Sherman [208] (in a weaker form than Tarski's) and by Emerson [67].

Marczewski was the first to investigate measures vanishing on the meager sets. The equivalences of Theorem 9.5 are due to Mycielski [166], [169]. The latter paper, however, deals with S^2 rather than $\mathbf{R}^3$ and assumes that Corollary 9.2 is valid for $\mathscr{B}$. This is not known, and so Theorem 9.5 is not known to be true as stated for S^2 (see remarks following 9.9). The trick of using similarities to sidestep the problem in $\mathbf{R}^n$ (Theorem 9.5, $(2) \Rightarrow (1)$) is due to Mycielski.

Lemma 9.7 is due to Tarski [235, p. 65], and has turned out to be important in recent work on characterizing Lebesgue measure (see remarks preceding Definition 11.9).

Proposition 9.10 was essentially known to Lebesgue [121].

Chuaqui [31, p. 74] conjectured that Tarski's Theorem is valid for countably additive measures and $\sim_\infty$, but then he discovered the counterexample of 9.12 [32]. (A counterexample was first discovered by Lang [120], but he assumed the Continuum Hypothesis.)

The notions of countable equidecomposability and almost countable equidecomposability were first investigated by Banach and Tarski [11], who proved Theorems 9.14 and 9.17. Their proof of 9.17 was more elementary and more complicated than the proof presented here, which is due to Mycielski. The extension of 9.17 to Haar measure is due to von Neumann (unpublished; see [34] and [136]).

Lemma 9.18 is due to Banach and Tarski [11, p. 277] for $\mathbf{R}^n$ (see also [31]), and the fact that the result is valid on spheres without having to use the Axiom of Choice was pointed out to the author by B. Grünbaum. The fact that 9.18 could be used to obtain the characterization of 9.19 is due to Chuaqui [31], although for open sets the result was known to Banach and Tarski. The possibility that 9.19 is valid for Haar measure, that is, Question 9.20, was first investigated by Chuaqui [34].

Chapter 10

Measures in Groups

In Part I we saw that the main idea in the construction of a paradoxical decomposition of a set was to first get such a decomposition in a group acting on the set, and then transfer it to the set. An almost identical theme pervades the construction of invariant measures on a set X acted upon by a group G. If there is a finitely additive, left-invariant measure defined on all subsets of G, then it can be used to produce a finitely additive, G-invariant measure defined on all subsets of X. Such measures on X yield that X (and certain subsets of X) are not G-paradoxical.

It was von Neumann [246] who realized that such a transference of measures was possible, and he began the job of classifying the groups that bear measures of this sort. In this chapter we first study some properties of the class of groups having measures and show that it is fairly extensive, containing all solvable groups. We then give the important application to the case of isometries acting on the line or plane, obtaining the nonexistence of Banach-Tarski-type paradoxes in these two dimensions.

Definition 10.1. *If, for a group G, μ is a finitely additive measure on $\mathscr{P}(G)$ such that $\mu(G) = 1$ and μ is left-invariant ($\mu(gA) = \mu(A)$ for $g \in G$, $A \subseteq G$), then μ will be called simply a measure on G. An amenable group is one that bears such a measure, and AG denotes the class of all amenable groups.*

If a group G is paradoxical with respect to left translations, then G cannot be amenable, since any measure would have to satisfy $\mu(G) = 2\mu(G)$.

146

Thus, for example, any group that has a free subgroup of rank 2 is not amenable (Corollary 1.11). In fact, an application of Tarski's Theorem (Corollary 9.2) to a group's action on itself yields that groups that are not paradoxical are necessarily amenable, and hence AG coincides with the class of paradoxical groups. While this characterization of amenability can be used to establish the amenability of some families of groups (e.g., the amenability of Abelian groups follows from Theorem 12.8(a)), we shall establish the existence of a measure more directly whenever possible, avoiding the rather difficult theorem of Tarski.

Note that if G is an infinite group there cannot be a left-invariant measure on $\mathscr{P}(G)$ that has total measure one and is countably additive. This is easily proved by the Vitali technique (1.5), using instead the equivalence relation based on any countable subgroup of G. In fact, there cannot even be a σ-finite, left-invariant, countably additive measure on $\mathscr{P}(G)$ (see [69, A2]).

The basic result on AG is that all elementary groups (see the end of Chapter 1) lie in this class, that is, $EG \subseteq AG$. No simple example of a nonelementary amenable group has yet been constructed, but it was recently shown that such groups do exist (see Theorem 12.17 and the discussion at the end of Chapter 1). Before proving that all elementary groups are amenable, we discuss a few simple consequences of the definition of amenability.

For a group G, let $B(G)$ denote the collection of bounded real-valued functions on G. If μ is a measure on G, then the standard construction of an integral from a measure (see, e.g., [196, Ch. 11]) can be applied to the triple $(G, \mathscr{P}(G), \mu)$. This construction, which proceeds by defining the integral first on simple functions, then, via limits, on all measurable functions, is somewhat simplified in this context because all sets, and hence all functions, are measurable. Thus, $\int f\, d\mu$ defines a linear functional on all of $B(G)$. On the other hand, because the measure is not necessarily countably additive, certain standard theorems, such as the monotone and dominated convergence theorems, cannot be proved. In summary then, a real number, $\int f\, d\mu$, is assigned to each $f \in B(G)$, and this integral satisfies the following properties:

(i) $\int af + bg\, d\mu = a \int f\, d\mu + b \int g\, d\mu$ if $a, b \in \mathbf{R}$;

(ii) $\int f\, d\mu \geqslant 0$ if, for all $g \in G$, $f(g) \geqslant 0$;

(iii) $\int \chi_G\, d\mu = 1$;

(iv) For each $g \in G$, $f \in B(G)$, $\int {}_g f\, d\mu = \int f\, d\mu$, where $({}_g f)(h) = f(g^{-1}h)$; that is, the integral is left-invariant.

It is easy to see that conditions (ii) and (iii) may be replaced by the single condition: $\inf\{f(g): g \in G\} \leqslant \int f\, d\mu \leqslant \sup\{f(g): g \in G\}$. Because of this, a linear functional on $B(G)$ that satisfies (i)–(iv) is called a *left-invariant mean* on G; therefore an amenable group always bears a left-invariant mean. Conversely, if $F: B(G) \to \mathbf{R}$ is a left-invariant mean, then defining $\mu(A) = F(\chi_A)$

yields a measure on G. Hence a group is amenable if and only if it bears a left-invariant mean.

While the existence of a measure is sufficient to prove some properties of amenable groups, in some proofs the corresponding mean is the more useful object. The following proposition is an example. Defining a right-invariant measure in the obvious way, one sees that a left-invariant measure, μ, on G yields a right-invariant measure, μ_0, defined by $\mu_0(A) = \mu(A^{-1})$. Note that the integral with respect to μ_0 is invariant under the right action of G on $B(G)$ given by $f_g(h) = f(hg^{-1})$. A (left-invariant) measure on an amenable group is not necessarily right-invariant (see [95, p. 230]), but at least some measures on an amenable group are two-sided invariant.

Proposition 10.2. *If G is amenable, then there is a measure ν on $\mathscr{P}(G)$ that is left-invariant and right-invariant.*

Proof. Let μ be a measure on G, and μ_0 the corresponding right-invariant measure. If $A \subseteq G$, define f_A in $B(G)$ by $f_A(g) = \mu(Ag^{-1})$; f_A is bounded by 1. Then define ν on $\mathscr{P}(G)$ by $\nu(A) = \int f_A \, d\mu_0$. It is easy to check that $\nu(G) = 1$ and ν is finitely additive ($f_{A \cup B} = f_A + f_B$ if $A \cap B = \varnothing$). Moreover, $f_{gA} = f_A$ and $f_{Ag} = (f_A)_g$, so ν is both left- and right-invariant. $\qquad\qquad\square$

A much more important application of means is the Invariant Extension Theorem, Theorem 10.8. The next result is a very simple special case. It shows why a set cannot be paradoxical if the group acting on it is amenable.

Theorem 10.3. *Suppose the amenable group G acts on X. Then there is a finitely additive, G-invariant measure on $\mathscr{P}(X)$ of total measure one; hence X is not G-paradoxical.*

Proof. Choose any $x \in X$ and, if μ is a measure on G, define $\nu : \mathscr{P}(X) \to [0, 1]$ by $\nu(A) = \mu(\{g \in G : g(x) \in A\})$. It is easy to check that ν, which gives measure one to the orbit of x, is as required. $\qquad\qquad\square$

Because of Tarski's Theorem, this result may be stated as: if G acts on X and G is not paradoxical, then X is not G-paradoxical. In fact, it is easy to give a direct proof of this latter result, which is a strong converse to Proposition 1.10.

We now turn our attention to proving that all elementary groups—in particular, all Abelian and solvable groups—are amenable. Note, however, that Theorem 10.3, together with the amenability of the solvable groups G_1 and G_2, yields only that $\mathbf{R}^n$ is not G_n-paradoxical if $n = 1$ or 2. We are more interested in the fact that no interval or square is paradoxical, and this requires a measure that does not vanish on intervals or squares; the measures provided by 10.3 necessarily vanish on all bounded sets. The proof that a measure that agrees with Lebesgue measure exists (if $n \leqslant 2$) requires a stronger form of 10.3, the Invariant Extension Theorem (10.8).

Theorem 10.4

(a) *Finite groups are amenable.*

(b) *(AC) Abelian groups are amenable.*

(c) *(AC) A subgroup of an amenable group is amenable.*

(d) *If N is a normal subgroup of the amenable group G, then G/N is amenable.*

(e) *If N is a normal subgroup of G, and each of N, G/N are amenable, then G is amenable.*

(f) *(AC) If G is the direct union of a directed system of amenable groups, $\{G_\alpha : \alpha \in I\}$, then G is amenable.*

Before proving this theorem, we note some consequences. By parts (b) and (e), every solvable group is amenable; in particular, O_1, G_1, and G_2 are amenable. Also, since any group is the direct union of its finitely generated subgroups, it follows from (c) and (f) that a group is amenable if and only if all of its finitely generated subgroups are. Thus amenability may be regarded as being primarily a property of finitely generated groups.

Proof of 10.4

(a) If $|G| = n < \infty$, then defining $\mu(A) = |A|/n$ yields the desired measure on G.

(b) This part is the most difficult, and will be proved after the others.

(c) Let μ be a measure on G, and suppose H is a subgroup of G. Let M be a set of representatives (choice set) for the collection of right cosets of H in G. Then define v on $\mathscr{P}(H)$ by $v(A) = \mu(\bigcup\{Ag : g \in M\})$. It is easy to check that v is a measure on H.

(d) If μ is a measure on G, then define $v : \mathscr{P}(G/N) \to [0, 1]$ by setting $v(A) = \mu(\bigcup A)$. Again, it is routine to check that v is as desired.

(e) Let v_1, v_2, be measures on N, G/N, respectively. For any $A \subseteq G$ let $f_A : G \to \mathbf{R}$ be defined by $f_A(g) = v_1(N \cap g^{-1} A)$. Then if g_1 and g_2 define the same coset of N in G, $f_A(g_1) = f_A(g_2)$. For if $g_2^{-1} g_1 = h \in N$, then $f_A(g_2) = v_1(N \cap g_2^{-1} A) = v_1(N \cap hg_1^{-1} A) = v_1(h(N \cap g_1^{-1} A)) = v_1(N \cap g_1^{-1} A) = f_A(g_1)$. This means that f_A may be regarded as a (bounded) real-valued function with domain G/N. Define $\mu(A)$ to be $\int f_A \, dv_2$. Since $f_G = \chi_G$, $\mu(G) = 1$, and if $A, B \subseteq G$, $A \cap B = \emptyset$, then for any $g \in G$, $g^{-1} A \cap g^{-1} B = \emptyset$ whence $f_{A \cup B}(g) = f_A(g) + f_B(g)$ (more precisely, $f_{A \cup B}(gN) = f_A(gN) + f_B(gN)$). This yields the finite additivity of μ. Finally, $f_{gA}(g_0) = v_1(N \cap g_0^{-1} gA) = f_A(g^{-1} g_0) = {}_g(f_A)(g_0)$, and so the left-invariance of the integral defined by v_2 yields that $\mu(gA) = \mu(A)$.

(f) We are given that $G = \bigcup\{G_\alpha : \alpha \in I\}$ where each G_α is amenable (with measure μ_α), and for each $\alpha, \beta \in I$ there is some $\gamma \in I$ such that G_α and G_β are each subgroups of G_γ. Consider the compact topological space $[0,1]^{\mathscr{P}(G)}$; this is where the Axiom of Choice, via Tychonoff's Theorem, is used. For each $\alpha \in I$, let $\mathscr{M}_\alpha$ consist of those finitely additive $\mu : \mathscr{P}(G) \to [0,1]$ such that $\mu(G) = 1$ and $\mu(gA) = \mu(A)$ whenever $g \in G_\alpha$. Then each $\mathscr{M}_\alpha$ is nonempty, as can be seen by defining $\mu(A) = \mu_\alpha(A \cap G_\alpha)$. And, as in the proof of 9.1, one may check that each $\mathscr{M}_\alpha$ is a closed subset of $[0,1]^{\mathscr{P}(G)}$. Since $\mathscr{M}_\alpha \cap \mathscr{M}_\beta \supseteq \mathscr{M}_\gamma$ if $G_\alpha, G_\beta \subseteq G_\gamma$, the collection $\{\mathscr{M}_\alpha : \alpha \in I\}$ has the finite intersection property. By compactness, then, there is some $\mu \in \bigcap\{\mathscr{M}_\alpha : \alpha \in I\}$, and such a μ witnesses the amenability of G.

(b) As pointed out before, any group is the direct union of its finitely generated subgroups; hence, by (f) it suffices to consider finitely generated Abelian groups. So suppose G is Abelian, with generating set $\{g_1, \ldots, g_m\}$. We claim that it suffices to show that for all $\varepsilon > 0$, there is a function $\mu_\varepsilon : \mathscr{P}(G) \to [0,1]$ such that

(1) $\mu_\varepsilon(G) = 1$,
(2) μ_ε is finitely additive, and
(3) μ_ε is almost invariant with respect to the generators in the sense that for each $A \subseteq G$ and generator g_k, $|\mu_\varepsilon(A) - \mu_\varepsilon(g_k A)| \leq \varepsilon$.

Once the existence of such a μ_ε is established, we may let $\mathscr{M}_\varepsilon$ denote the set of functions from $\mathscr{P}(G)$ to $[0,1]$ satisfying (1)–(3). Then each $\mathscr{M}_\varepsilon$ is nonempty and closed, since if a function fails to lie in $\mathscr{M}_\varepsilon$, then that failure is evident from finitely many values of the function (see 9.1's proof). Furthermore, the collection of the sets $\mathscr{M}_\varepsilon$ has the finite intersection property: $\bigcap \mathscr{M}_{\varepsilon_i} = \mathscr{M}_{\min \varepsilon_i}$, which is nonempty. Therefore, by the compactness of $[0,1]^{\mathscr{P}(G)}$, there is some μ lying in each $\mathscr{M}_\varepsilon$. Such a μ is left-invariant with respect to each g_k, and hence is left-invariant with respect to any member of G, as desired.

The idea behind the construction of a single μ_ε is really very simple. For example, consider the case where G has the single generator g_1. Choose N so large that $2/N \leq \varepsilon$ and let $\mu_\varepsilon(A) = |\{i : 1 \leq i \leq N$ and $g_1^i \in A\}|/N$. Then μ_ε differs from $\mu_\varepsilon(gA)$ by no more than $2/N \leq \varepsilon$. For the general case, choose N as before and let $\mu_\varepsilon(A)$ be $|\{(i_1, \ldots, i_m) : 1 \leq i_1, \ldots, i_m \leq N$ and $g_1^{i_1} g_2^{i_2} \cdots g_m^{i_m} \in A\}|/N^m$. Then $\mu_\varepsilon(G) = 1$, μ is finitely additive, and because g_k commutes with the other generators, $\mu_\varepsilon(g_k A)$ differs from $\mu_\varepsilon(A)$ by no more than $|\{(i_1, \ldots, i_m) : 1 \leq i_1, \ldots, i_{k-1}, i_{k+1}, \ldots, i_m \leq N$ and $i_k = 1$ or $N+1\}|/N^m = 2N^{m-1}/N^m = 2/N \leq \varepsilon$, as desired. $\qquad\Box$

It is an immediate consequence of this theorem that all elementary groups are amenable, and therefore $EG \subseteq AG \subseteq NF$. The work of Grigorchuk and Ol'shanskii mentioned at the end of Chapter 1 shows that $EG \neq AG$ and $AG \neq NF$. But there are still several questions remaining, such as

Conjecture 12.27 and the following question, which asks about the location of a specific group.

Question 10.5.   Is $B(2, 665)$ amenable? In other words (see the end of Chapter 1), is $B(2, 665)$ in $NF \setminus AG$, or is it in $AG \setminus EG$?

A theorem of Tits yields that, when restricted to a certain class of groups, the three classes EG, AG, and NF do, in fact, coincide. Moreover, this class includes all groups of isometries of $\mathbf{R}^n$, and hence the result is applicable to the central examples of this book. If a group has a normal subgroup of finite index with a certain property, we shall say that the group *almost* has the property.

Theorem 10.6.   *Let G be any group of $n \times n$ nonsingular matrices (under multiplication) with entries in a field K; that is, G is a subgroup of $GL_n(K)$.*

(a) *If K has characteristic 0, then either G has a free subgroup of rank 2 or G is almost solvable.*

(b) *If K has nonzero characteristic, then either G has a free subgroup of rank 2 or G has a normal solvable subgroup H, such that G/H is locally finite (meaning: every finite subset generates a finite subgroup).*

The interested reader is referred to [237] for a proof of this powerful result, and some further applications. More recent expositions can be found in [60] and [255], and an illuminating discussion appears in [A6]. A refinement of part (a) of the theorem appears in [254]. Since a locally finite group is the direct limit of its finite subgroups and hence is elementary, this result yields that a matrix group without a free subgroup of rank 2 is elementary and hence amenable. Note that the extra complication of G/H in part (b) is necessary. If K is infinite and $n \geqslant 2$, then $GL_n(K)$ is not almost solvable, and if, in addition, K is an algebraic extension of a finite field, then $GL_n(K)$ is locally finite and hence has no free subgroup of rank 2. Thus, for example, $GL_n(K)$, where K is the algebraic closure of a finite field, shows that the result of part (a) is not valid in the case of nonzero characteristic. Now, the Euclidean affine group A_n is isomorphic to a subgroup of GL_{n+1} (see Appendix A); it follows that any group of isometries of $\mathbf{R}^n$ either has a free subgroup of rank 2, or is almost solvable.

Tits's Theorem is an algebraic one. A topological approach was used by Balcerzyk and Mycielski [5] to show that any locally compact, connected topological group either is solvable or contains a free subgroup of rank the continuum. It follows that the three classes, EG, AG, and NF, coincide when restricted to this collection of topological groups. The work of Balcerzyk and Mycielski builds upon the earlier investigation into free subgroups of Lie groups by Kuranishi [A9].

For applications to equidecomposability theory, the most important question is whether or not a group is amenable. But there are many interesting questions concerning the types of measures on an amenable group. For instance, it is clear that the counting measure is the only possible measure on a finite group, and it is now known (Granirer [78]; see also [80, Appendix 1]) that any infinite amenable group carries more than one measure. Indeed, if G is infinite and amenable, then there are as many measures on G as there are real-valued functions on $\mathscr{P}(G)$ (Chou [29]). However, it is unknown whether there are always at least two G-invariant measures on $\mathscr{P}(X)$, when G is an amenable group acting on an infinite set X. The unique measure on a finite group is also inverse-invariant $(\mu(A) = \mu(A^{-1}))$, and it is easy to see that an amenable group always bears an inverse-invariant measure: simply let $v(A) = (\mu(A) + \mu(A^{-1}))/2$ where μ is both left- and right-invariant (use 10.2). It may happen that all measures are necessarily inverse-invariant, however. For instance, if G is Abelian, then all measures are inverse-invariant if and only if $2G (= \{g + g : g \in G\})$ is finite ([189]).

In order to prove several results about the existence of invariant measures in as uniform a way as possible, it is most convenient to work in the context of Boolean algebras. There are other approaches, however, and we shall indicate after the proof of 10.11 how the important Corollaries 10.9 and 10.10 can be proved using linear functionals and the Hahn-Banach Theorem. But Boolean algebras are more natural for some of the other applications of amenability. We now summarize the basic facts about Boolean algebras that will be needed. The reader who is familiar with the Hahn-Banach Theorem and desires a short proof that analogs of the Banach-Tarski Paradox do not exist in $\mathbf{R}^1$ or $\mathbf{R}^2$ might skip ahead to Theorem 10.11 $((1) \Rightarrow (5))$. For a more detailed treatment of Boolean algebras, see [221].

A *Boolean algebra* $\mathscr{A}$ is a nonempty set (also denoted by $\mathscr{A}$) together with three operations defined on elements of $\mathscr{A}: a \vee b$ (join), $a \wedge b$ (meet), and a' (complement). These operations, which are viewed as generalizations of the set-theoretic operations, union, intersection, and complement, must satisfy the following axioms:

(i) join and meet are commutative and associative,
(ii) $(a \wedge b) \vee b = b, (a \vee b) \wedge b = b$,
(iii) $a \wedge (b \vee c) = (a \wedge b) \vee (a \wedge c), a \vee (b \wedge c) = (a \vee b) \wedge (a \vee c)$,
(iv) $(a \wedge a') \vee b = b, (a \vee a') \wedge b = b$.

We assume that Boolean algebras are nondegenerate, that is, have at least two elements. It follows that there are two distinguished elements, $\mathbf{0}$ and $\mathbf{1}$, defined by $\mathbf{0} = a \wedge a'$ and $\mathbf{1} = a \vee a'$ (the choice of a is immaterial). The *complement of* b *in* a, $a \wedge b'$, is denoted simply by $a - b$. If A is a finite subset of $\mathscr{A}$, $A = \{a_0, \ldots, a_n\}$, then $\sum A$ denotes $a_0 \vee a_1 \vee \cdots \vee a_n$. Two elements $a, b \in \mathscr{A}$ are called *disjoint* if $a \wedge b = \mathbf{0}$. There is a natural partial ordering in any

Boolean algebra, defined by $a \leqslant b$ if $a \wedge b = a$; $\mathbf{1}$ is the greatest element and $\mathbf{0}$ the least element with respect to this ordering.

A subset $\mathscr{A}_0$ of $\mathscr{A}$ is a *subalgebra* if $\mathscr{A}$ is closed under $\vee$, $\wedge$, and $'$. For any subset X of $\mathscr{A}$, there is a smallest subalgebra of $\mathscr{A}$ containing X. It is called the subalgebra of $\mathscr{A}$ generated by X. See [221, p. 14] for an explicit description of the subalgebra generated by a set; an important fact is that the subalgebra generated by a finite set is finite. An equally important concept is that of relativization—the restriction of a Boolean algebra to elements smaller than a fixed $b \in \mathscr{A}$. More precisely, let $\mathscr{A}_b = \{a \in \mathscr{A} : a \leqslant b\}$, with the same join as in $\mathscr{A}$, but with $a_1 \wedge a_2$ in $\mathscr{A}_b$ defined to be $a_1 \wedge a_2 \wedge b$ (in $\mathscr{A}$), and with a' in $\mathscr{A}_b$ defined to be $a' \wedge b$ (in $\mathscr{A}$). Then $\mathscr{A}_b$ is a Boolean algebra whose zero is the same as the zero of $\mathscr{A}$, but whose unit is b. Note that $\mathscr{A}_b$ is not a subalgebra of $\mathscr{A}$. An element $b \in \mathscr{A}$ is called an *atom* if there are no elements below b except for $\mathbf{0}$ and b, that is, if $\mathscr{A}_b = \{\mathbf{0}, b\}$, a two-element Boolean algebra. Any finite Boolean algebra is isomorphic to the algebra of all subsets of a finite set. It follows that any nonzero element b of a finite Boolean algebra satisfies $b = \sum \{a : a \leqslant b$ and a is an atom$\}$; in particular, there is at least one atom a such that $a \leqslant b$.

An *automorphism* of a Boolean algebra $\mathscr{A}$ is a bijection $g : \mathscr{A} \to \mathscr{A}$ that preserves the three Boolean operations. For instance, if a group G acts on a set X, then each $g \in G$ induces an automorphism of the Boolean algebra $\mathscr{P}(X)$ by $g(A) = \{g(x) : x \in A\}$. A *measure* on a Boolean algebra $\mathscr{A}$ is a function $\mu : \mathscr{A} \to [0, \infty]$ such that μ is finitely additive (meaning: $\mu(a \vee b) = \mu(a) + \mu(b)$ if $a \wedge b = \mathbf{0}$) and $\mu(\mathbf{0}) = 0$. If G is a group of automorphisms of $\mathscr{A}$, then a measure μ on $\mathscr{A}$ is *G-invariant* if $\mu(g(b)) = \mu(b)$ for all $b \in \mathscr{A}, g \in G$.

Finally, we shall need the notion of a subring of a Boolean algebra. A nonempty subset $\mathscr{A}_0$ of a Boolean algebra $\mathscr{A}$ is called a *subring* if $a_1 \vee a_2$ and $a_1 - a_2$ are in $\mathscr{A}_0$ whenever $a_1, a_2 \in \mathscr{A}_0$. For instance, the collection of bounded subsets of $\mathbf{R}^n$ is a subring of $\mathscr{P}(\mathbf{R}^n)$. Note that a subalgebra is always a subring. A *measure* on a subring $\mathscr{A}_0$ means a finitely additive function $\mu : \mathscr{A}_0 \to [0, \infty]$ with $\mu(\mathbf{0}) = 0$.

The next theorem is fundamental to the study of measures in Boolean algebras. For instance, letting $\mathscr{A}_0 = \{\mathbf{0}, \mathbf{1}\}$ and $\mu(\mathbf{0}) = 0$, $\mu(\mathbf{1}) = 1$, the theorem yields that every Boolean algebra admits a measure of total measure one. Or, letting $\mathscr{A}_0 = \mathscr{L}$, the subalgebra of $\mathscr{P}(\mathbf{R})$ consisting of all Lebesgue measurable sets, and $\mu = \lambda$, one gets a finitely additive extension of Lebesgue measure to all sets. Such a measure will not necessarily have any invariance properties, but Theorem 10.8 will show how invariance can be guaranteed, provided the group in question is amenable.

Theorem 10.7 (Measure Extension Theorem) (AC). *Suppose $\mathscr{A}_0$ is a subring of the Boolean algebra $\mathscr{A}$, and μ is a measure on $\mathscr{A}_0$. Then there is a measure, $\bar{\mu}$ on $\mathscr{A}$ that extends μ.*

Proof. We first prove the theorem under the additional assumption that $\mathscr{A}$ is finite, proceeding by induction on the number of atoms in $\mathscr{A}$. If $\mathscr{A}$ has one atom, then $\mathscr{A} = \{\mathbf{0}, \mathbf{1}\}$, and the assertion is trivial. In general, choose b to be a minimal (with respect to $\leqslant$) element of $\mathscr{A}_0 \setminus \{\mathbf{0}\}$ (if $\mathscr{A}_0$ consists only of $\mathbf{0}$, simply let $\bar{\mu}$ be identically 0). Let $c = b'$ and consider the relativised algebra $\mathscr{A}_c$ with associated subring $\mathscr{A}_0 \cap \mathscr{A}_c$ and measure $\mu \upharpoonright \mathscr{A}_0 \cap \mathscr{A}_c$. Let a_0 be an atom of $\mathscr{A}$ such that $a_0 \leqslant b$. Then $a_0 \notin \mathscr{A}_c$, whence $\mathscr{A}_c$ has fewer atoms than $\mathscr{A}$ and the induction hypothesis can be used to obtain a measure v on $\mathscr{A}_c$ that extends $\mu \upharpoonright \mathscr{A}_0 \cap \mathscr{A}_c$. Now, define $\bar{\mu}$ on $\mathscr{A}$ by first defining $\bar{\mu}$ on all atoms of $\mathscr{A}$, and then extending to all of $\mathscr{A}$ by $\bar{\mu}(e) = \sum \{\mu(a) : a \leqslant e, a \text{ an atom}\}$; $\bar{\mu}$ will then automatically be finitely additive on $\mathscr{A}$. Let a be any atom of $\mathscr{A}$ and define $\bar{\mu}(a)$ by $\bar{\mu}(a) = v(a)$ if $a \leqslant c$, $\bar{\mu}(a) = \mu(b)$ if $a = a_0$, and $\bar{\mu}(a) = 0$ if $a \leqslant b$ but $a \neq a_0$. Note that $\bar{\mu}$ must agree with v on all of $\mathscr{A}_c$; it remains to show that $\bar{\mu}$ agrees with μ on $\mathscr{A}_0$. Observe that by the minimality of b, if $d \in \mathscr{A}_0$, then either $d \wedge b = \mathbf{0}$ or $d \geqslant b$ (otherwise, $b - d \in \mathscr{A}_0 \setminus \{\mathbf{0}\}$). If the former, then $d \leqslant c$, so $\bar{\mu}(d) = v(d) = \mu(d)$. If, instead, $d \geqslant b$, then $d = b \vee (d - b)$ and $\bar{\mu}(d) = \bar{\mu}(b) + \bar{\mu}(d - b) = \mu(b) + v(d - b) = \mu(b) + \mu(d - b) = \mu(d)$.

The result will be extended to infinite Boolean algebras by the usual compactness technique applied to the product space, $[0, \infty]^{\mathscr{A}}$, viewed as consisting of functions from $\mathscr{A}$ to $[0, \infty]$. For each finite subalgebra, $\mathscr{C}$, of $\mathscr{A}$, let $\mathscr{M}(\mathscr{C}) = \{v \in [0, \infty]^{\mathscr{A}} : v \upharpoonright \mathscr{C} \text{ is a measure extending } \mu \upharpoonright \mathscr{C} \cap \mathscr{A}_0\}$. Each $\mathscr{M}(\mathscr{C})$ is easily seen to be closed and, as shown, nonempty. Moreover, finitely many finite subalgebras of $\mathscr{A}$ generate a finite subalgebra; it follows that the family of all $\mathscr{M}(\mathscr{C})$ has the finite intersection property. By compactness then, the intersection of all the $\mathscr{M}(\mathscr{C})$ over all finite subalgebras $\mathscr{C}$ is nonempty, and any element of this intersection satisfies the conclusion of the theorem. □

As explained in Chapter 9 (see 9.4), the compactness of $[0, \infty]^{\mathscr{A}}$ does not require the Axiom of Choice if $\mathscr{A}$ is countable; hence neither does the Measure Extension Theorem in this case. Some form of choice is definitely required for the general Measure Extension Theorem just proved, however. This is because if the theorem is applied to $\mathscr{A} = \mathscr{P}(\mathbf{N})$, $\mathscr{A}_0 = \{A \in \mathscr{A} : A \text{ or } \mathbf{N} \setminus A \text{ is finite}\}$, and μ, the measure assigning measure 0 to finite sets and measure one to cofinite sets, one obtains a finitely additive measure on $\mathscr{P}(\mathbf{N})$ that has total measure one. In fact, by using $\{0, 1\}^{\mathscr{A}}$ instead of $[0, \infty]^{\mathscr{A}}$ in the latter part of the proof, one obtains a $\{0, 1\}$-valued measure on $\mathscr{P}(\mathbf{N})$ of total measure one that vanishes on singletons. Such a measure (more precisely, the sets having measure one with respect to such a measure) is called a *nonprincipal ultrafilter* on $\mathbf{N}$. In Chapter 13 (see 13.4, 13.5) it will be indicated why the existence of such an ultrafilter, or even of a finitely additive measure on $\mathscr{P}(\mathbf{N})$ that vanishes on singletons, cannot be proved in ZF alone. It will also be shown in that chapter why parts (c), (f), and, with an additional hypothesis about inaccessible cardinals, (b) of Theorem 10.4 cannot be proved in ZF.

However, the Measure Extension Theorem is not strong enough to prove the full Axiom of Choice. This is because the proof of the Measure Extension Theorem used the fact that a product of compact Hausdorff spaces is compact, rather than the full Tychonoff Theorem. This Hausdorff version of the Tychonoff Theorem is equivalent to the Boolean Prime Ideal Theorem (every Boolean algebra has a $\{0,1\}$-valued measure of total measure one; see [98]), and it is known that these assertions do not imply the Axiom of Choice. In fact, the Measure Extension Theorem is equivalent (in ZF) to the Hahn-Banach Theorem on extending linear functionals ([129]), and it is known that these statements are strictly weaker than the Boolean Prime Ideal Theorem; see Figure 13.1.

Now, it is a slightly more complex situation that interests us. Namely, if G is a group of automorphisms of $\mathscr{A}$, $\mathscr{A}_0$ is a G-invariant subring ($a \in \mathscr{A}_0$ and $g \in G$ imply $g(a) \in \mathscr{A}_0$), and μ is a G-invariant measure on $\mathscr{A}_0$, can we be sure that there is a G-invariant extension of μ to all of $\mathscr{A}$? This can be done if G is amenable: first use the preceding theorem to obtain some extension v, and then use a left-invariant mean on G to average out the noninvariance of v.

Theorem 10.8 (Invariant Extension Theorem) (AC). *If in the Measure Extension Theorem G is an amenable group of automorphisms of $\mathscr{A}$, and $\mathscr{A}_0$ and μ are G-invariant, then $\bar{\mu}$ can be chosen to be G-invariant as well.*

Proof. Use the Measure Extension Theorem to get a measure v on $\mathscr{A}$ extending μ. Let θ be a measure on the amenable group G. Now, if $b \in \mathscr{A}$, define $f_b: G \to \mathbf{R}$ by $f_b(g) = v(g^{-1}(b))$. Then define $\bar{\mu}$ by $\bar{\mu}(b) = \int f_b \, d\theta$, if $f_b \in B(G)$, that is, if f_b is bounded, and $\bar{\mu}(b) = \infty$ if f_b is unbounded (where a function that takes on the value ∞ is considered to be unbounded.) It is easy to see that $\bar{\mu}$ is a G-invariant extension of μ. $\qquad\square$

In fact, it is easy to see that the amenable groups are the only ones for which the Invariant Extension Theorem holds. If the theorem holds for G, apply it to $\mathscr{A} = \mathscr{P}(G)$, $\mathscr{A}_0 = \{\varnothing, G\}$, $\mu(\varnothing) = 0$, and $\mu(G) = 1$, with G acting on $\mathscr{A}$ by left translation to obtain a measure on G.

The Invariant Extension Theorem can be applied to extend Lebesgue measure to an invariant, finitely additive measure defined on all subsets of $\mathbf{R}^1$ or $\mathbf{R}^2$. Hence the Banach-Tarski Paradox has no analog using isometries of the line or plane. Because of the Banach-Tarski Paradox (more precisely, by Theorem 2.6), such invariant extensions of Lebesgue measure do not exist in $\mathbf{R}^3$ and beyond; but if invariance with respect to an amenable group of isometries, rather than the full group, is desired, such extensions exist in all dimensions.

Corollary 10.9 (AC).   *If G is an amenable group of isometries of* $\mathbf{R}^n$ *(or of* S^n*), then there is a finitely additive, G-invariant extension of Lebesgue measure,* λ, *to all subsets of* $\mathbf{R}^n$ *(or* S^n*). In particular, Lebesgue measure on* S^1, $\mathbf{R}^1$, *or* $\mathbf{R}^2$ *has an isometry-invariant, finitely additive extension to all sets.*

Proof.   Apply the Invariant Extension Theorem to $\mathscr{A} = \mathscr{P}(\mathbf{R}^n)$, $\mathscr{A}_0 = \mathscr{L}$, the subalgebra of Lebesgue measurable sets, and $\mu = \lambda$. The second assertion follows from the fact that the corresponding isometry groups are solvable (see Appendix A).   □

Recall from Corollary 1.6 that there is no countably additive, translation-invariant measure on all subsets of $\mathbf{R}^n$ that normalizes a cube. But the translation group is Abelian, so Corollary 10.9 shows that if only finite rather than countable additivity is desired, then such measures do exist.

Corollary 10.10 (AC).   *If G is an amenable group of isometries of* $\mathbf{R}^n$, *then no bounded subset of* $\mathbf{R}^n$ *with nonempty interior is G-paradoxical. In particular, no bounded subset of* $\mathbf{R}^1$ *or* $\mathbf{R}^2$ *with nonempty interior is paradoxical.*

Proof.   Use the preceding corollary to obtain μ, a finitely additive, G-invariant extension of λ to all subsets of $\mathbf{R}^n$. Then $0 < \mu(A) < \infty$ for any bounded set A with nonempty interior, so A is not G-paradoxical.   □

Corollary 10.10 actually applies to a wider class of sets. For if A has positive inner Lebesgue measure and bounded outer Lebesgue measure, then if μ is as in the proof of 10.10, it is easy to see that $0 < \mu(A) < \infty$, whence A is not paradoxical. But some conditions on A are necessary, because of the Sierpiński-Mazurkiewicz Paradox (1.7), which provides an example of a paradoxical subset of $\mathbf{R}^2$. Recall that the set of that paradox is countable and unbounded. It is not known whether there can be a bounded paradoxical subset of the plane; however, no bounded subset of $\mathbf{R}^2$ can be paradoxical using two pieces (see 12.14).

The situation in $\mathbf{R}^1$ is quite different, because the isometry group is much simpler. In fact, in $\mathbf{R}^1$ Corollary 10.10 is valid without any conditions on the set, that is, no nonempty subset of the line is paradoxical. This will follow from 12.8 and 12.11, which show that this is the case whenever the group acting on the set is almost Abelian.

Although paradoxes using isometries are missing in the plane, recall that the Von Neumann Paradox (7.3) shows that a square is paradoxical using area-preserving affine transformation. Of course, this enlargement of the isometry group is nonsolvable and nonamenable.

In the next chapter we shall investigate the necessity of amenability in these two corollaries. It turns out that in the case of spheres, amenability is a necessary condition for the existence of a measure as in 10.10. But amenability is not necessary for the absence of paradoxes, that is, the existence of a finitely additive, G-invariant measure on $\mathscr{P}(\mathbf{R}^n)$ that normalizes the unit cube, but does not necessarily agree with Lebesgue measure.

We have used the Axiom of Choice to prove that the analog of the Banach-Tarski Paradox fails in $\mathbf{R}^1$ and $\mathbf{R}^2$. Thus Choice sits on both sides of the fence—it is used to construct the paradoxes in the higher dimensions, and to get rid of them in the lower dimensions. If one desires only the nonexistence of paradoxes rather than the stronger (in the absence of Choice) existence of total measures, however, one can get by without using the Axiom of Choice (see 12.12 and 13.9). Thus a more accurate portrayal of the axiom's role, at least in $\mathbf{R}^1$ and $\mathbf{R}^2$, is that it is used to destroy countably additive measures, but to construct finitely additive measures. In the next chapter we shall give some further applications of amenability to the construction of measures.

CHARACTERIZATIONS OF AMENABILITY

We conclude this chapter with some more properties of amenable groups, including an alternate proof of Corollary 10.10 based on the use of linear functionals rather than Boolean algebras. The following result gives several characterizations of amenability, showing how natural this notion is.

Theorem 10.11 (AC). *For a group G, the following are equivalent:*

(1) *G is amenable.*
(2) *There is a left-invariant mean on G.*
(3) *G is not paradoxical.*
(4) *G satisfies the Invariant Extension Theorem: A G-invariant measure on a subring of a Boolean algebra may be extended to a G-invariant measure on the entire algebra.*
(5) *G satisfies the Hahn-Banach Extension Property: Suppose*
    (a) *G is a group of linear operators on a real vector space V;*
    (b) *F is a G-invariant linear functional on V_0, a G-invariant subspace of V; and*
    (c) *$F(v) \leqslant p(v)$ for all $v \in V_0$, where p is some real-valued function on V such that $p(v_1 + v_2) \leqslant p(v_1) + p(v_2)$ for $v_1, v_2 \in V$, $p(\alpha v) = \alpha p(v)$ for $\alpha \geqslant 0$, $v \in V$, and $p(g(v)) \leqslant p(v)$ for $g \in G$, $v \in V$.*
    Then there is a G-invariant linear functional $\bar{F}$ on V that extends F and is dominated by p.
(6) *G satisfies Følner's Condition: For any finite subset W of G and every $\varepsilon > 0$, there is another finite subset W^* of G such that for any $g \in W, |gW^* \triangle W^*|/|W^*| \leqslant \varepsilon$.*
(7) *G satisfies Dixmier's Condition: If $f_1, \ldots, f_n \in B(G)$ and $g_1, \ldots, g_n \in G$, then for some $h \in G$, $\sum f_i(h) - f_i(g_i^{-1}h) \leqslant 0$.*
(8) *G satisfies the Markov-Kakutani Fixed Point Theorem: Let K be a compact convex subset of a locally convex linear topological space X,*

and suppose G acts on K in such a way that each transformation $g: K \to K$ is continuous and affine $\big(g(\alpha x + (1 - \alpha)y) = \alpha g(x) + (1 - \alpha)g(y)$ whenever $x, y \in K$ and $0 \leqslant \alpha \leqslant 1\big)$. Then there is some x in K that is fixed by each $g \in G$.

Proof. The equivalence of (1)–(4) follows from previous work. The equivalence of (1) and (2) was discussed at the beginning of this chapter; that of (1) and (3) follows from Tarski's Theorem (9.2); and that of (1) and (4) follows from Theorem 10.8 and the remarks following its proof. We shall prove that $(1) \Leftrightarrow (5)$ and $(1) \Rightarrow (8) \Rightarrow (2) \Rightarrow (7) \Rightarrow (2)$, yielding the equivalence of all statements but (6). That (6) implies (1) will be proved, but the proof of the converse, which uses techniques of functional analysis, will be omitted. However, we shall sketch an elementary proof of the special case that Abelian groups satisfy Følner's Condition.

$(1) \Rightarrow (5)$. Use the standard Hahn-Banach Theorem [196, p. 187] to obtain a linear functional F_0 on V that extends F and is dominated by p. Then, for any $v \in V$, define $f_v: G \to \mathbf{R}$ by $f_v(h) = F_0(h^{-1}(v))$. Since $F_0(h^{-1}(v)) \leqslant p(h^{-1}(v)) \leqslant p(v)$, f_v is bounded by $p(v)$. Hence, choosing a measure μ on G, we may define $\bar{F}(v)$ to be $\int f_v \, d\mu$. Then $\bar{F}(v) \leqslant p(v)$ and $\bar{F}$ is a linear functional on V. Moreover, $\bar{F}$ extends F, and because $f_{g(v)} = {}_g(f_v)$, the G-invariance of $\bar{F}$ follows from that of μ.

$(5) \Rightarrow (1)$. Let $V = B(G)$, with V_0 taken to be the subspace of constant functions. The action of g on $B(G)$ by $f \mapsto {}_g f$ is linear, and V_0 is G-invariant. Letting $F(\alpha \chi_G) = \alpha$ and $p(f) = \sup\{f(g): g \in G\}$, we see that the hypotheses of (5) are satisfied. Hence there is a left-invariant linear functional $\bar{F}$ on $B(G)$ with $\bar{F}(\chi_G) = 1$. To prove that $\bar{F}$ is a left-invariant mean, it remains to show that $\bar{F}(f) \geqslant 0$ if $f(g) \geqslant 0$ for each $g \in G$. But $p(-f) \leqslant 0$ for such f, whence $\bar{F}(-f) \leqslant p(-f) \leqslant 0$ and $\bar{F}(f) = -\bar{F}(-f) \geqslant 0$.

$(2) \Rightarrow (7)$. If F is a left-invariant mean on $B(G)$, then $F(\sum f_i - {}_{gi}f_i) = 0$, and hence for some $h \in G$, $f_i(h) - ({}_{g_i}f_i)(h) \leqslant 0$.

$(7) \Rightarrow (2)$. Let V_0 be the subspace of $B(G)$ generated by the constant functions and all functions of the form $f - {}_g f$, where $f \in B(G)$ and $g \in G$. An element of V_0 has the form $f - {}_g f + \alpha \chi_G$ for some real α that, by the hypothesis that each $f - {}_g f$ takes on a nonpositive value, is unique; hence we can define a linear functional on V_0 by $F(f - {}_g f + \alpha \chi_G) = \alpha$. We claim that $F(v) \leqslant \sup v$ for each $v \in V_0$. For if $v = f - {}_g f + \alpha \chi_G$, then since $-(f - {}_g f) = -f - {}_g(-f)$ takes on a nonpositive value, $v = \alpha \chi_G - \big(-(f - {}_g f)\big)$ takes on a value no smaller than $\alpha = F(v)$. So, letting $p(v) = \sup v$ for $v \in B(G)$, we may apply the Hahn-Banach Theorem to obtain a linear functional $\bar{F}$ on $B(G)$ that extends F and is dominated by p. The definition of F guarantees that $\bar{F}$ is left-invariant and normalizes χ_G. And if $f(h) \geqslant 0$ for all $h \in G$, then $\bar{F}(f) = -\bar{F}(-f) \geqslant -p(-f) \geqslant 0$. Hence $\bar{F}$ is a left-invariant mean on $B(G)$.

(8) $\Rightarrow$ (2). Turn $B(G)$ into a normed linear space by using the sup norm, $\|f\| = \sup\{|f(g)|: g \in G\}$ and let X be the dual space of $B(G)$ (all bounded linear functionals on $B(G)$), equipped with the weak* topology. With this topology, X is a locally convex linear topological space. Let K be the subset of X consisting of all functionals F satisfying $\inf(f) \leqslant F(f) \leqslant \sup(f)$. Note that each $F \in K$ satisfies $|F(f)|/\|f\| \leqslant 1$; hence K is contained in the unit ball of X. Since this ball is compact (Alaoglu's Theorem) and K is closed, K is a compact subset of X. Moreover, K is convex. Now, each $g \in G$ acts on X by $(_gF)(f) = F(_gf)$, and the transformation of X induced by g is linear (hence affine) and continuous. Moreover, since $\inf(_gf) = \inf(f)$ and $\sup(_gf) = \sup(f)$, each g maps K into K. So, by (8), there is some $F \in K$ that is fixed by each $g \in G$; such an F is a left-invariant mean on $B(G)$.

(1) $\Rightarrow$ (8). Let μ be a measure on G and choose any $x \in K$. The idea is to use μ to define an average value (integral) of the function $f: G \to K$ defined by $f(g) = g(x)$; this average will be the desired fixed point. Let D be the directed set consisting of all finite open covers, $\pi = \{U_i\}$, of K, ordered by refinement, where it is assumed that each $U_i \cap K$ is nonempty. If V is a neighborhood of the origin, then π will be called *V-fine* if each set in π can be translated to fit into V. Define a net $\phi: D \to K$ as follows: given $\pi \in D$, choose points $s_i \in U_i \cap K$ and let $\phi(\pi) = \sum \mu(E_i)s_i$, where $E_i = f^{-1}(U_i)\backslash \bigcup \{E_j : j < i\}$. Then $\phi(\pi)$ is a convex combination of the s_i, and therefore $\phi(\pi) \in K$.

Claim 1. If V is a convex symmetric ($V = -V$) neighborhood of the origin, π is $V/2$-fine, and π' refines π, then $\phi(\pi') - \phi(\pi) \in V$.

Proof. We give the details for the simple case where π consists of a single open set U, with designated point s, and leave the general case as an exercise. Let $\pi' = \{U'_1\}$ with selected points $s_i \in U'_1$; then $\phi(\pi) = s$ and $\phi(\pi') = \sum \alpha_i s_i$ where the reals α_i sum to one. Now, $K \subseteq U \subseteq t + V/2$ for some $t \in X$, so for each s_i there is some $v_i \in V$ such that $s_i = t + v_i/2$. Also, $s = t + v/2$ for some $v \in V$. Now, by convexity, $\sum \alpha_i s_i = t + v'/2$, with $v' \in V$, and so $(\sum \alpha_i s_i) - t = v'/2$. But $t - s = -v/2$, and since $-v \in V$, summing these equations yields that $(\sum \alpha_i s_i) - s \in V$, as desired.

Claim 2. There is a unique $x \in K$ such that the net ϕ converges to x.

Proof. The compactness of K yields that ϕ has at least an accumulation point x in K. To prove that $\phi \to x$ it suffices to show that if U is any convex symmetric neighborhood of the origin, then there is some $\pi \in D$ such that $\phi(\pi') \in x + U$ whenever π' refines π. To this end, let π be any $U/4$-fine cover in D (such exists by compactness), and using the fact that ϕ accumulates at x to refine π if necessary, assume $\phi(\pi) \in x + U/2$. Now, if π' refines π, then $\phi(\pi') - \phi(\pi) \in U/2$ by Claim 1. Since $\phi(\pi) - x \in U/2$, this yields that $\phi(\pi') - x \in U$, as required. Since X is a Hausdorff space, limits of nets are unique.

The proof that x, the limit of ϕ, is the desired fixed point requires the following claim.

Claim 3. If ρ is a net on D defined in the same way as ϕ, but with different designated points s_i, then ρ converges to x.

Proof. Suppose U is a convex symmetric neighborhood of the origin, and choose π to be $U/4$-fine so that all refinements π' of π satisfy $\phi(\pi') \in x + U/2$. Suppose $\phi(\pi')$ is defined from points s_i, while $\rho(\pi')$ uses r_i. Since π' is $U/4$-fine, there are points t_i such that $r_i, s_i \in t_i + U/4$. Since $U = -U$, it follows that $r_i - s_i \in U/2$, and because U is convex, this implies that $\rho(\pi') - \phi(\pi') \in U/2$. Now, $\phi(\pi') - x \in U/2$ so $\rho(\pi') - x \in U$. Hence $\rho(\pi') \in x + U$ for any π' refining π, proving $\rho \to x$.

Now, g is continuous, so the net $g\phi$ defined by $(g\phi)(\pi) = g(\phi(\pi))$ converges to $g(x)$. To complete the proof we shall show that $g\phi \to x$; uniqueness of limits then yields $x = g(x)$. Note that each $g \in G$ induces an order-preserving map from D to D by $g(\pi) = \{\{y : g(y) \in U_i\} : U_i \in \pi\}$. Use this mapping to define a net ψ on D by $\psi(\pi) = \phi(g(\pi))$; it is easy to prove that if $\psi \to y$, then $\phi \to y$ too. We claim that the net $g\psi$ converges to x, the limit of ϕ. For $(g\psi)(\pi) = g(\psi(\pi)) = g(\phi(g(\pi))) = g(\sum \alpha_i r_i)$ where r_i is some point on the ith set of the cover $g(\pi)$ and α_i is the μ-measure of the appropriate subset of G. Since $g : X \to X$ is affine, $g(\sum \alpha_i r_i) = \sum \alpha_i g(r_i)$ and the left-invariance of μ yields that α_i is the measure of the subset of G arising from π. Since $g(r_i) \in U_i$, this means that the net $g\psi$ satisfies the condition of Claim 3. Hence $g\psi \to x$. This yields that $\psi \to g^{-1}(x)$, which implies by the preceding remark that $\phi \to g^{-1}(x)$. Therefore $g\phi \to x$, as desired. (For a rather different proof of the Markov-Kakutani Theorem in the case that G is Abelian (but without the assumption that X is locally convex) see [63, p. 456].)

Følner's Condition is a very interesting property that abstracts the essential fact about Abelian groups that makes them amenable. Loosely speaking, it states that for any finite subset of the group there is another finite subset that is almost invariant with respect to translation on the left by elements of the first set. It is not hard to see (and we shall give proofs) that Abelian groups satisfy Følner's Condition and that Følner's Condition yields amenability. (The proof of 10.4(b) used a more direct route, via N^m.) The surprising fact is that all amenable groups satisfy Følner's Condition. This was discovered by Følner [71], but the reader interested in a proof is referred to Namioka [172], where a much simplified proof is presented. Or, see [80] where a topological version of Følner's Condition (involving Haar measure) is proved equivalent to topological amenability for locally compact topological groups (see the remarks at the end of this chapter). Since Haar measure on a discrete group is just the counting measure, the equivalence for abstract groups is a consequence of the topological generalization.

(6) $\Rightarrow$ (1). The proof of amenability from Følner's Condition is similar to the proof of 10.4(b). For each finite $W \subseteq G$ and $\varepsilon > 0$, let $\mathscr{M}_{W,\varepsilon}$ consist of those finitely additive functions $\mu : \mathscr{P}(G) \to [0, 1]$ such that $\mu(G) = 1$, and for each $g \in W$ and $A \subseteq G$, $|\mu(A) = \mu(gA)| \leqslant \varepsilon$. Then $\mathscr{M}_{W,\varepsilon}$ is a closed subset of $[0, 1]^{\mathscr{P}(G)}$, and, as usual, a compactness argument will complete the proof once it is shown that each $\mathscr{M}_{W,\varepsilon}$ is nonempty. For this, simply define $\mu(A)$ to be $|A \cap W^*|/|W^*|$ where W^* is as provided by Følner's Condition; then μ is in $\mathscr{M}_{W,\varepsilon}$.

(1) $\Rightarrow$ (6). (Abelian case). We shall sketch a proof that Abelian groups satisfy Følner's Condition that proceeds by induction on m, where $W = \{g_1, \ldots, g_m\}$. One could also prove this using the Fundamental Theorem of Finitely Generated Abelian Groups. For $m = 1$, let $W^* = \{1, g_1, g_1^2, \ldots, g_1^{r-1}\}$ where r is the order of g_1, or if g_1 does not have finite order, any integer such that $2/r \leqslant \varepsilon$. For the general case, let $V = \{g_1, \ldots, g_{m-1}\}$ and let H be the subgroup of G generated by V. If no positive power of g_m lies in H, simply choose r so that $2/r \leqslant \varepsilon$ and let $V^* \subseteq H$ be as obtained by the induction hypothesis on V and ε/r. Then let $W^* = V^* \cup g_m V^* \cup \cdots \cup g_m^{r-1} V^*$. The hypothesis on g_m yields that the sets in this union are pairwise disjoint, so $|W^*| = r|V^*|$, and it is easy to see that W^* is as desired with respect to W and ε. If $g_m^r \in H$ for some $r \geqslant 1$, with g_m^r equal to a word of length s in $g_1^{\pm 1}, \ldots, g_{m-1}^{\pm 1}$, then use the induction hypothesis to choose V^* with respect to V and ε/rs. Let $W^* = V^* \cup g_m V^* \cup \cdots \cup g_m^{r-1} V^*$. Since V^* works for $V \cup \{g_m^r\}$, ε/r, it follows that W^* works for W, ε. $\qquad \square$

There are several other properties of a group that are equivalent to amenability. One such is due to Kesten [103, 104]. He proved, using Følner's Condition, that amenability is equivalent to an assertion about recurrence for random walks in a group with respect to symmetric probability distributions on the group. See [45] for a different, shorter proof of Kesten's characterization. More recently, Cohen [35] has obtained a very useful characterization of amenability in terms of a growth condition on the group; this characterization is discussed at the end of Chapter 12.

Now, by considering the problem of extending Lebesgue measure to all sets as one of extending the Lebesgue integral to all functions, we can use the Hahn-Banach Extension Property for amenable groups to give an alternate proof of Corollaries 10.9 and 10.10. The only part of the previous theorem that is required is (1) $\Rightarrow$ (5).

Another Proof of Corollary 10.9. Suppose G is an amenable group of isometries of $\mathbf{R}^n$. Let V_0 be the space of all Lebesgue integrable real-valued functions on $\mathbf{R}^n$, and V be the space of all functions $f : \mathbf{R}^n \to \mathbf{R}$ such that for some $g \in V_0$, $f(x) \leqslant g(x)$ for all $x \in \mathbf{R}^n$. Then any group of isometries of $\mathbf{R}^n$ acts on V and V_0 in the obvious way, and we may let F be the G-invariant linear functional on V_0 defined by the Lebesgue integral, that is, $F(f) = \int f \, d\lambda$. Finally, define a G-invariant pseudo-norm p on V that dominates F by

$p(f) = \inf\{F(g): g \in V_0 \text{ and } g(x) \geqslant f(x) \text{ for all } x\}$; recall that each $f \in V$ is dominated by some $g \in V_0$.

Now, use the Hahn-Banach Extension Property to obtain a G-invariant linear functional $\bar{F}$ on V, and use $\bar{F}$ to define the desired measure, μ, on $\mathscr{P}(\mathbf{R}^n)$ by $\mu(A) = \bar{F}(\chi_A)$ if $\chi_A \in V$; $\mu(A) = \infty$ otherwise. It is easy to see that μ is finitely additive and G-invariant. And because $\bar{F}(\chi_A) = \int \chi_A \, d\lambda = \lambda(A)$ if A has finite Lebesgue measure, μ extends λ. Finally, note that $\mu(A) \geqslant 0$; for if $f(x) \geqslant 0$ for all $x \in \mathbf{R}^n$, then $-f(x) \leqslant 0$, whence $\bar{F}(-f) \leqslant p(-f) \leqslant 0$ and $\bar{F}(f) = -\bar{F}(-f) \geqslant 0$. $\square$

Much of the preceding discussion of amenability is valid in semigroups. The definition is the same and many of the applications remain valid. For instance, the Hahn-Banach Extension Property and Følner's Condition are valid for precisely the amenable semigroups. See [43] for more on amenable semigroups.

TOPOLOGICAL AMENABILITY

An important generalization of amenability is a topological one. In the summary that follows all topological groups will be assumed to be locally compact and Hausdorff; some of the results are valid without the assumption, but the theory is much more coherent in this class of groups. A topological group is called (topologically) amenable if there is a finitely additive, left-invariant measure on the Borel subsets of the group that has total measure one. In what follows we shall use the word amenable in this topological sense. To say that G is amenable as a discrete group (i.e., with the discrete topology, which is locally compact) is the same as saying that G is amenable in the usual, nontopological sense: in the discrete case, all sets are Borel. Note that a topological group that is amenable as an abstract group is topologically amenable: simply restrict the measure on $\mathscr{P}(G)$ to the Borel sets. If G is compact, then there is a countably additive, left-invariant Borel measure with total measure one, namely Haar measure; hence compact groups are topologically amenable.

The book by Greenleaf [80] contains a detailed account of the theory of topological amenability, which is much richer than the theory of amenability in abstract groups and has many applications to diverse areas of analysis. Theorem 10.4 is valid after making appropriate topological modifications; for example, a closed subgroup of an amenable locally compact group is amenable. It is not surprising that G is topologically amenable if and only if there is a left-invariant mean on the bounded, Borel measurable functions from G to $\mathbf{R}$, but in fact it is sufficient to have a mean on the bounded, continuous real-valued functions on G. The following version of Følner's Condition is equivalent to amenability. If K is a compact subset of G and

$\varepsilon > 0$, then there is a Borel set $K^* \subseteq G$ with $0 < \theta(K^*) < \infty$ such that $\theta(gK^* \triangle K^*)/\theta(K^*) \leqslant \varepsilon$ for all $g \in K$ (here θ denotes left-invariant Haar measure). Also, there is a topological version of the Markov-Kakutani Fixed Point Theorem that is equivalent to amenability in locally compact groups. One simply adds the condition that the function from $G \times K$ to K induced by the action is continuous, that is, the map $(g, x) \rightarrow g(x)$ is jointly continuous. Many more characterizations of topological amenability are given in [80].

If a locally compact group G has a closed subgroup H that is a free group on two generators, then G is not amenable. This is because H is countable and Hausdorff, so all of its subsets are Borel; hence the standard paradox of a free group of rank 2 yields that H is not topologically amenable, and therefore, since H is closed, G is not amenable. Just as with abstract groups, this leads to the question whether a locally compact group is amenable if and only if it has no closed subgroup that is free of rank 2. This optimistic characterization is false, since by Theorem 1.12, it is false already in the discrete case. But yet, just as in the discrete case, it may be that this characterization is valid for large classes of interesting locally compact groups (just as Tits's Theorem yields the discrete version for a large class of groups). In fact, Rickert [191] has shown that if the locally compact group G is almost connected (i.e., G/N is compact where N is the component of the identity), then G is amenable if and only if G does not have a closed subgroup that is free of rank 2.

Clearly a group cannot be amenable if it is paradoxical using Borel pieces. But the converse is not clear. Tarski's Theorem (9.3) yields that a group is amenable if and only if, for each n, it is not possible to pack $n + 1$ copies of the group into n copies using Borel pieces (and left multiplication). But, because no cancellation law for Borel equidecomposability is known, it is not clear that the assertion referring to all n is equivalent to the one for $n = 1$, that is, the assertion that G is paradoxical.

Question 10.12. Is it true that a locally compact group that is not paradoxical using Borel sets is amenable?

NOTES

The material of this chapter has its origin in the seminal paper of Banach [8]. In that paper Banach developed the main ideas of the Hahn-Banach Theorem and showed that the isometry groups of $\mathbf{R}^1$ and $\mathbf{R}^2$ satisfy the Hahn-Banach Extension Property of Theorem 10.11. Thus he proved Corollary 10.9 for $\mathbf{R}^1$ and $\mathbf{R}^2$, using the technique of the proof given after the proof of Theorem 10.11. Von Neumann [246] realized that Banach's results could be formulated more abstractly. He introduced the definition of an amenable group and proved Theorem 10.4 (a), (b), and (e). Von Neumann was the first to state explicitly that a free group of rank 2 is not amenable, and the problem whether $AG = NF$ (see 1.12) has often been attributed to him. Agnew and Morse [3] showed how von Neumann's ideas applied to the Hahn-Banach Theorem in its modern form. They showed that all solvable groups satisfied the Hahn-Banach Extension Property, and obtained corresponding results about extensions of Lebesgue measure.

The class of amenable groups was studied extensively by Day [42, 43]. He proved Proposition 10.2; parts (c), (d), and (f) of Theorem 10.4 (parts (c) and (d) were obtained independently by Følner [71]); and many other results about amenable groups and semigroups.

The Measure Extension Theorem (10.7) is due to Horn and Tarski [36]. See [129, 182, 183, 184] for various results related to the axiomatic strength of the Measure Extension Theorem and comparisons with the Hahn-Banach Theorem; foundational considerations are also discussed in Chapter 13. The idea of the Invariant Extension Theorem (10.8) goes back to Banach and von Neumann; the formulation given here, using subrings, appears in Mycielski [168].

Theorem 10.11 contains only a sample of the many properties of a group now known to be equivalent to amenability. The Hahn-Banach Extension Property was considered by Agnew and Morse [3], Klee [106], and Silverman [222, 223]. Silverman proved that the property implies amenability and investigated generalizations of the property to vector-valued linear functionals. Følner's Condition was introduced by Følner [71], whose proof of the sufficiency of amenability was substantially simplified by Namioka [172]. Dixmier's Condition was introduced in [59]; see [66] and [80]. The fact that the Markov-Kakutani Fixed Point Theorem is valid for amenable groups was proved by Day [44], and this paper also contains the converse, which is due to Granirer. Chen [28] presents a weaker version of the Markov-Kakutani Fixed Point Property that holds for groups without a free non-Abelian subgroup.

There has been a large amount of work on amenability done in the Soviet Union. For a summary, see [A10].

A variety of results, applications, and references on the subject of topological amenability is contained in Greenleaf [80]. The equivalence of topological amenability with a topological version of the Markov-Kakutani Fixed Point Theorem can be found in [191] (see also [45]). Emerson [67], unaware of Tarski's work, formulated Question 10.12 and obtained the characterization of topological amenability discussed just before 10.12.

Chapter 11

Applications of Amenability: Marczewski Measures and Exotic Measures

This chapter contains some refinements and further applications of the basic technique of constructing finitely additive measures that was introduced in the previous chapter. After constructing certain "exotic" measures in $\mathbf{R}^1$ and $\mathbf{R}^2$, we shall survey recent work on an old problem about the uniqueness of Lebesgue measure that shows that such strange measures do not exist in $\mathbf{R}^3$ or beyond. Much of the emphasis here is on measures on the algebra of measurable sets rather than measures on all sets. Finally, we shall discuss the problem of characterizing groups of Euclidean isometries (and more general group actions) for which paradoxes exist.

Corollary 10.9 showed how to construct finitely additive, invariant extensions of Lebesgue measure. Such measures perhaps seem a bit unnatural because they mix the two types of additivity; λ is countably additive, but the extension is finitely additive. In order to prove that paradoxical decompositions do not exist, all that is required is a finitely additive, isometry-invariant measure that normalizes the unit cube. Of course, if we are considering all isometries, then we know (by 10.9 and the Banach-Tarski Paradox) that such measures exist in $\mathbf{R}^n$ if and only if $n \leqslant 2$.

As we shall see, it is worthwhile to study the properties of a finitely additive, isometry-invariant measure that normalizes the cube. Recall from Proposition 9.6 that such a measure must agree with Jordan measure, v, on the Jordan measurable sets, $\mathscr{J}$. A famous problem, solved for all $\mathbf{R}^n$ only recently, is whether such a measure must agree with Lebesgue measure. If $\mathscr{B}$ denotes the sets with the Property of Baire, then $\mathscr{J} \subseteq \mathscr{L} \cap \mathscr{B}$ (see Appendix B). If $A \in \mathscr{J}$, then $v(A) = 0$ if and only if A is nowhere dense (or meager) if and only if $\lambda(A) = 0$. So, Jordan measure is unbiased regarding measure and category: the Jordan null sets are precisely those sets in $\mathscr{J}$ that are small in the sense of both measure and category. Lebesgue measure is an extension of v from $\mathscr{J}$ to $\mathscr{L}$ that is biased toward the measure side: λ gives positive measure to some nowhere dense sets (consider the complement in $[0, 1]$ of a small open cover of the rationals). We may also view λ as a finitely additive, invariant Borel measure that vanishes on the (Borel) sets of Lebesgue measure zero.

Marczewski raised the question of the existence of finitely additive, invariant Borel measures that are biased in the opposite direction, that is, measures that vanish on all meager sets (and hence necessarily disagreeing with λ on some closed, nowhere dense sets). Such measures may be viewed as measures on $\mathscr{B}$ (since if $A \in \mathscr{B}$, then $A = C \vartriangle E$ where C is Borel and E is meager); hence this question is equivalent to Question 9.9 about Marczewski measures). Note that the restriction to finite additivity is essential, since a countably additive measure is forced to agree with λ (see 9.10), and hence cannot vanish on all nowhere dense sets. The techniques of Chapter 10 can be used to produce Marczewski measures in $\mathbf{R}^1$ and $\mathbf{R}^2$; their existence in higher dimensions, and in S^2, is still unresolved. We shall construct these measures, and then examine the consequences of their existence for questions about Lebesgue measure's uniqueness.

The Marczewski construction and some of the other results to be presented in this chapter are most easily understood in the context of quotient Boolean algebras with respect to an ideal. An *ideal* in a Boolean algebra, $\mathscr{A}$, is a nonempty subset, I, of $\mathscr{A}$ such that $a \vee b \in I$ whenever $a, b \in I$, and $a \leqslant b \in I$ implies $a \in I$. (Any Boolean algebra can be turned into a ring using $a \vartriangle b$ as addition and $a \wedge b$ as multiplication (see [221, §17]), and the definition of ideal just given corresponds to the usual notion in rings.) Two prominent examples are the collection of meager sets, to be denoted by M, and the collection of sets of Lebesgue measure zero, to be denoted by N. In $\mathscr{P}(\mathbf{R}^n)$, both M and N are ideals, while M is an ideal in $\mathscr{B}$ (sets with the Property of Baire) and N is an ideal in $\mathscr{L}$. We shall also use the notion of ideal in a subring $\mathscr{C}$ of a Boolean algebra $\mathscr{A}$, defined by: $a \vee b \in I$ whenever $a, b \in I$, and $a \leqslant b \in I$ implies $a \in I$ whenever $a, b \in \mathscr{C}$. Note that if I is an ideal in $\mathscr{A}$ and $\mathscr{C}$ is a subring of $\mathscr{A}$, then $I \cap \mathscr{C}$ is an ideal in $\mathscr{C}$. Whenever μ is a measure on a Boolean algebra, the collection of μ-measure zero sets is an ideal.

If I is an ideal in $\mathscr{A}$, a quotient algebra can be defined using the equivalence relation: $a_1 \sim a_2$ if $a_1 \vartriangle a_2 \in I$ $\left(a_1 \vartriangle a_2 = (a_1 - a_2) \vee (a_2 - a_1)\right)$. Let $\mathscr{A}/I$ denote the collection of equivalence classes and define the Boolean operations in the obvious way using representatives, that is, $[a_1] \vee [a_2] = [a_1 \vee a_2]$, and so on. These operations are well-defined (see [22], §10]), and turn $\mathscr{A}/I$ into a Boolean algebra.

The next theorem shows that in the Invariant Extension Theorem, the property of "vanishing on an ideal" can be preserved.

Theorem 11.1 (AC).   *Suppose $\mathscr{A}$ is a Boolean algebra, G is a group of automorphisms of $\mathscr{A}$, I is a G-invariant ideal in $\mathscr{A}$, $\mathscr{C}$ is a G-invariant subring of $\mathscr{A}$, and μ is a G-invariant measure on $\mathscr{C}$ that vanishes on $\mathscr{C} \cap I$. Suppose further that G is amenable. Then there is a G-invariant extension of μ to a measure on $\mathscr{A}$ that vanishes on I.*

Proof. Form the quotients $\mathscr{A}/I$ and $\mathscr{C}/\mathscr{C} \cap I$. Because I is G-invariant $(g(a) \in I$ whenever $a \in I)$, G may be regarded as acting on these two quotients, and $\mathscr{C}/\mathscr{C} \cap I$ is a G-invariant subring of $\mathscr{A}/I$. Moreover, the hypothesis on μ implies that μ may be viewed as a G-invariant measure on $\mathscr{C}/\mathscr{C} \cap I$ by $\mu([c]) = \mu(c)$. The Invariant Extension Theorem now yields ν, a G-invariant measure on $\mathscr{A}/I$ that extends μ. Now just let $\bar{\mu}(a) = \nu([a])$ to obtain the desired extension of μ.   □

Corollary 11.2 (AC).   *If G is an amenable group of isometries of $\mathbf{R}^n$ (or S^n), then there is a finitely additive, G-invariant measure on $\mathscr{P}(\mathbf{R}^n)$ $(\mathscr{P}(S^n))$ that normalizes the unit cube (S^n) and vanishes on all meager sets.*

Proof. Apply the previous theorem with $\mathscr{A} = \mathscr{P}(\mathbf{R}^n)$, $\mathscr{C} = \mathscr{J}$, $\mu = \nu$, and $I = M$, the ideal of meager sets. It is shown in Appendix B that ν vanishes on all Jordan measurable, meager sets.   □

Corollary 11.3 (AC).   *Marczewski measures exist in $\mathbf{R}^1, \mathbf{R}^2$, and S^1.*

Proof. The isometry groups of these spaces are solvable, hence amenable. If μ is the total measure that exists by the previous corollary, then $\mu \restriction \mathscr{B}$ is a Marczewski measure.   □

Because of the Banach-Tarski Paradox (really, the Hausdorff Paradox), measures as in Corollary 11.2 cannot exist if G is the group of all isometries of $\mathbf{R}^n (n \geqslant 3)$ or $S^n (n \geqslant 2)$. But Marczewski's Problem concerns measures on $\mathscr{B}$ and, except for the equivalences of Theorem 9.5, almost nothing is known about whether such measures exist in $\mathbf{R}^3$, S^2, or beyond. Note that a Marczewski measure cannot be countably additive (see Theorem 9.15).

Corollary 11.3 was required to prove one of the implications in Theorem 9.5 $((3) \Rightarrow (7))$, and so the proof of that theorem is now complete. The existence of Marczewski measures has an interesting geometric consequence. Let A be an open dense subset of $[0, 1]$ with $\lambda(A) < 1$; just choose a small open covering of the rationals in $[0, 1]$. Then $[0, 1] \setminus A$ is nowhere dense;

therefore, if μ is an isometry-invariant measure on all sets as in Corollary 11.2, $\mu([0, 1]\setminus A) = 0$ and $\mu(A) = 1$. It follows that A is not equidecomposable with any subset of a proper subinterval of $[0, 1]$, even using arbitrary pieces. (However, if countably many pieces are allowed, then A is Borel equidecomposable with $(0, \lambda(A))$; see 9.18). Now, if A is the analogous subset of the unit cube, J, in $\mathbf{R}^3$, then it is not known whether A can be packed into a proper subcube of J using finitely many pieces having the Property of Baire (a packing with arbitrary pieces exists because of the Banach-Tarski Paradox). A proof that such a packing exists would yield a negative solution to the problem of the existence of a Marczewski measure in $\mathbf{R}^3$, because of $(3) \Rightarrow (9)$ in Theorem 9.5.

If in the two corollaries above, one uses the ideal of nowhere dense sets rather than M, one gets measures vanishing on the nowhere dense sets. A natural question is whether such a measure necessarily vanishes on the meager sets. The answer is no. Here is a sketch of the proof for S^1. Let A be a meager F_σ subset of S^1 with $\lambda(A) = 1$. (See [181, p. 5]; A is obtained as the complement of the intersection of a sequence of successively smaller open covers of a countable dense set). Let $\mathscr{C}$ be the subalgebra of $\mathscr{P}(S^1)$ generated by $\{\rho(A) : \rho \in O_2\}$; then any element of $\mathscr{C}$ is a finite union of terms, each of which is a finite intersection of terms of the form $\rho(A)$ or $S^1\setminus\rho(A)$ (see [221, p. 14]). It follows that $\mathscr{C}$ is contained in the algebra of Borel sets, and so λ may be viewed as a measure on $\mathscr{C}$. The representation of elements of $\mathscr{C}$ implies that λ is $\{0, 1\}$-valued on $\mathscr{C}$, and it follows that λ vanishes on any nowhere dense set in $\mathscr{C}$. Now, Theorem 11.1 may be applied to get an invariant extension of $\lambda \restriction \mathscr{C}$ to $\mathscr{P}(S^1)$ that vanishes on all nowhere dense sets. This extension, however, gives A measure 1.

There is another property that the standard measures (Jordan and Lebesgue) have that we can try to preserve in our extensions; namely, these measures react to similarities by multiplying the measure by the nth power of the magnifying factor, where n is the dimension. The following generalization of the Invariant Extension Theorem can be used to show how this property can be preserved.

Theorem 11.4 (AC). *Suppose G is an amenable group of automorphisms of a Boolean algebra $\mathscr{A}$, $\mathscr{C}$ is a G-invariant subring of $\mathscr{A}$, and μ is a measure on $\mathscr{C}$. Suppose further that $\pi : G \to (0, \infty)$ is a homomorphism into the multiplicative group of positive reals such that whenever $c \in \mathscr{C}$ and $\sigma \in G$, then $\mu(\sigma(c)) = \pi(\sigma)\mu(c)$. Then there is an extension $\bar\mu$ of μ to all of $\mathscr{A}$ that satisfies $\bar\mu(\sigma(a)) = \pi(\sigma)\bar\mu(a)$.*

Proof. The proof is identical to that of the Invariant Extension Theorem (10.8), except that for $a \in \mathscr{A}$, $f_a : G \to \mathbf{R}$ is defined by $f_a(\sigma) = \pi(\sigma)\nu(\sigma^{-1}(a))$. Then $\bar\mu$, which is defined by integrating f_a over the amenable group G, has the desired property, since $f_{\tau(a)} = \pi(\tau)\,_\tau(f_a)$. $\qquad\Box$

Now, let H_n be the group of similarities of $\mathbf{R}^n$; any element of H_n has a unique representation as $d\sigma$ where $\sigma \in G_n$ and d is a magnification through the origin: $d(x_1, \ldots, x_n) = (\alpha x_1, \ldots, \alpha x_n)$ for some positive real α. Let the homomorphism $\pi : H_n \to (0, \infty)$ be defined by $\pi(d\sigma) = \alpha^n$, where α is the magnification factor of d.

Corollary 11.5 (AC). *If $n = 1$ or 2, then there is a finitely additive extension μ of λ to all of $\mathscr{P}(\mathbf{R}^n)$ such that $\mu(g(A)) = \alpha^n \mu(A)$ whenever g is a similarity with magnification factor α.*

Proof. The mapping $d\sigma \mapsto \sigma$ is a homomorphism from H_n onto G_n whose kernel is the subgroup consisting of pure magnifications. Since this subgroup is Abelian, H_n is solvable if G_n is. But G_1 and G_2 are solvable, whence H_1 and H_2 are amenable, and the corollary is now an immediate consequence of the preceding theorem, using the homomorphism $\pi : H_n \to (0, \infty)$. $\square$

Of course, this corollary is valid for larger n, provided H_n is replaced by some amenable group of similarities. Also, Theorems 11.4 and 11.1 can be combined, so that one can get a Marczewski measure in $\mathbf{R}^1$ or $\mathbf{R}^2$ that behaves properly with respect to similarities.

Corollary 11.6*. *Suppose $A \subseteq \mathbf{R}^2$ is Lebesgue measurable, with $0 < \lambda(A) < \infty$. Suppose K is a square such that for any $\varepsilon > 0$, A is H_2-equidecomposable with K using similarities that are ε-magnifying (meaning: $1 - \varepsilon \leqslant \sqrt{\pi(g)} \leqslant 1 + \varepsilon$). Then K has area $\lambda(A)$.*

Proof. Let μ be a measure on all subsets of the plane as in 11.5. Suppose $\lambda(K) \neq \lambda(A)$ and choose any $\varepsilon < |1 - \sqrt{\lambda(K)/\lambda(A)}|$. Then, because A is H_2-equidecomposable with K using ε-magnifying similarities, $(1 - \varepsilon)^2 \mu(A) \leqslant \mu(K) \leqslant (1 + \varepsilon)^2 \mu(A)$, whence $1 - \varepsilon \leqslant \sqrt{\lambda(K)/\lambda(A)} \leqslant 1 + \varepsilon$, contradicting the choice of ε. $\square$

This corollary means that Theorem 7.8, which showed how for any $\varepsilon > 0$ the circle could be squared using ε-magnifying similarities, is not valid if the square's area is different than that of the circle.

Marczewski measures were a refinement of some measures produced by Banach to answer a question of Ruziewicz and Lebesgue regarding the uniqueness of Lebesgue measure as a finitely additive measure. We have already seen (9.10) that λ is the only countably additive, translation-invariant measure on $\mathscr{L}$ normalizing the unit cube. As stated earlier, this is false if λ is viewed as a finitely additive measure: simply let $\mu(A) = \lambda(A)$ if $A \in \mathscr{L}$ is bounded, and $\mu(A) = \infty$ if $A \in \mathscr{L}$ is unbounded. This led Ruziewicz to pose

* The Axiom of Choice can be eliminated friom this corollary in the same way that it is eliminated from Corollary 10.10 when $n \leqslant 2$; see Corollary 13.9.

the problem of λ's uniqueness as a finitely additive, invariant measure on $\mathscr{L}_b$, the bounded Lebesgue measurable subsets of $\mathbf{R}^n$ (on S^n, $\mathscr{L}_b$ denotes $\mathscr{L}$). Call a finitely additive measure μ on $\mathscr{L}_b$ *exotic* if μ is isometry-invariant and normalizes the unit cube (or, S^n), but $\mu \neq \lambda$. Ruziewicz's Problem asks whether exotic measures exist in $\mathbf{R}^n$ or S^n.

This problem is closely related to questions of amenability and paradoxical decompositions. First of all, note that Banach's solution is an easy consequence of the construction given in Corollary 11.2. If μ is a finitely additive, isometry-invariant measure on $\mathscr{P}(\mathbf{R}^1)$, $\mathscr{P}(\mathbf{R}^2)$, or $\mathscr{P}(S^1)$ that normalizes the unit interval, square, or sphere, respectively, but vanishes on all meager sets (such measures exist by Corollary 11.2), then μ vanishes on a closed nowhere dense set of positive Lebesgue measure. Hence $\mu \upharpoonright \mathscr{L}_b$ is an exotic measure. The solution of the Ruziewicz Problem in higher dimensions, which makes use of the Banach-Tarski Paradox, was completed only recently and will be discussed. First we examine Banach's solution for S^1 more closely.

The fact that an exotic measure exists for S^1 relies heavily on the amenability of O_2. How general is this phenomenon? Suppose $(X, \mathscr{A}, m)$ is an arbitrary nonatomic measure space with $m(X) = 1$ (i.e., $\mathscr{A}$ is a σ-algebra, m is a countably additive measure on $\mathscr{A}$, and any set of positive measure splits into two sets, each having positive measure). Suppose further that a group G acts on X so that $\mathscr{A}$ and m are G-invariant (in short, G is measure-preserving). Then a finitely additive, G-invariant measure v on $\mathscr{A}$ with $v(X) = 1$ and $v \neq m$ is called *exotic*. If, in addition, v vanishes on all sets in $\mathscr{A}$ having m-measure zero, then v is called an *absolutely continuous exotic measure*.

Now, Corollary 11.2 yields an exotic measure for O_2's action on $(S^1, \mathscr{L}, \lambda)$. Indeed, one can get an absolutely continuous exotic measure by letting I be the O_2-invariant ideal in $\mathscr{L}$ generated by N together with a single nowhere dense set of positive measure; because a finite union of nowhere dense sets is nowhere dense and hence has Lebesgue measure less than one, I is a proper ideal. The desired measure arises by applying Theorem 11.1 with $\mathscr{A} = \mathscr{L}$, $m = \lambda$, $\mathscr{C} = I \cup \{S^1 \backslash A : A \in I\}$, and μ the $\{0, 1\}$-valued measure on $\mathscr{C}$ that vanishes on I. In its use of a nowhere dense set, these arguments use a bit of the topology of S^1. J. Rosenblatt has raised the following question, an affirmative answer to which would be a wide generalization of Banach's classical result for the circle.

Question 11.7 (Rosenblatt's Problem). Let $(X, \mathscr{A}, m)$ be a non-atomic measure space with $m(X) = 1$ and let G be a measure-preserving action of a group G on X. Is the amenability of G sufficient to yield an exotic measure on $\mathscr{A}$? Does amenability always yield an absolutely continuous exotic measure?

Before discussing a partial solution to this problem, we explain the importance of the absolute continuity condition. This is best understood in the context of functional analysis rather than measure theory. The measure m on $\mathscr{A}$ induces a G-invariant linear functional on $L^\infty(X)$ (which denotes m-equivalence classes of bounded, measurable, real-valued functions on X) that assigns nonnegative values to nonnegative (almost everywhere) functions, and normalizes the constant function with value 1, namely the integral with respect to m. Such a functional is called an *invariant mean* on $L^\infty(X)$. The integral with respect to an exotic measure on $\mathscr{A}$ will be a new invariant mean on $L^\infty(X)$ only if the exotic measure is absolutely continuous with respect to m, for otherwise the new integral will not be well defined on the m-equivalence classes of functions. Thus the existence of an absolutely continuous exotic measure on $\mathscr{A}$ is equivalent to the nonuniqueness of the m-integral as an invariant mean. Most of the results in this area have been motivated by the uniqueness question for invariant means, and the proofs use techniques of functional analysis applied to $L^\infty(X)$.

In the case of countable groups, Rosenblatt's Problem has been solved. Moreover, for arbitrary groups it is known that the amenability condition in Rosenblatt's Problem is necessary.

Theorem 11.8 (AC).   Let $(X, \mathscr{A}, m)$ and G be as in Question 11.7. Then (2) implies (1) and, if G is countable, (1) implies (2):

(1) *G is amenable.*
(2) *There is an absolutely continuous exotic measure on $\mathscr{A}$; that is, the m-integral is not the unique G-invariant mean on $L^\infty(X)$.*

Proof

(2) $\Rightarrow$ (1). (Sketch; see [127, 195] for more details.) Suppose G fails to be amenable. Let X denote $\mathbf{Z}_2^G$, a compact topological group with Haar measure m on its Borel sets. There is a natural action of G on X with respect to which m is invariant. But it can be shown that the m-integral is the unique G-invariant mean on $L^\infty(X)$, contradicting (2).

(1) $\Rightarrow$ (2). This direction uses the notion of an asymptotically invariant sequence, by which is meant a sequence of sets A_n such that $m(A_n) > 0$, and for every $g \in G$,

$$\lim_{n \to \infty} m(A_n \bigtriangleup g A_n)/m(A_n) = 0.$$

For an example of such a sequence in a slightly different context, see the proof of Theorem 11.15. This notion, with *net* replacing *sequence*, also plays a role in the omitted details of the proof that (2) implies (1). Now, the proof splits into two cases. If the action of G on X fails to be ergodic, that is, there is some measurable set A such that $0 < m(A) < 1$ and $m(gA \bigtriangleup A) = 0$ for all $g \in G$, then (2) is obviously true. The measure v defined by $v(B) = m(B \cap A)/m(A)$ is absolutely continuous and exotic. On the other hand, del Junco and

Rosenblatt [58] proved that if G is countable and amenable and the action is ergodic, then an asymptotically invariant sequence $\{A_n\}$ with the additional property that $\lim_{n \to \infty} m(A_n) = 0$ exists. Such a sequence yields a new invariant mean as follows. Thinning the sequence if necessary, assume $m(A_n) \leqslant 1/2^{n+2}$, and then define m_n by $m_n(A) = m(A \cap A_n)/m(A_n)$. Now, let U be a nonprincipal ultrafilter on $\mathbf{N}$ (see remarks after 10.7's proof) and define $v(A) = \int f_A \, dU$, where $f_A(n) = m_n(A)$. Because $\lim_{n \to \infty} f_A(n) - f_{gA}(n) = 0$ and finite sets have U-measure zero, v is G-invariant. Clearly, v is absolutely continuous with respect to m, and v is exotic because $v(\bigcup A_n) = 1$ while $m(\bigcup A_n) \leqslant 1/4 + 1/8 + \cdots = 1/2$. $\qquad\qquad\square$

The notion of an asymptotically invariant sequence, which plays a central role in the preceding proof and in the solution to Ruziewicz's Problem to be discussed, is intimately related to Følner's Condition (see Theorem 10.11). Indeed, another way of phrasing Følner's Condition for a countable group G is: There is a sequence $\{A_n\}$ of subsets of G that is asymptotically invariant with respect to the counting measure on $\mathscr{P}(G)$ and is such that $0 < |A_n| < \infty$.

For details of the preceding proofs, and various related results on asymptotically invariant sequences and the problem of uniqueness for invariant means, see [38, 39, 40, 58, 127, 137, 138, 195, 204, 205, 228].

Question 11.7, or at least the part that asks about absolutely continuous exotic measures, can be given another interpretation. Let us, for the moment, call a measure-preserving action of G on $(X, \mathscr{A}, m)$ *superergodic* if for any set A of positive measure, there are $g_1, \ldots, g_n \in G$ such that $m(\bigcup g_i A) = 1$; a superergodic action on a nonatomic space is necessarily ergodic. Now, for absolutely continuous measures, Question 11.7 is equivalent to asking whether any measure-preserving action of an amenable group on a finite, nonatomic measure space fails to be superergodic. For if v is exotic and absolutely continuous, then as shown by Rosenblatt [195, Prop. 1.1], there is another absolutely continuous exotic measure v' such that v' vanishes on a set of positive m-measure; this set witnesses the nonsuperergodicity of the action. Conversely, if A is such that $m(A) > 0$ and $m(g_1 A \cup \cdots \cup g_n A) < 1$ for any finite subset $\{g_i\}$ of G, then I, the collection of sets contained in a set of the form $E \cup \bigcup g_i A$, where $m(E) = 0$ and $\{g_i\}$ is a finite subset of G, is a proper G-invariant ideal in $\mathscr{A}$. Then Theorem 11.1, with $\mathscr{C} = I \cup \{X \setminus B : B \in I\}$ and μ equal to the $\{0, 1\}$-valued measure on $\mathscr{C}$ that vanishes on I, yields an absolutely continuous finitely additive measure on $\mathscr{A}$ that vanishes on I and hence is exotic. In short, a measure-preserving action of an amenable group G on $(X, \mathscr{A}, m)$ is superergodic if and only if the m-integral is the unique invariant mean on $L^\infty(X)$.

For about 50 years, no progress was made on Ruziewicz's Problem in higher dimensions. All that was known was Tarski's observation (9.7) that because of the Banach-Tarski Paradox, any exotic measure necessarily vanishes on the bounded sets of Lebesgue measure zero, that is, is absolutely continuous with respect to Lebesgue measure on $\mathscr{L}_b$. This means that to prove the nonexistence of an exotic measure, it suffices to prove that the Lebesgue integral is the unique invariant mean on $L^\infty(\mathbf{R}^n), n \geqslant 3$, or $L^\infty(S^n), n \geqslant 2$. Then, in 1980–81, a solution for S^n, $n \geqslant 4$, was obtained simultaneously in Russia (by G. Margulis) and in the United States (by D. Sullivan and J. Rosenblatt). The solution leans heavily on a property of groups diametrically opposed to amenability called Property T, first introduced by Kazhdan [102]. We state the definition for countable groups, although the property can be defined more generally for locally compact topological groups.

Definition 11.9. *A countable group G has* Property T *if whenever a unitary representation π of G on a complex Hilbert space H admits an asymptotically invariant sequence of nonzero vectors (i.e., a sequence of nonzero $v_n \in H$ such that for all $g \in G$,*

$$\lim_{n \to \infty} \|v_n - \pi(g)v_n\|/\|v_n\| = 0),$$

then there is a nonzero G-invariant vector (i.e., a nonzero $v \in H$ such that for all $g \in G$, $\pi(g)v = v$).

The key step in the solution of the Ruziewicz Problem in S^4 and beyond was the identification, using techniques of algebraic group theory, of a countable dense subgroup of $SO_n(n \geqslant 5)$ that has Property T. The relevance of Property T to the uniqueness of invariant means is given by the following theorem.

Theorem 11.10. *Suppose a countable group G with Property T acts on X in a way that is measure-preserving and ergodic, where $(X, \mathscr{A}, m)$ is a nonatomic measure space and $m(X) = 1$. Then the m-integral is the unique G-invariant mean on $L^\infty(X)$.*

Proof. Suppose the m-integral is not the unique mean. Then by a result of del Junco and Rosenblatt ([58]; see [195] for a proof), there is an asymptotically invariant sequence $\{A_n\}$ of sets of positive measure such that $\lim_{n \to \infty} m(A_n) = 0$. Let $L^2(X)$ denote the Hilbert space of all square-integrable complex functions on X and let H be the subspace consisting of those f such that $\int f \, dm = 0$. Let π be the natural representation of G as isometries of H: $(\pi(g)f)(x) = f(g^{-1}x)$. Let $f_n(x) = \chi_{A_n}(x) - m(A_n)$; then $f_n \in H$, $\|f_n\|^2 = m(A_n)(1 - m(A_n))$, and $\|f_n - \pi(g)f_n\|^2 = m(A_n \triangle gA_n)$. It follows that $\{f_n\}$ is an asymptotically invariant sequence in H, so because G has Property T, some nonzero $f \in H$ satisfies $\pi(g)f = f$ for all $g \in G$. But then set $E_r = \{x \in X : |f(x)| \geqslant 1/r\}$; E_r has positive measure for some m, and $m(E_r \triangle gE_r) = 0$ for all $g \in G$, which contradicts the ergodicity of G's action. $\qquad\square$

The converse of the previous theorem is valid too. If invariant means are unique for all actions of a countable group G as in the theorem, then G has Property T ([39, 205]). Property T can also be characterized in terms of cohomological properties of G [205, 252].

Now, the action of a countable dense subgroup of SO_n on S^{n-1} is ergodic, like the action of all of SO_n. Hence Theorem 11.10, together with the existence, mentioned above, of a countable dense subgroup of SO_n, $n \geqslant 5$, having Property T, implies that the Lebesgue integral is the unique invariant mean on $L^\infty(S^n)$, $n \geqslant 4$, and hence by Tarski's observation, there is no exotic measure on S^n, $n \geqslant 4$.

The preceding technique can be adapted to $\mathbf{R}^5$ and beyond, but the lower-dimensional cases, $\mathbf{R}^3$, $\mathbf{R}^4$, S^2, and S^3, proved more troublesome. Nevertheless, Margulis [138] has obtained a solution for all $\mathbf{R}^n$, $n \geqslant 3$, and Drinfeld has reportedly solved the remaining two cases, S^2 and S^3, using some deep mathematics, including Deligne's proof of the Peterson Conjecture. To summarize:

Theorem 11.11 (AC).   *An exotic measure in S^n or $\mathbf{R}^n$ exists only in the cases* $\mathbf{R}^1$, $\mathbf{R}^2$, *and* S^1.

The role of absolute continuity in the proof of Theorem 11.11, via Tarski's observation, shows how the Banach-Tarski Paradox, which is usually interpreted negatively (invariant extensions of Lebesgue measure to all sets do not exist), can be looked upon more positively if the focus is shifted from all of $\mathscr{P}(\mathbf{R}^n)$ to the measurable sets. The paradox yields that exotic measures must be absolutely continuous, which by the work just outlined, yields a characterization of Lebesgue measure as a finitely additive measure.

Theorem 11.11 points to a close connection between the existence of exotic measures and the amenability of the group in question, since exotic measures exist only in the cases that the isometry group is amenable. How closely are these two properties related? We shall see that for a somewhat trivial reason, exotic measures on S^3, for example, can exist even when the group in question is a nonamenable group of rotations. Thus it is more appropriate to formulate a question in terms of invariant means on $L^\infty(S^n)$.

Question 11.12.   Does the nonamenability of a subgroup G of O_{n+1} guarantee the uniqueness of the Lebesgue integral as a G-invariant mean on $L^\infty(S^n)$? Can the mean's uniqueness be proved under the stronger assumption that S^n is G-paradoxical?

An affirmative answer to the first part of this question would yield (in fact, is equivalent to) the assertion that if G is a nonamenable subgroup of O_{n+1} and F is a G-invariant mean on $L^\infty(S^n)$, then F is, in fact, O_{n+1}-invariant. Note that if G is an amenable subgroup of O_{n+1} and n is at least 2, then there is a

G-invariant mean on $L^\infty(S^n)$ that is not O_{n+1}-invariant: any G-invariant mean that differs from the Lebesgue integral (such exist by the method preceding Question 11.7) is not O_{n+1}-invariant because of Theorem 11.11.

All the known results about exotic measures have been obtained by studying the related question about the uniqueness of invariant means on L^∞. But these techniques do not apply if one is considering the family of Borel sets rather than the collection of measurable sets. Lemma 9.7 cannot be used because the proof of that lemma uses subsets of sets of measure zero, a technique that can introduce non-Borel sets. Thus we have the following question, which is completely unresolved. Note that a Marczewski measure on the collection of subsets of S^2 having the Property of Baire, if it exists, yields an exotic measure when restricted to the Borel sets. Thus an affirmative answer to the following question would imply that Marczewski measures do not exist.

Question 11.13. Is it true that in S^n, $n \geqslant 2$ (or $\mathbf{R}^n$, $n \geqslant 3$), Lebesgue measure is the only finitely additive, isometry-invariant measure on the Borel sets that has total measure one (or, normalizes the unit cube)?

PARADOXES MODULO AN IDEAL

The nonexistence of exotic measures leads to a new type of paradoxical decomposition, one that uses measurable pieces. Of course, J, the unit cube in $\mathbf{R}^3$, cannot be paradoxical using pieces in $\mathscr{L}$. But it is possible that J is paradoxical in $\mathscr{L}$ provided sets in some ideal I of $\mathscr{L}$ are ignored. We have already seen the usefulness of considering equidecomposability modulo an ideal; see Definition 9.16 and the results following it. In order to fit this generalization into the machinery already developed, we digress briefly to discuss equidecomposability in arbitrary Boolean algebras.

If G is a group of automorphisms of a Boolean algebra $\mathscr{A}$, and $a, b \in \mathscr{A}$, then a and b are G-equidecomposable ($a \sim b$) if there are two pairwise disjoint collections of n elements, $\{a_i\}$, $\{b_i\}$, and $g_i \in G$ such that $a = a_1 \vee \cdots \vee a_n$, $b = b_1 \vee \cdots \vee b_n$, and $g_i(a_i) = b_i$. An element a in $\mathscr{A}$ is G-paradoxical if there are disjoint $b, c \leqslant a$ such that $b \sim a$ and $c \sim a$. Most of the algebras considered so far have been fields of sets, but Ruziewicz's Problem leads to the consideration of paradoxical decompositions in quotient algebras of the form $\mathscr{L}/I$. One consequence of this increased generality is that the proof of the Banach-Schröder-Bernstein Theorem (3.5) breaks down. The two axioms for equidecomposability used in that proof, (a) and (b), remain valid if suitably reformulated for Boolean algebras: (a) if $a \sim b$, then there is a Boolean isomorphism $g: \mathscr{A}_a \to \mathscr{A}_b$ such that $c \sim g(c)$ for each $c \leqslant a$, and (b) if $a_1 \wedge a_2 = \mathbf{0} = b_1 \wedge b_2$, $a_1 \sim b_1$, and $a_2 \sim b_2$, then $a_1 \vee a_2 \sim b_1 \vee b_2$; for (a), let $g(c) = g_1(c \wedge a_1) \vee \cdots \vee g_n(c \wedge a_n)$, where a_i, g_i witness $a \sim b$. But the proof of 3.5 uses an infinite union at one point, and the Boolean counterpart of that operation does not always exist. A Boolean algebra $\mathscr{A}$ is called *countably complete* if sups and infs (with respect to $\leqslant$) of countable subsets of $\mathscr{A}$ exist in $\mathscr{A}$. It is easy to see that in countably complete Boolean algebras, the proof of 3.5 yields the Banach-Schröder-Bernstein Theorem for G-equidecomposability.

An example showing that the Banach-Schröder-Bernstein Theorem is not valid in all Boolean algebras can be constructed as follows. Kinoshita [105] found countable Boolean algebras $\mathscr{A}$, $\mathscr{C}_1$, and $\mathscr{C}_2$ such that $\mathscr{A} \cong \mathscr{A} \times \mathscr{C}_1 \times \mathscr{C}_2$, but $\mathscr{A} \not\cong \mathscr{A} \times \mathscr{C}_1$ ($\cong$ denotes isomorphism of Boolean algebras). This result was improved by Hanf [88], who showed that $\mathscr{C}_1$ could be taken isomorphic to $\mathscr{C}_2$. Now, let $\mathscr{B} = \mathscr{A} \times \mathscr{C}_1 \times \mathscr{C}_2 \times \mathscr{A}$ and let G be the group of all automorphisms of $\mathscr{B}$. Let g be the automorphism taking the first factor of $\mathscr{B}$ to the last three, and vice versa, and let h be the automorphism that switches the first and last coordinates. Finally, let $a = (1, 0, 0, 0)$ and $b = (1, 1, 0, 0)$. Then $a \leqslant b$ and $b \sim_G gh(b) = g(0, 1, 0, 1) \leqslant g(0, 1, 1, 1) = a$. But a and b are not G-equidecomposable since a piecewise automorphism would yield an isomorphism of $\mathscr{A}$ with $\mathscr{A} \times \mathscr{C}_1$.

The generalization of equidecomposability to arbitrary Boolean algebras provides a convenient context for studying equidecomposability in quotient algebras, but strictly speaking this can all be carried out within the framework used earlier, that of G-equidecomposability in an algebra of subsets of some set, where G acts on the set. This is because by the Stone Representation Theorem ([221, p. 24]), any Boolean algebra $\mathscr{A}$ is isomorphic to an algebra of subsets of X for some set X (X is the set of ultrafilters on $\mathscr{A}$), and a group of automorphisms of $\mathscr{A}$ may be viewed as a group acting on X. Because of these remarks, we may talk of equidecomposability of objects na where $n \in \mathbf{N}$ and $a \in \mathscr{A}$; simply apply the semigroup construction preceding Question 8.12 to the algebra of sets corresponding to $\mathscr{A}$. Moreover, Tarski's Theorem in the form of Corollary 9.3 applies to equidecomposability in Boolean algebras. Because of the lack of a general cancellation law, we cannot be sure that the stronger Corollary 9.2 is valid.

With all these preliminaries, we can now derive a new type of paradoxical decomposition of a cube in $\mathbf{R}^n$ ($n \geqslant 3$) or of S^n ($n \geqslant 2$) from the nonexistence of exotic measures. Let I be any isometry-invariant ideal in $\mathscr{L}$, the measurable subsets of $\mathbf{R}^n$, $n \geqslant 3$, or of S^n, $n \geqslant 2$, and let $\mathscr{A}$ be the Boolean algebra $\mathscr{L}/I$; let $[A]$ denote the I-equivalence class of a measurable set A. If I consists only of measure zero sets, then Lebesgue measure induces an isometry-invariant measure on $\mathscr{A}$. But in all other cases, that is, whenever I is an isometry-invariant ideal in $\mathscr{L}$ that contains a set of positive measure, there is no isometry-invariant measure on $\mathscr{A}$ normalizing $[J]$ (or, normalizing $[S^n]$). This is a consequence of Theorem 11.11, since such a measure would induce an exotic measure on $\mathscr{L}_b$. Note that such ideals are easy to obtain; for example, choose any closed nowhere dense set A of positive measure and let $I = \{B \in \mathscr{L}:$ for some $\sigma_1, \ldots, \sigma_m \in G_n,$

$B \subseteq \sigma_1 A \cup \cdots \cup \sigma_m A\}$. Now, an application of Corollary 9.3 yields the following result.

Corollary 11.14 (AC). *Let $\mathscr{L}$ denote the class of measurable subsets of $\mathbf{R}^n$, $n \geqslant 3$, or S^n, $n \geqslant 2$; let I be any proper isometry-invariant ideal in $\mathscr{L}$ that contains a set of positive measure; let $\mathscr{A} = \mathscr{L}/I$; and let α be the equivalence class of the unit cube in $\mathbf{R}^n$ or of the sphere S^n. Then there is some positive integer m such that $(m + 1)\alpha \leqslant m\alpha$ in $\mathscr{A}$; that is, $m + 1$ copies of the cube or sphere can be packed into m copies using measurable pieces, but ignoring sets in I.*

This corollary is noteworthy because it applies to every ideal satisfying the hypothesis; but the conclusion is weaker than one might hope because of the usual problems with the cancellation law requiring 9.3 to be used rather than 9.2. Nevertheless, even before Ruziewicz's Problem for S^2 had been solved, Rosenblatt [194] had constructed a specific isometry-invariant ideal I in the algebra of measurable subsets of S^2 such that S^2 is paradoxical modulo I using measurable pieces. Note also that the case of $\mathbf{R}^n$ in Corollary 11.14 can be improved if I is assumed, in addition, to be closed under countable unions and invariant under similarities. Then $\mathscr{L}/I$ is countably complete and satisfies the Banach-Schröder-Bernstein Theorem, whence the proof of (2) $\Rightarrow$ (1) of Theorem 9.5 can be applied to obtain that the unit cube is paradoxical in $\mathscr{L}/I$. In other words, the extra hypotheses on I in the case of $\mathbf{R}^n$ allow m to be taken to be 1.

HOW TO ELIMINATE EXOTIC MEASURES IN R^2

Recall from Chapter 7 that even though the square in $\mathbf{R}^2$ is not paradoxical using isometries, the square is paradoxical using a larger group, one that contains some area-preserving linear transformations. Let G be the group generated by G_2 and σ, where $\sigma(x, y) = (x + y, y)$. Then $\begin{bmatrix} 1 & 2 \\ 0 & 1 \end{bmatrix} = \sigma^2$ and $\begin{bmatrix} 1 & 0 \\ 2 & 1 \end{bmatrix} = \begin{bmatrix} 0 & -1 \\ 1 & 0 \end{bmatrix} \sigma^{-1} \begin{bmatrix} 0 & 1 \\ -1 & 0 \end{bmatrix}$, so A and B of Proposition 7.1 lie in G, and by 7.3 the square is G-paradoxical. It follows as in Lemma 9.7 that any G-invariant finitely additive measure on $\mathscr{L}_b$, the Lebesgue measurable subsets of $\mathbf{R}^2$, that normalizes the unit square must be absolutely continuous with respect to λ.

Now, Rosenblatt [195] has shown that despite the existence of absolutely continuous exotic measures in $\mathbf{R}^2$, no such measure can be σ-invariant. But, by the preceding remarks, the absolute continuity condition is redundant in the presence of σ-invariance. Thus any finitely additive measure on $\mathscr{L}_b$ that normalizes the unit square and is G-invariant must coincide with Lebesgue measure. In other words, if one modifies the original Ruziewicz Problem by considering area-preserving affine transformations instead of just isometries, then one obtains a characterization of Lebesgue measure as a finitely additive measure that is valid in all $\mathbf{R}^n$ except $\mathbf{R}^1$.

Rosenblatt's result was obtained as a corollary to results on the n-dimensional torus, T^n. He proved that the Lebesgue integral on $L^\infty(T^n)$, where $n \geq 2$, is the unique mean that is invariant under the natural action of $SL_n(\mathbf{Z})$; hence there is a unique mean on $L^\infty(T^n)$ that is invariant under all topological automorphisms of T^n. Therefore absolutely continuous $SL_n(\mathbf{Z})$-exotic measures on the measurable subsets of T^n do not exist. As pointed out in [195], however, finitely additive, $SL_n(\mathbf{Z})$-invariant measures differing from Lebesgue measure do exist. Thus the toroidal case shows that the uniqueness of invariant means can be a different question than the nonexistence of exotic measures. For a further analysis of invariant, finitely additive measures both on the family of measurable subsets of T^n and on $\mathscr{P}(T^n)$, see [40].

It follows from Theorems 11.8 and 11.10 that for countable groups, amenability and Property T are mutually exclusive. Suppose G is countable and amenable. Let X be as in the proof of $(2) \Rightarrow (1)$ of Theorem 11.8. By $(1) \Rightarrow (2)$ of that theorem, the m-integral is not the unique G-invariant mean on $L^\infty(X)$. But the action of G on X is ergodic, so if G had Property T, Theorem 11.10 would yield a contradiction.

These two properties are not exhaustive, however. For example, $SL_1(\mathbf{Z})$ is amenable and $SL_n(\mathbf{Z})$ has Property T if $n \geq 3$ (see [137]), but $SL_2(\mathbf{Z})$, which is nonamenable (Proposition 7.1), fails to have Property T (see [205, Example 3.7]). Theorem 11.17 presents an application of the fact that $SL_n(\mathbf{Z})$ has Property T if $n \geq 3$. Also, a free group of rank n, $n \geq 2$, fails to have Property T (see [4, 102]). Of course, such groups are nonamenable.

PARADOXES USING MEASURABLE PIECES

Recall from Figure 5.3 that the hyperbolic plane (and also H^n, $n \geq 2$) is paradoxical using measurable pieces. The measure-theoretic consequence of this is that there is no finitely additive measure on the measurable subsets of H^2 that has total measure one and is invariant under hyperbolic isometries. Analogous paradoxical decompositions of $\mathbf{R}^n$ using measurable pieces do not exist. This is not too difficult to prove geometrically (see [165, §5]), but it can also be derived from the construction of a measure of the type that cannot exist in hyperbolic space. Part of the interest in the following theorem is that it is valid even in the cases where the isometry group is nonamenable. Moreover, it yields the somewhat surprising result that paradoxes using measurable pieces do not exist even if one allows similarities, which can change areas greatly.

Theorem 11.15 (AC).   For any n, there is a finitely additive, isometry-invariant measure μ on $\mathscr{L}$ such that $\mu(\mathbf{R}^n) = 1$. Moreover, μ can be chosen so that $\mu(A) = \mu(g(A))$ for any similarity g.

Proof. Let U be a nonprincipal ultrafilter on $\mathbf{N}$ (see discussion following Theorem 10.7). Let B_m denote the ball of radius m centered at the origin and, for $A \in \mathcal{L}$, define $f_A: \mathbf{N} \to [0, \infty)$ by $f_A(m) = \lambda(A \cap B_m)/\lambda(B_m)$. Then let $\mu(A) = \int f_A \, dU$. It is clear that μ is finitely additive, $\mu(\mathbf{R}^n) = 1$, and μ is O_n-invariant. Since $O_n \cup T$ generates G_n, it must be shown that for any $v \in \mathbf{R}^n$ and $A \in \mathcal{L}$, $\mu(A) = \mu(A + v)$. But $\lim_{m \to \infty} \lambda(B_m \triangle (B_m + v))/\lambda(B_m) = 0$, which implies that $\lim_{m \to \infty} f_A(m) - f_{A+v}(m) = 0$. Using the fact that finite subsets of $\mathbf{N}$ have U-measure zero, it follows that $\int f_A \, dU = \int f_{A+v} \, dU$, as required.

For the part of the theorem dealing with H_n-invariance, where H_n is the group of similarities of $\mathbf{R}^n$, modify the measure of the preceding paragraph as follows. Let M be the group of magnifications from the origin; then (see remarks preceding 11.5) $H_n/G_n \cong M$ and M is Abelian, hence amenable. Let θ be a left-invariant measure on M and define μ^* by $\mu^*(A) = \int f_A \, d\theta$, where $f_A: M \to \mathbf{R}$ is given by $f_A(d) = \mu(d^{-1}(A))$. Then, as usual, μ^* is finitely additive and M-invariant, and $\mu^*(\mathbf{R}^n) = 1$. But because G_n is a normal subgroup of H_n, μ^* is G-invariant too: $f_{\sigma(A)}(d) = \mu(d^{-1}\sigma(A)) = \mu(\sigma_0 d^{-1}(A)) = \mu(d^{-1}(A)) = f_A(d)$, where $\sigma_0 = d^{-1}\sigma d \in G_n$. Since H_n is generated by $G_n \cup M$, μ^* is H_n-invariant. $\square$

Corollary 11.16. $\mathbf{R}^n$ *is not paradoxical using measurable pieces and isometries (or even similarities).*

Note that the key point of the preceding proof is that any unbounded increasing sequence of concentric balls in $\mathbf{R}^n$ is asymptotically G_n-invariant.

Now, consider the action of SA_n, the group of affine transformations with determinant 1, on $\mathbf{R}^n$. For $\mathbf{R}^1$ this adds nothing new, but in $\mathbf{R}^2$ and beyond, SA_n is substantially larger than the isometry group. The technique of Chapter 7 easily yields that if $n \geqslant 2$, $\mathbf{R}^n$ is SA_n-paradoxical. Indeed, if $n \geqslant 3$, this follows already from Corollary 5.8, while for $\mathbf{R}^2$ this follows from the local commutativity of $SL_2(\mathbf{Z})$ in its action on $\mathbf{R}^2 \setminus \{0\}$. But can it happen that $\mathbf{R}^n$ is SA_n-paradoxical using Lebesgue measurable pieces? In other words, does Theorem 11.15 extend to the case of an SA_n-invariant measure?

Theorem 11.17 (AC). *If $n \geqslant 2$, then there does not exist a finitely additive, SA_n-invariant measure of total measure one on $\mathcal{L}$, the Lebesgue measurable subsets of $\mathbf{R}^n$.*

Proof. We shall sketch the proof in the case $n > 2$; the $\mathbf{R}^2$ case is due to Kallman (unpublished) and uses completely different methods. Suppose, by way of contradiction, that μ is an SA_n-invariant measure with $\mu(\mathbf{R}^n) = 1$. Then μ must vanish on any set E with $\lambda(E) = 0$. Because $\mathbf{R}^n$ is SA_n-paradoxical, arbitrarily many copies of $\mathbf{R}^n$ may be packed into one copy of $\mathbf{R}^n$; and because a subset of E is necessarily Lebesgue measurable, this implies that for any $m \in \mathbf{N}$, $m\mu(E) \leq \mu(\mathbf{R}^n) = 1$, that is, $\mu(E) = 0$. This means that it is sufficient to prove that there is no SA_n-invariant mean on $L^\infty(\mathbf{R}^n)$ that normalizes the constant function with value 1.

Suppose there is such a mean. We shall appeal to the following theorem*, which is a σ-finite version of the theorem mentioned at the beginning of 11.10's proof; it can be proved by the same techniques used by del Junco and Rosenblatt in [58]. If G is countable and acts in a measure-preserving way on the σ-finite nonatomic measure space $(X, \mathscr{A}, m)$ where $m(X) = \infty$, and if there is a G-invariant mean on $L^\infty(X)$ that normalizes χ_X, then there is an asymptotically invariant sequence of sets in $\mathscr{A}$ having positive, finite measure. Now, an SA_n-invariant mean is obviously $SL_n(\mathbf{Z})$-invariant, so an application of this theorem yields an asymptotically $SL_n(\mathbf{Z})$-invariant sequence $\{A_k\} \subseteq \mathscr{L}$ with $0 < \lambda(A_k) < \infty$. But then $\{\chi_{A_k}\}$ is asymptotically invariant in $L^2(\mathbf{R}^n)$, and so because $SL_n(\mathbf{Z})$ has Property T (see page 178), there must be some nonzero $f \in L^2(\mathbf{R}^n)$ such that $_\sigma f = f$ for all $\sigma \in SL_n(\mathbf{Z})$. But then for any real r, $E_r = \{x : |f(x)| \geq r\}$ is $SL_n(\mathbf{Z})$-invariant up to a set of measure zero. Note that $r^2\lambda(E_r) \leq \int |f|^2 \, d\lambda < \infty$, so $\lambda(E_r) < \infty$. Now suppose there is some positive r such that $\lambda(E_r) > 0$. The action of $SL_n(\mathbf{Z})$ on $\mathbf{R}^n$ is easily seen to have the property that no set with positive finite measure is $SL_n(\mathbf{Z})$-invariant up to a set of measure zero. This means that $\lambda(E_r) = 0$ for all positive r, whence $f = 0$ almost everywhere, a contradiction. $\square$

Applying Corollary 9.3 (and the Banach-Schröder-Bernstein Theorem) to the previous result yields a type of paradoxical decomposition of $\mathbf{R}^n$ using measurable pieces and measure-preserving transformations.

Corollary 11.18 (AC).   *If $n \geq 2$, there is a positive integer m and $m + 1$ partitions $\{\{A_1^j, \ldots, A_{r_j}^j\} : 1 \leq j \leq m + 1\}$ of $\mathbf{R}^n$ where each A_i^j is measurable such that, for suitable $\sigma_i^j \in SA_n$, the sets $\sigma_i^j(A_i^j)$ can be regrouped to form m partitions of $\mathbf{R}^n$. In short, $m + 1$ copies of $\mathbf{R}^n$ are SA_n-equidecomposable with m copies of $\mathbf{R}^n$, using Lebesgue measurable pieces.*

Since the assertion of this corollary for any $n \geq 2$ follows from the assertion for $\mathbf{R}^2$, the same m can be used in all cases. It would be nice to have a more explicit paradoxical decomposition, that is, one where $m = 1$ and the measurable pieces are given explicitly in some way. One could get such an improvement if one had a measurable subset of $\mathbf{R}^2$ that met every orbit of G's action in one point, where G is the subgroup of $SL_2(\mathbf{Z})$ generated by $\left[\begin{smallmatrix} 2 & 1 \\ 1 & 1 \end{smallmatrix}\right]$ and $\left[\begin{smallmatrix} 5 & 2 \\ 2 & 1 \end{smallmatrix}\right]$. This is because G, the Magnus-Neumann group, acts without nontrivial

---

* Note that this theorem immediately yields that countable amenable groups (and therefore all amenable groups) satisfy Følner's Condition. Just consider the left action of G on the measure space $(G, \mathscr{P}(G), m)$ where m is the counting measure.

fixed points on $\mathbf{R}^2 \setminus \{0\}$, whence by the proof of 1.10, a measurable choice set for G's action would immediately yield a paradoxical decomposition of $R^2 \setminus \{0\}$ using measurable pieces. By the usual absorption technique, this would yield a paradoxical decomposition of $\mathbf{R}^2$ using SA_2 and measurable pieces.

Although no finitely additive measure on the measurable subsets of $\mathbf{R}^2$ having total measure one is SA_2-invariant, Belley and Prasad [15] have shown that there are interesting subalgebras of the algebra of Borel subsets of $\mathbf{R}^n$ that (a) are invariant under all affine transformations, and (b) bear a finitely additive measure of total measure one that is invariant under all affine transformations.

CHARACTERIZING GROUPS OF EUCLIDEAN
ISOMETRIES THAT YIELD PARADOXES

We now turn our attention to a finer analysis of paradoxical decompositions, and attempt to characterize the groups of isometries of $\mathbf{R}^n$ (or S^n) with respect to which the cube (or S^n) is paradoxical. In other words, for which subgroups of G_n or O_n are Corollaries 10.9 and 10.10 valid? This leads to two distinct problems. For the first, we are asking to characterize the groups, G, of isometries for which λ has a G-invariant, finitely additive extension to all sets. For the second, we are asking about groups with respect to which no paradox exists, which by Tarski's Theorem is equivalent to asking about groups G for which there exists a finitely additive, G-invariant measure on all sets that normalizes the unit cube, or the sphere.

The following theorem treats the first question for spheres; the case of $\mathbf{R}^n$ follows.

Theorem 11.19 (AC). *Let G be a subgroup of O_{n+1}. Then the following are equivalent to each other and to the statements of Theorem 10.11.*

(1) *G is amenable.*
(2) *G has no free subgroup of rank 2.*
(3) *There is a finitely additive, G-invariant extension of λ to all subsets of S^n.*
(4) *There is a finitely additive, G-invariant measure μ on $\mathscr{P}(S^n)$ with $\mu(S^n) = 1$ and $\mu(E) = 0$ if $\lambda(E) = 0$.*

Proof. As indicated following the statement of Tits's Theorem (10.5), that theorem implies the equivalence of (1) and (2) for groups of Euclidean (and hence spherical) isometries. Since (1) $\Rightarrow$ (3) is just Corollary 10.9, and (3) $\Rightarrow$ (4) is trivial, it remains only to show that (4) implies (1). Assume μ is as in (4), but G is not amenable. Then by 10.4(f) and the fact that any group is the union of its finitely generated subgroups, there is some finitely generated, and hence countable, subgroup, H, of G that fails to be amenable. Let $D = \{x \in S^n : \sigma(x) = x \text{ for some } \sigma \in H \backslash \{1\}\}$. Now, the fixed point set (in $\mathbf{R}^{n+1}$) of a single $\sigma \in H \backslash \{1\}$ is a linear subspace of dimension at most n. Such a subspace has Lebesgue measure 0 in $\mathbf{R}^{n+1}$, and hence its intersection with S^n has measure zero with respect to surface Lebesgue measure on S^n. Since H is countable, this implies that $\lambda(D) = 0$, and hence $\mu(D) = 0$ and $\mu(S^n \backslash D) = 1$. But H acts on $S^n \backslash D$ without nontrivial fixed points, and H, being nonamenable, is paradoxical by Tarski's Theorem. Therefore, by Proposition 1.10, $S^n \backslash D$ is H-paradoxical, which contradicts $\mu(S^n \backslash D) = 1$. (Note: The appeal to Tarski's Theorem here can be avoided by using an easy variation of 1.10 to transfer the measure, μ, on $S^n \backslash D$ directly to H, contradicting H's nonamenability.) $\square$

We now use the previous theorem to obtain a characterization of subgroups of G_n for which invariant extensions of Lebesgue measure on $\mathbf{R}^n$ exist.

Theorem 11.20 (AC).   *Let G be a subgroup of G_n. Then the following are equivalent.*
   (1) *G is amenable.*
   (2) *G has no free subgroup of rank 2.*
   (3) *There is a finitely additive, G-invariant extension μ of λ to all subsets of $\mathbf{R}^n$.*

Proof.   As in 11.19, only (3) $\Rightarrow$ (2) requires proof. Suppose F is a rank-2 free subgroup of G. Consider first the case that F is locally commutative on $\mathbf{R}^n$. Then, by 4.5, $\mathbf{R}^n$ is F-paradoxical. But this contradicts the existence of a finitely additive, G-invariant measure v on $\mathscr{P}(\mathbf{R}^n)$ having total measure one, which can be obtained from μ as in Theorem 11.15. More precisely, let U and B_m be as in 11.15 and define $v(A)$ to be $\int \mu(A \cap B_m)/\mu(B_m) \, dU$.
   If F is not locally commutative, choose $\sigma_1, \sigma_2 \in F$ and $P \in \mathbf{R}^n$ such that $\sigma_1\sigma_2 \neq \sigma_2\sigma_1$ and $\sigma_1(P) = P = \sigma_2(P)$. Let τ be the translation by $-P$ and define v on $\mathscr{P}(\mathbf{R}^n)$ by $v(A) = \mu(\tau^{-1}(A))$; let $\sigma_i' = \tau\sigma_i\tau^{-1} \in O_n$. Then v is a finitely additive extension of λ that is $\langle \sigma_1', \sigma_2' \rangle$-invariant. But $\langle \sigma_1', \sigma_2' \rangle$ is isomorphic to $\langle \sigma_1, \sigma_2 \rangle$, which, because σ_1 and σ_2 do not commute, is freely generated by σ_1, σ_2. Since v induces a $\langle \sigma_1', \sigma_2' \rangle$-invariant extension of λ on $\mathscr{P}(S^{n-1})$ by adjunction of radii, this contradicts (3) $\Rightarrow$ (2) of Theorem 11.19. $\square$

A question related to the previous characterization problem asks: which groups of isometries can arise as the group with respect to which a finitely additive extension of λ to $\mathscr{P}(\mathbf{R}^n)$ is invariant? If μ is such a measure, let $\text{Inv}(\mu)$ denote the subgroup of G_n containing those isometries, σ, such that μ is

σ-invariant. The following result shows that precisely the amenable groups arise as $\text{Inv}(\mu)$.

Theorem 11.21 (AC). *Suppose G is a subgroup of G_n. Then there is a finitely additive extension μ of λ to $\mathscr{P}(\mathbf{R}^n)$ such that $\text{Inv}(\mu) = G$ if and only if G is amenable.*

Thus, for example, there are total extensions of λ in $\mathbf{R}^1$ that are invariant under all translations, but not under any reflections, or that are invariant under all rational translations, but no others. In fact, in $\mathbf{R}^1$ and $\mathbf{R}^2$, any group of isometries is realizable as $\text{Inv}(\mu)$. Note that in other contexts things may turn out quite differently in that invariance with respect to one group is sufficient to imply invariance with respect to a larger group. For instance, any countably additive, translation-invariant measure on $\mathscr{L}$ that normalizes $[0,1]$ must coincide with Lebesgue measure, and hence is necessarily invariant under reflections. (But see [94] for a variation on this problem.) Or, letting G be the countable dense subgroup of SO_5 having Property T (see remarks preceding Theorem 11.10), any G-invariant mean on $L^\infty(S^4)$ is necessarily O_5-invariant (because it must equal the Lebesgue integral).

The necessity of amenability in Theorem 11.21 follows from Theorem 11.20. The proof of sufficiency is due to Wagon [247]. The main idea is to find a set A such that $B = \{\chi_{\sigma(A)} : \sigma \in G_n\}$ is linearly independent over $L^1(\mathbf{R}^n)$, the space of Lebesgue integrable, real-valued functions on $\mathbf{R}^n$. Then the Hahn-Banach Theorem is used to extend the Lebesgue integral to a linear function F on the space generated by $L^1(\mathbf{R}^n) \cup B$, in such a way that $F(\chi_{\sigma(A)}) = 1$ or 0 according as σ is or is not in G. Then F, which is precisely G-invariant, is extended by 10.11(5) to a G-invariant linear functional on the space of all real-valued functions on $\mathbf{R}^n$ that are bounded by a function in $L^1(\mathbf{R}^n)$. This extension induces a measure μ on $\mathscr{P}(\mathbf{R}^n)$ such that μ extends λ and $\text{Inv}(\mu) = G$.

We now consider the rather different problem of characterizing the groups of isometries of S^n for which an invariant, finitely additive measure on $\mathscr{P}(S^n)$ having total measure one exists. By Corollary 9.2 this is equivalent to asking for the subgroups, G, of O_n with respect to which S^n is paradoxical. For the case of S^2, the local commutativity of SO_3 yields an easy solution as follows.

Theorem 11.22 (AC). *For a subgroup G of O_3 the following are equivalent:*

(1) *G is amenable.*
(2) *G has no free subgroup of rank 2.*
(3) *S^2 is not G-paradoxical.*
(4) *There is a finitely additive, G-invariant measure on $\mathscr{P}(S^2)$ having total measure one.*

Proof. $(1) \Leftrightarrow (2)$ follows from 11.19 and $(3) \Leftrightarrow (4)$ follows from 9.2. Moreover, $(1) \Rightarrow (4)$ by Theorem 10.3. Finally, Theorem 4.5 (see also 8.6) and the local commutativity of SO_3 yield $(3) \Rightarrow (2)$ for subgroups of SO_3. The result for O_3 follows, since if σ, τ are independent in O_3, then σ^2 and τ^2 are independent in SO_3. $\qquad\square$

Property (4) is much more sensitive to an action's fixed points than the corresponding property (3) of Theorem 11.19. Because of this, Theorem 11.22 does not extend to $\mathbf{R}^3$ or to S^3. If $\sigma, \rho \in SO_3$ are independent, then G, the group they generate, satisfies (4) with respect to $\mathbf{R}^3$: just let $\mu(A)$ be 1 or 0 according as the origin is or is not in A. But G does not satisfy (2). For a similar example in S^3, use the same σ, ρ, but extend them to $\sigma', \rho' \in SO_4$ by fixing the new coordinate. Then σ', ρ' are still free generators of G, the group they generate, so G fails to satisfy (2). But there is a G-invariant measure on S^4; just let μ be the principal measure determined by the point $(0, 0, 0, 1)$, which is fixed by the action of G.

The point of the examples of the preceding paragraph is that if a group G acting on X fixes some x in X, then the principal measure determined by x is G-invariant. In fact, it is not necessary that G fix a point, but only that a certain subgroup of G fix a point. For suppose H is a normal subgroup of G such that G/H is amenable and H has a common fixed point x. Then let E be the G-orbit of x and consider the natural action of G/H on E. The amenability of G/H yields a G/H-invariant measure on E that is, in fact, G-invariant. Giving $X \backslash E$ measure zero yields a G-invariant measure on $\mathscr{P}(X)$. A recent result of Dani (who used the technique introduced in [40]) confirms a conjecture of Wagon [247] that this condition is necessary as well as sufficient for G-invariant measures on $\mathscr{P}(S^n)$ to exist, where G is a subgroup of O_{n+1}.

Theorem 11.23 (AC). *For a subgroup G of O_{n+1} the following are equivalent:*

(1) *S^n is not G-paradoxical.*
(2) *There is a finitely additive, G-invariant measure on $\mathscr{P}(S^n)$ having total measure one.*
(3) *There is a normal subgroup H of G such that G/H is amenable and some point in S^n is fixed by all isometries in H.*

Of course, (1) and (2) are equivalent by Tarski's Theorem, and as proved above, (3) implies (2). Dani's proof that (2) implies (3) combines Theorem 11.25 with the fact that G's action on S^n has the following property. If $x \in S^n$, then G_x denotes the subgroup of G that fixes x; if H is a subgroup of G, then F_H denotes the set of points in S^n left fixed by each member of H. Then Dani's property is that every intersection of members of $\{G_x : x \in S^n\}$ (and of $\{F_H : H \text{ a subgroup of } G\}$) is in fact an intersection of finitely many members of the family.

If G, a subgroup of O_{n+1}, contains a locally commutative free subgroup of rank two, then by 4.5 or 8.6, S^n is G-paradoxical. Thus each of (1)–(3) of 11.23 imply that G has no such subgroup. This leads to the following question.

Question 11.24.   Does O_4 have a subgroup G such that for any normal subgroup H of G with $G/H \in NF$, H has no common fixed point on S^3, but yet G does not have a free locally commutative subgroup of rank two? Equivalently, is there a subgroup G of O_4 such that S^3 is G-paradoxical, but S^3 is not G-paradoxical using four pieces?

The preceding discussion for spheres brings us to the general question of what conditions on a group action are necessary and sufficient for the existence of an invariant measure. Condition 11.23(3) is sufficient but not necessary. Van Douwen [241] has shown that F, a free group of rank two, can act on an infinite set X in such a way that the action is transitive and each element of F fixes only finitely many points in X (it follows that condition 11.23(3) fails), but yet a finitely additive, F-invariant measure on $\mathscr{P}(X)$ exists. Another counterexample was presented by Promislow [186]: let G be the group with generators $\{g_1, g_2, \ldots, h_1, h_2, \ldots\}$ and subject to the relations that h_i and h_j commute for all i, j and that g_i and h_j commute if $i \leqslant j$. Let G act on $G\backslash\{1\}$ by conjugation. Then there is an invariant measure on all subsets of $G\backslash\{1\}$: for any finite subset S of G let μ_S be the principal measure determined by h_n, where n is so large that $i \leqslant n$ for all $g_i \in S$. Then μ_S is S-invariant, and applying the usual compactness technique to subsets $\mathscr{M}_S$ of $[0, 1]^{\mathscr{P}(G)}$ (see the proof of Theorem 10.4), one gets a G-invariant measure on $\mathscr{P}(G)$. But it is easy to see that condition 11.23(3) fails for this action.

For some positive results, recall that a locally commutative action of a free group of rank two is paradoxical. In fact, a somewhat stronger result is true; the following result shows that the hypothesis can be weakened to nonamenability.

Theorem 11.25 (AC).   *If a nonamenable group G acts on X in such a way that for every $x \in X$, $G_x = \{g \in G : gx = x\}$ is amenable, then there is no finitely additive, G-invariant measure on X (and hence X is G-paradoxical).*

Proof.   The proof is a generalization of that of 10.4(e). Suppose μ is a finitely additive, G-invariant measure on X. Let $M \subseteq X$ be a set of representatives for the orbits of G's action, and for each $x \in M$, let v_x be a finitely additive, left-invariant measure on the amenable group G_x. Now, for any $A \subseteq G$, define $f_A : X \to \mathbf{R}$ by setting $f_A(y) = v_x(G_x \cap h^{-1}A)$ where x is the unique element of M such that y is in x's orbit and h is any element of G such that $hx = y$. To see that f_A is well defined, note that if $h_1 x = h_2 x$, then $f_A(h_1 x) = v_x(G_x \cap h_1^{-1}A) = v_x(h_2^{-1}h_1(G_x \cap h_1^{-1}A)) = f_A(h_2 x)$. Now define v on $\mathscr{P}(G)$ by $v(A) = \int f_A \, d\mu$. Since $f_{gA} = {}_g(f_A)$, v is a left-invariant finitely additive measure, contradicting the nonamenability of G.   □

In the case that G is a free group, the preceding result has been strengthened by Promislow [186]: if m is a positive integer, and G, a free group of rank $m + 1$ or greater, acts on X in such a way that each G_x, $x \in X$, is a free group of rank at most m, then there is no finitely additive, G-invariant measure on X. Still, these results leave us quite short of a general solution to the question posed earlier, which we restate in the possibly simpler case of a transitive action.

Question 11.26.   Suppose a nonamenable group G acts transitively on a set X. Is there a condition on the action that is both necessary and sufficient that no finitely additive, G-invariant measure on $\mathscr{P}(X)$ exists?

NOTES

The existence of Marczewski measures in $\mathbf{R}^1$, $\mathbf{R}^2$, and S^1 was proved by E. Marczewski in the 1930s (see [168, 169]); [168] contains the explicit formulation of Theorem 11.1. The modification of the Invariant Extension Theorem to obtain a measure that scales according to a fixed homomorphism from G to $(0, \infty)$ (Theorem 11.4) is due to Klee [106] for solvable groups; as a corollary he proved 11.5. For amenable groups in general the result is given in [168].

The Ruziewicz Problem is stated in [8], and in that paper, Banach solved the problem in $\mathbf{R}^1$, $\mathbf{R}^2$, and S^1 (p. 170). In fact, the problem had been posed earlier by Lebesgue, in the case of $\mathbf{R}^1$. Lebesgue had realized [121, p. 106] that the Lebesgue integral was the unique invariant mean on $L^\infty[0, 1]$ that satisfied the Monotone Convergence Theorem, and he asked whether this last, rather complicated, condition was necessary for the characterization. Because the Monotone Convergence Theorem is equivalent to countable additivity of the underlying measure, and because countable additivity does characterize λ as an invariant measure, Lebesgue was really posing the same problem as Ruziewicz.

Rosenblatt's problem (11.7), which is an attempt to generalize Banach's negative solution of the Ruziewicz Problem, is posed in [195]. Theorem 11.8 (1) $\Rightarrow$ (2), which solves this problem for countable groups, is due to del Junco and Rosenblatt [58]. The other implication of 11.8 is due to Losert and Rindler [127, Ex. (d)]; a related result was proved independently by Rosenblatt [195, Th. 3.8].

The paper by del Junco and Rosenblatt turned out to be the impetus for the solution of the higher-dimensional cases of Ruziewicz's Problem, a problem that had been dormant for 50 years. Using their result that the nonuniqueness of the invariant mean yields an asymptotically invariant sequence, Margulis [137] discovered the argument of 11.10 and constructed a countable dense subgroup of SO_n, $n \geqslant 5$, that has Property T. Simultaneously and independently, Rosenblatt [195] used the work of [58] to prove that if $n \geqslant 2$, then the Lebesgue integral is the unique mean on $L^\infty(\mathbf{R}^n)$ that normalizes χ_J, is invariant under isometries, and is invariant under the shear: $\sigma(x_1, \ldots, x_n) = (x_1 + x_2, x_2, \ldots, x_n)$. After seeing Rosenblatt's paper, Sullivan [228] saw how to solve Ruziewicz's Problem in S^n, $n \geqslant 4$, using essentially the same technique as Margulis. The solution in $\mathbf{R}^n$, $n \geqslant 3$, is due to Margulis [138], and V. Drinfeld (unpublished) is reported to have settled the two remaining cases, S^2 and S^3. It should be mentioned that the fundamental result of [58] on uniqueness of means and asymptotically invariant sequences was inspired by Namioka's proof [172] that amenable groups satisfy Følner's Condition.

Many others have studied Property T, and some of their results have a bearing on the topics considered here. See [4, 39, 102, 205, 228, 251, 252, 253].

The fact that Theorem 11.11 could be used to obtain a paradoxical decomposition modulo an ideal was pointed out by Rosenblatt [195], although that paper asserts that m can be taken equal to 1 in Corollary 11.14, which is not known unless I is closed under countable unions and invariant under similarities. However, Corollary 11.14 for a particular ideal of measurable subsets of S^2, and with $m = 1$, was proved by Rosenblatt earlier [194].

Theorem 11.15 is due to Mycielski [165, 168], although a geometric proof of Corollary 11.16 for isometries that avoids the Axiom of Choice was found by Davies [165]. Some results related to Theorem 11.15 may be found in [A3]. Theorem 11.17 is due to Rosenblatt for $n \geqslant 3$, and to Kallman (unpublished) for $n = 2$.

Theorem 11.19 (4) $\Rightarrow$ (1) is due to Wagon [247]; a version of this result for the action of $SL_n(\mathbf{Z})$ on T^n is given by Dani [40]. The proof of 11.20 is due to Mycielski. Theorem 11.21 is due to Wagon [247] and that paper also contains Theorem 11.22, although, as indicated in 11.22's proof, this characterization follows from known results. Theorem 11.23 is due to Dani, exploiting the ideas of [40], and answers a question of Wagon [247]. Question 11.24 was suggested by A. Borel.

The general problem of characterizing those actions of a free group of rank 2 that yield a paradox was posed by Greenleaf [80, p. 18]. The paper of van Douwen [241] contains several relevant results. Theorem 11.25, which, because AG has turned out to be strictly smaller than NF (Theorem 1.12), is a substantial strengthening of Corollary 8.6, is due to Rosenblatt [195, Th. 3.5]. It is interesting to note that the proof of Theorem 11.25 is the third proof presented here of the fact that a locally commutative action of a free group of rank 2 is paradoxical (see also 4.5 and 8.6). The result was discovered again, with yet a different proof, by Akemann [4]. Promislow's result mentioned at the end of the chapter (see [186]) is noteworthy because its proof combines methods of graph theory and combinatorial group theory.

Chapter 12

Growth Conditions in Groups and Supramenability

In this chapter we shall present some interesting connections between the amenability of a group and the rate of growth of a group, that is, the speed at which new elements appear when one considers longer and longer words using letters from a fixed finite subset of the group. This approach to amenability sheds light on a basic difference between Abelian and solvable groups. Both types of groups are amenable, but their growth properties can be quite different. This will explain why there is a paradoxical subset of the plane (Sierpiński-Mazurkiewicz Paradox, 1.7), but no similar subset of $\mathbf{R}^1$.

The study of growth conditions also elucidates the amenability of Abelian groups. One may prove quite simply (without the Axiom of Choice) that an Abelian group is not paradoxical (12.8), but the proof that Abelian groups are amenable (10.4(b)) is more complicated. One can then obtain an alternate proof of 10.4(b) by applying Tarski's Theorem (9.2).

We shall also discuss some recent work on the cogrowth of a group, a notion that refines the idea of the growth of a group. This leads to a striking and important characterization of amenable groups, a characterization that is central to the proof of Theorem 1.12, which constructs a group in $NF \setminus AG$.

The notion of amenability of a group is based on the existence of a measure of total measure one. But we are often interested in invariant measures that assign specific subsets measure one. The following definition is the appropriate strengthening of amenability that guarantees the existence of such measures for any nonempty subset of a set on which the group acts.

Definition 12.1. *A group G is* supramenable *if for any nonempty $A \subseteq G$, there is a finitely additive, left-invariant measure $\mu : \mathscr{P}(G) \to [0, \infty]$ with $\mu(A) = 1$.*

Of course, if G is supramenable, then no nonempty subset of G is paradoxical. By Tarski's Theorem, this latter condition is equivalent to supramenability. Before discussing some simple properties and examples of supramenable groups, we give the main result on actions of such groups.

Theorem 12.2. *Suppose a supramenable group G acts on X, and A is a nonempty subset of X. Then there is a finitely additive, G-invariant measure $\mu : \mathscr{P}(X) \to [0, \infty]$ such that $\mu(A) = 1$. Therefore no nonempty subset of X is G-paradoxical.*

Proof.   Fix any point $x \in A$ and then assign a subset B^* of G to every $B \subseteq X$ as follows: $B^* = \{g \in G : g(x) \in B\}$. Note that the identity of G lies in A^*. Therefore, by G's supramenability, there is a finitely additive, left-invariant $v : \mathscr{P}(G) \to [0, \infty]$ with $v(A^*) = 1$. Now, define the desired measure μ on $\mathscr{P}(X)$ by setting $\mu(B) = v(B^*)$. Then $\mu(A) = v(A^*) = 1$ and the finite additivity of v easily yields the same for μ. Finally, if $h \in G$, then $(h(B))^* = h(B^*)$, whence the G-invariance of μ follows from the left-invariance of v.   □

Of course, any nonamenable group fails to be supramenable, but the Sierpiński-Mazurkiewicz Paradox (1.7) provides an example of an amenable group that is not supramenable. The planar isometry group, G_2, is solvable and hence amenable, but the existence of a paradoxical subset of the plane, together with the previous theorem, implies that G is not supramenable. This can be shown more directly by using Theorem 1.8, which constructed a free subsemigroup of rank 2 in G_2. In fact, free semigroups of rank 2 play much the same role for supramenability that free groups of rank 2 do for amenability. As we now show, a group that contains a free subsemigroup of rank 2 cannot be supramenable. Recall that a semigroup S is free with free generating set A if S is the semigroup generated by A and distinct words using elements of A as letters yield distinct elements of S.

Proposition 12.3. *If G contains σ, ρ, which are free generators of a free subsemigroup of G, then G is not supramenable.*

Proof.   Let S be the subsemigroup of G generated by σ and ρ. It was shown in Proposition 1.3 that S is G-paradoxical: $S \supseteq \sigma S, \rho S$ and $S = \sigma^{-1}(\sigma S) = \rho^{-1}(\rho S)$.   □

Let SG denote the class of supramenable groups and NS the class of groups with no free subsemigroup of rank 2; then 12.3 states that $SG \subseteq NS$.

While no simple example of a group in $NS \setminus SG$ is known, the recent work on the analogous problem for amenable groups shows that such examples exist. Recall from Theorem 1.12 that there is a nonamenable periodic group G. Then G is not supramenable, but because all elements of G have finite order, $G \in NS$. Another possibility for a characterization of supramenability, discussed further in Question 12.7, comes from the observation that $SG \subseteq AG \cap NS$. It is quite possible that these two classes coincide, that is, that $SG = NS$ when restricted to amenable groups. Indeed, it is known that $SG = NS$ when restricted to the subclass, EG, of AG; see remarks following Theorem 12.18.

The next theorem details the known closure properties of SG, and it is easy to see that NS satisfies all of them as well. (For (f) use the fact that if σ, τ freely generate a subsemigroup of G, then for each $n \geqslant 1$, so does the pair σ^n, τ^n. But for some n, σ^n and τ^n both lie in H.) Furthermore, neither class is closed under general group extension since the solvable group G_2 is not amenable. But, unlike the case of the classes AG, NF, and EG, it is not clear that SG and NS share the same closure properties (see Question 12.5).

Theorem 12.4

(a) *Finite groups are supramenable.*

(b) *(AC) Abelian groups are supramenable.*

(c) *A subgroup of a supramenable group is supramenable.*

(d) *If N is a normal subgroup of a supramenable group G, then G/N is supramenable.*

(e) *(AC) If G is the direct union of a directed system of supramenable groups, $\{G_\alpha : \alpha \in I\}$, then G is supramenable.*

(f) *If H, a subgroup of G, is supramenable and H has finite index in G, then G is supramenable.*

Proof

(a) If $A \subseteq G$ is nonempty, let μ be defined by $\mu(B) = |B|/|A|$.

(b) This will be deduced from the more general result, to be proved later (12.8), that groups that do not grow too fast are necessarily supramenable.

(c) If A is a nonempty subset of H which, in turn, is a subgroup of a supramenable group G, then simply restrict a measure on $\mathscr{P}(G)$ that normalizes A to $\mathscr{P}(H)$.

(d) If A is a nonempty set of N-cosets, let μ be a left-invariant measure on $\mathscr{P}(G)$ that normalizes $\bigcup A$. Then define ν on $\mathscr{P}(G/N)$ by $\nu(B) = \mu(\bigcup B)$.

(e) Let a nonempty $A \subseteq G$ be given. Since each G_α is contained in a G_β that intersects A, we may assume that each G_α intersects A; simply delete from the system any subgroups that miss A. Consider the topological space $[0, \infty]^{\mathscr{P}(G)}$, which is compact. For each $\alpha \in I$, let $\mathscr{M}_\alpha$ consist of those finitely additive $\mu:\mathscr{P}(G) \to [0, \infty]$ such that $\mu(A) = 1$ and μ is G_α-invariant. Each G_α is supramenable, so if μ_α is a G_α-invariant measure on $\mathscr{P}(G_\alpha)$ with $\mu_\alpha(A \cap G_\alpha) = 1$, then the measure defined by $\mu(B) = \mu_\alpha(B \cap G_\alpha)$ lies in $\mathscr{M}_\alpha$. As in 10.4, each $\mathscr{M}_\alpha$ is closed, and because $\mathscr{M}_\alpha \cap \mathscr{M}_\beta \supseteq \mathscr{M}_\gamma$ if $G_\alpha, G_\beta \subseteq G_\gamma$, the collection $\{\mathscr{M}_\alpha : \alpha \in I\}$ has the finite intersection property. Compactness yields $\mu \in \bigcap \mathscr{M}_\alpha$, and such a μ is a G-invariant measure normalizing A.

(f) Suppose A is a nonempty subset of G and let $\{g_1, \ldots, g_m\}$ $(m < \infty)$ represent the right cosets of H in G. Since H is supramenable, we may apply Theorem 12.2 to the action of H on G by left multiplication to obtain an H-invariant measure v on $\mathscr{P}(G)$ with $v(\bigcup g_i A) = 1$. Note that $0 < \sum v(g_i A) < \infty$, for otherwise some $v(g_j A) = \infty$, contradicting $v(\bigcup g_i A) = 1$.

Now, let $a = \sum v(g_i A)$ and define μ on $\mathscr{P}(G)$ by $\mu(B) = (1/a) \sum v(g_i B)$. Then μ is finitely additive and $\mu(A) = 1$. Moreover, for any $g \in G$, the set $\{g_i g\}$ represents the right cosets of H. Hence $\mu(gB) = (1/a) \sum_i v(g_i g B) = (1/a) \sum_k v(h_k g_k B)$, where $h_k \in H$, and the H-invariance of v then yields that $\mu(gB) = \mu(B)$. $\qquad\square$

It can be shown that NS is closed under (finite) direct products (see [193]); therefore the same is true of $NS \cap AG$. But it is not known whether the corresponding closure property for SG holds.

Question 12.5. If G and H are supramenable, is $G \times H$ supramenable?

We now discuss growth rates in groups and how they relate to supramenability and the proof of 12.4(b). Recall that the length of a reduced word $g_1^{m_1} \cdots g_r^{m_r}$ (g_i not necessarily distinct, $g_i \neq g_{i+1}$, $m_i \in \mathbf{Z} \setminus \{0\}$) is $\sum m_i$; the identity is assumed to have length 0.

Definition 12.6. *If S is a finite subset of a group G, then the* growth function *(with respect to S) $\gamma_S : \mathbf{N} \to \mathbf{N}$ is defined by setting $\gamma_S(n)$ equal to the number of elements of G obtainable as a reduced word of length at most n using elements of $S \cup S^{-1}$ as letters.*

Of course, γ_S is nondecreasing. Also, $\gamma_S(0) = 1$ and $\gamma_S(1) = |\{1\} \cup S \cup S^{-1}|$. Since $\gamma_S(n + m) \leqslant \gamma_S(n)\gamma_S(m)$, it follows that $\gamma_S(n) \leqslant \gamma_S(1)^n$, so γ_S is always bounded by an exponential function. We are interested in whether or not γ_S really exhibits exponential growth. If G contains a free subsemigroup of rank 2, and S contains two free generators of such a semigroup, then $\gamma_S(n) \geqslant 2^n$, the number of words in S with only positive exponents and with length exactly n. Hence if G contains a free subsemigroup (or free subgroup) of rank 2, then the growth function exhibits exponential growth, with respect to some choice of S. At the other extreme lie the Abelian groups. For suppose

$S = \{g_1, \ldots, g_r\} \subseteq G$, which is Abelian. Any word in S is equal, in G, to a word of the form $g_1^{m_1} g_2^{m_2} \cdots g_r^{m_r}$ where $m_i \in \mathbf{Z}$. Hence the number of group elements that arise from words of length n is at most $(2n + 1)^r$ (since each m_i must lie in $[-n, n]$). Hence $\gamma_S(n)$ is certainly dominated by $n(2n + 1)^r$, a polynomial of degree $r + 1$, which shows that Abelian groups have slow, that is, nonexponential, growth for any choice of S. (A more careful word count shows that in fact $\gamma_S(n)$ is dominated by a polynomial in n of degree r; see [258].)

Definition 12.7.   *A group G is* exponentially bounded *if for any finite $S \subseteq G$ and any $b > 1$, there is some $n_0 \in \mathbf{N}$ such that $\gamma_S(n) < b^n$ whenever $n > n_0$; equivalently, $\lim_{n \to \infty} \gamma_S(n)^{1/n} = 1$. If G fails to be exponentially bounded, G is said to have* exponential growth. *The class of exponentially bounded groups is denoted by EB.*

We shall give more examples of exponentially bounded groups, but first we point out the important connection with supramenability.

Theorem 12.8

   (a) *(AC) If G is exponentially bounded, then G is supramenable.*
   (b) *If G is exponentially bounded, G acts on X, and A is a nonempty subset of X, then A is not G-paradoxical.*

Proof.   We first prove (b) (without the Axiom of Choice). Tarski's Theorem then yields a finitely additive, G-invariant measure on $\mathscr{P}(X)$ that normalizes A; this in turn yields (a), if one considers the left action of G on itself.

Suppose, to get a contradiction, that A is G-paradoxical. Then there are two one-one piecewise G-transformations, $F_1: A \to A$ and $F_2: A \to A$, such that $F_1(A) \cap F_2(A) = \varnothing$. Let $S = \{g_1, \ldots, g_r\}$ be all the elements of G occurring as multipliers in F_1 and F_2. Since G is exponentially bounded, there is an integer n such that $\gamma_S(n) < 2^n$. Consider the 2^n functions H_i obtainable as compositions of a string of n F_1's and F_2's. Each $H_i: A \to A$, and if $i \neq j$, then $H_i(A) \cap H_j(A) = \varnothing$. To see this, let p be the first (i.e., leftmost) of the n positions where H_i and H_j differ. Since F_1 and F_2 map A into disjoint sets, and since the function obtained by restricting H_i and H_j to the first $p - 1$ positions is one-one, H_i and H_j must map A into disjoint sets too. If we now choose any $x \in A$, it must be that the set $\{H_i(x): 1 \leqslant i \leqslant 2^n\}$ has 2^n elements. But each $H_i(x)$ has the form wx where w is a word of length n composed of elements of S, and this contradicts the fact that $\gamma_S(n) < 2^n$. $\square$

We have explained why Abelian groups are exponentially bounded, and therefore by the theorem just proved, such groups are supramenable. In particular, this yields a proof that Abelian groups are amenable that is completely different from the proof of 10.4(b). This new proof is somewhat more informative, because in addition to proving the stronger conclusion regarding supramenability, it shows quite clearly and effectively why the existence of a G-paradoxical set in any action of G means that G has exponential growth.

By 12.3 and 12.8 we have the containments $EB \subseteq SG \subseteq AG \cap NS$. When restricted to the elementary groups, these three classes coincide: by Theorem 12.18 they all coincide with the class of groups with the property that all finitely generated subgroups have a nilpotent subgroup of finite index. But it is not known whether either of the two equalities holds in general.

Question 12.9. (a) Is every supramenable group exponentially bound? That is, does $EB = SG$? (b) **(Rosenblatt's Conjecture).** Is every amenable group with no free subsemigroup of rank 2 necessarily supramenable? That is, does $SG = AG \cap NS$?

Figure 12.1 illustrates the relationships between various classes of groups.

The proof of 12.8 used only the fact that for all finite $S \subseteq G$, $\gamma_S(n) < 2^n$ for n sufficiently large, but there was no loss of generality in assuming the

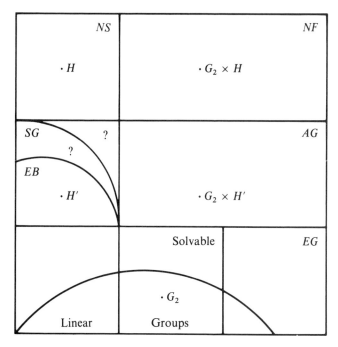

Figure 12.1. Various subclasses of *NF*, the class of groups with no free subgroup of rank 2. All of the labels describe the area below and to the left. G_2 is the isometry group of $\mathbf{R}^2$ and H, H' are the groups of Theorems 1.12, 12.17, respectively. The position of the linear groups without a free subgroup of rank 2 is a consequence of Tits's Theorem (10.6).

ostensibly stronger hypothesis that for all finite S and all $b > 1$, $\gamma_S(n) < b^n$ for n sufficiently large. The equivalence of these two conditions can be proved quite simply as follows.

Proposition 12.10

(a) For any finite subset S of a group G, the sequence $\{\gamma_S(n)^{1/n}\}$ converges.
(b) If G is not exponentially bounded, then for any $c > 1$, there is some finite $S \subseteq G$ such that $\gamma_S(n) \geqslant c^n$ for arbitrarily large n.

Proof

(a) Let γ denote γ_S, and fix a positive integer m. Then, for any n, let $k = [n/m] + 1$. It follows that $\gamma(n) \leqslant \gamma(km) \leqslant \gamma(m)^k \leqslant \gamma(m) \cdot \gamma(m)^{n/m}$, and since $\gamma(m)^{1/n} \to 1$ as $n \to \infty$, this means that $\limsup \gamma(n)^{1/n} \leqslant \gamma(m)^{1/m}$. Now, $\limsup \gamma(n)^{1/n} \leqslant \inf\{\gamma(m)^{1/m}\} \leqslant \liminf \gamma(n)^{1/n}$, yielding the existence of a limit of $\gamma(n)^{1/n}$.

(b) If G is not exponentially bounded, then for some finite $S \subseteq G$. $\gamma_S(n)^{1/n} \to \ell > 1$. Choose r such that $\ell^r > c$ and let S' consist of the identity of G together with all words of length exactly r from S. Any word of length at most rn from S may be viewed (using $1 \in S'$) as a word of length at most n from S'. Hence $\gamma_{S'}(n) \geqslant \gamma_S(rn)$, and since $\gamma_S(rn)^{1/n} \to \ell^r$ as $n \to \infty$, this means that $\gamma_{S'}(n) > c^n$ for arbitrarily large n. $\square$

The next proposition gives some simple closure properties of the class of exponentially bounded groups, which are sufficient for some geometric applications on the line and in the plane. But there are more exponentially bounded groups than the ones provided by Proposition 12.11; we shall return to this class later in the chapter, and mention a possible characterization of exponentially bounded groups.

Proposition 12.11

(a) Finite and Abelian groups are exponentially bounded.
(b) A subgroup or a homomorphic image of an exponentially bounded group is exponentially bounded.
(c) If H is an exponentially bounded subgroup of G and H has finite index in G, then G is exponentially bounded.
(d) A direct union of exponentially bounded groups is exponentially bounded; in particular, a group is exponentially bounded if (and only if) all of its finitely generated subgroups are.

Proof. The case of finite groups is trivial, and the Abelian case has already been discussed. The subgroup case is also trivial, and if S is a finite subset of H, a homomorphic image of G, let S' be a finite subset of G whose image is S. Then $\gamma_S(n)$ in H is bounded by $\gamma_{S'}(n)$ in G. For (c), choose a set of representatives $g_1, \ldots, g_m$ of the right cosets of H in G, with $g_1 = 1$. Then, given a finite $S \subseteq G$, let S' consist of the finitely many $h \in H$ that arise when each $g_i s$, where $s \in S$ and $1 \leqslant i \leqslant m$, is written in the form $h g_k$. We claim that each $w \in G$ that arises as a word of length at most n from S may be represented as $w' g_i$, where $1 \leqslant i \leqslant m$ and w' is a word of length at most n from S'. Suppose $w = s_1 s_2 \cdots$ with $s_i \in S$. Since $s_1 = 1 s_1 = g_1 s_1$, there is some $h_1 \in H$ and $k_1 \in \{1, \ldots, m\}$ such that $s_1 = h_1 g_{k_1}$. Similarly, there are h_2 and k_2 such that $g_{k_1} s_2 = h_2 g_{k_2}$. Continuing to the end of w and substituting into w transforms w to $h_1 \cdots h_r g_{k_r}$, where r is the length of w, as required. This claim yields that $\gamma_S(n)$ in G is at most $m \cdot \gamma_{S'}(n)$ in H. Therefore $\lim \gamma_S(n)^{1/n} \leqslant \lim m^{1/n} \gamma_{S'}(n)^{1/n} = 1$, so γ_S, and hence G, is exponentially bounded. Finally, suppose G is the union of the directed system of subgroups $\{G_i : i \in I\}$ and G fails to be exponentially bounded. If this failure is witnessed by $\gamma_{S'}$ choose $i \in I$ such that $S \subseteq G_i$. Then, since γ_S only refers to elements in the group generated by S, G_i must have exponential growth too.   □

Corollary 12.12.   *The isometry group of the line G_1 is exponentially bounded. Therefore no nonempty subset of $\mathbf{R}$ is paradoxical.*

Proof.   If T is the group of translations of $\mathbf{R}$, then T is a normal subgroup of G_1, $G_1/T \cong \mathbf{Z}_2$ and T is Abelian. It follows from 12.11(a) and (c) that G_1 is exponentially bounded. The rest of the corollary follows from the supramenability of G_1 (Theorem 12.8(a)), but this would use the Axiom of Choice. Instead, use 12.8(b).   □

Note that this Corollary, which shows why there is no Sierpiński-Mazurkiewcz Paradox in $\mathbf{R}^1$, also yields Corollary 10.10 for $\mathbf{R}^1$ without using the Axiom of Choice. The elimination of Choice from the $\mathbf{R}^2$ case is more complicated (see Corollary 13.9).

We have already mentioned that G_2 is not supramenable; indeed, the Sierpiński-Mazurkiewicz Paradox yields a paradoxical subset of the plane. But an interesting geometric problem arises if we consider only bounded subsets of $\mathbf{R}^2$. For $\mathbf{R}^3$ this restriction would add nothing new, since the bounded set S^2 is paradoxical. For the plane, however, there is the following unsolved problem.

Question 12.13.   Is there a nonempty paradoxical subset of $\mathbf{R}^2$ that is bounded?

The only partial result is the following, which shows that at least the strong sort of paradoxical decomposition given by the Sierpiński-Mazurkiewicz Paradox, that is, one using only two pieces, cannot be constructed for any bounded set.

Theorem 12.14.   *No nonempty bounded subset E of $\mathbf{R}^2$ contains two disjoint subsets, each of which is congruent to E.*

Proof.   Suppose $A, B \subseteq E$, $A \cap B = \emptyset$, and σ_1, σ_2 are isometries such that $\sigma_1(E) = A$ and $\sigma_2(E) = B$. Each σ_i^n maps E into a proper subset and therefore is not equal to the identity; hence σ_1 and σ_2 have infinite order. This means that σ_1 and σ_2 cannot be reflections, but in fact they cannot be translations or glide reflections either. For if σ_i is a translation by $\vec{v}$, choose P to be a point in $\bar{E}$ such that any other point of $\bar{E}$ is to the left of P in the direction determined by $\vec{v}$. Then $P + \vec{v} \notin \bar{E}$, contradicting $\sigma_i(\bar{E}) \subseteq \bar{E}$, which follows from $\sigma_i(E) \subseteq E$. An identical argument works if σ_i is a glide reflection; use the fact that $\vec{v}$, the translation vector, may be assumed to be parallel to the line of reflection.

So σ_1 and σ_2 must both be rotations; denote their fixed points by P_1, P_2, respectively. We first prove that $P_1 = P_2$. Let C be the smallest circle such that E is contained in C and its interior. Since $A \subseteq E$ and $\sigma^{-1}A = E$, it follows that E is also contained in $\sigma_1^{-1}C$ and its interior. But by the choice of C this means that $\sigma_1^{-1}C = C$, and therefore σ_1 leaves the center of C invariant; that is, P_1 is the center of C. Similarly, the same is true of P_2. Hence $P_1 = P_2$. This implies that σ_1 commutes with σ_2. Now choose any $Q \in E$ and obtain a contradiction by observing that $\sigma_1 \sigma_2 Q$, which is in A, equals $\sigma_2 \sigma_1 Q$, which is in B. $\square$

We now return to the problem of classifying the exponentially bounded groups. Groups with a free subsemigroup of rank 2 are the only examples of groups having exponential growth presented so far, but there are more. Adian [2] showed not only that the Burnside groups $B(r, e)$ with $r \geqslant 2$ and $e \geqslant 665$ are infinite, but also that they have exponential growth. Indeed, the number of group elements in $B(2, 665)$ corresponding to words in the two given generators of length exactly n is at least $4(2.9)^{n-1}$, while the free group of rank 2 has $4 \cdot 3^{n-1}$ such words. Now, since all elements of $B(r, e)$ have finite order, the group has no free subsemigroup. Hence the class of exponentially bounded groups is a proper subclass of the class of groups without a free subsemigroup of rank 2. To study the class of exponentially bounded groups in more detail, it is useful to delineate those groups that, like Abelian groups, have growth functions bounded by a polynomial.

Definition 12.15.   *A finitely generated group G has polynomial growth if for any finite $S \subseteq G$, there are positive constants c, d such that for all n, $\gamma_S(n) \leqslant cn^d$.*

It is easy to see that in order for a finitely generated group to have polynomial growth, it is sufficient that γ_S have polynomial growth for a single finite generating set S. For if S' is another finite subset of G, then there is an integer k such that $\gamma_{S'}(n) \leqslant \gamma_S(kn)$. (This observation also shows that if a finitely generated group has exponential growth, then $\lim \gamma_S(n)^{1/n} > 1$ for *any* finite set S that generates G.) The proof given earlier that Abelian groups are exponentially bounded shows that a finitely generated Abelian group has polynomial growth. This was extended by Wolf [258], who showed that a finitely generated nilpotent* group has polynomial growth; hence by 12.11(c) (modified for polynomial growth), any finitely generated almost nilpotent group has polynomial growth. The converse was proved first for solvable groups (Milnor and Wolf [148, 258]), then for elementary groups (Chou; see 12.18), and finally Gromov [82] showed that it is true in general, thus yielding the following beautiful characterization of groups with polynomial growth.

Theorem 12.16.   *A finitely generated group has polynomial growth if and only if it is almost nilpotent.*

Note that the definition of polynomial growth asserts only that $\gamma_S(n)$ is bounded above by a polynomial. It is not clear that there is a single d such that for appropriate positive constants c_i, $c_1 n^d \leqslant \gamma_S(n) \leqslant c_2 n^d$. However, it does follow from the remark after Definition 12.15 that the lowest degree of a polynomial bound on $\gamma_S(n)$ is independent of the choice of a generating set S. Bass [13] refined Wolf's work to show that if G is finitely generated and almost nilpotent, then there is an integer d such that for any finite generating set S, there are constants c_1 and c_2 so that $c_1 n^d \leqslant \gamma_S(n) \leqslant c_2 n^d$ for all n. Thus the degree of polynomial growth is a well-defined invariant of all groups having polynomial growth. For a finitely generated Abelian group, the degree of growth d is simply the torsion-free rank of the group.

The characterization of 12.16 was conjectured by Milnor [146], but Wolf [258] conjectured even more. A function of n that dominates all polynomials is not necessarily greater than some nontrivial exponential function $c^n, c > 1$; for example, consider $2^{\sqrt{n}}$ or $n^{\log n}$. Wolf conjectured that if a group's growth function dominates all polynomials, then it does dominate a nontrivial exponential function (this has become known as the *Milnor-Wolf Conjecture*). This conjecture is true for many interesting groups, but Grigorchuk [A8] has recently shown that it is not true in general.

Theorem 12.17.   *There exist finitely generated periodic groups that are exponentially bounded but do not have polynomial growth.*

Milnor and Wolf [148, 258] had proved their conjecture for solvable groups, and their work was strengthened by Rosenblatt [193] and Chou [30], who proved the following theorem.

---

* A group G is nilpotent if there is a finite sequence of normal subgroups $\{1\} = N_0 \lhd N_1 \lhd \cdots \lhd N_n = G$ such that for each i, N_{i+1}/N_i is contained in the center of G/N_i. (G is solvable if each N_{i+1}/N_i is Abelian; therefore nilpotent groups are solvable.)

Theorem 12.18. *If G is a finitely generated elementary group, then either G is almost nilpotent (and hence has polynomial growth) or G has a free subsemigroup of rank 2 (and hence has exponential growth).*

This theorem shows that the Milnor-Wolf Conjecture is valid in *EG*. Therefore the example of Theorem 12.17 is a nonelementary amenable (in fact, supramenable) group.

By Theorems 12.8 and 12.11(d), the preceding theorem shows that the strongest possible characterization of supramenability is valid in *EG*, the class of elementary groups: *G* is supramenable if and only if *G* has no free subsemigroup of rank 2. It is also known that *NS* = *SG* is valid when restricted to connected, locally compact topological groups (see [193, Cor. 4.20]). Theorem 12.18 also settles the Milnor-Wolf Conjecture in *EG* in the strongest possible way, since it yields that a finitely generated group in *EG* fails to have polynomial growth if and only if it has a free subsemigroup of rank 2. Even before Theorem 12.17 was proved, it was known that this strong form of the Milnor-Wolf Conjecture is not valid, since $B(2, 665)$ is a counterexample.

To summarize, we have that each of the following statements implies the next (and (1) and (1') are equivalent):

(1)   All finitely generated subgroups of *G* have polynomial growth;
(1') All finitely generated subgroups of *G* are almost nilpotent;
(2)   *G* is exponentially bounded;
(3)   *G* is supramenable;
(4)   *G* has no free subsemigroup of rank 2.

Moreover, (4) implies (1) in *EG*, and so these statements are all equivalent for elementary groups. In general, (4) does not imply (3) (by 1.12) and (2) does not imply (1) (by 12.17). Whether (3) implies (2) is Question 12.9(a).

Another large class of groups for which the four statements are equivalent is the class of all linear groups, that is, subgroups of $GL_n(K)$ for some positive integer *n* and field *K*. For any group *G* that satisfies (4) can have no free subgroup of rank 2, and hence by Tits's Theorem (10.6), a linear group that satisfies (4) is elementary. Hence for linear groups—in particular, all groups of isometries of $\mathbf{R}^n$—both the Milnor-Wolf Conjecture and Rosenblatt's Conjecture are valid in their strong forms. Note that, by the preceding remarks about elementary groups, the lower left square in Figure 12.1 consists of precisely the groups all of whose finitely generated subgroups have polynomial growth (equivalently, all such subgroups are almost nilpotent).

In light of Theorem 12.17, the following result pertaining to the Milnor-Wolf Conjecture is quite interesting. As a consequence of their investigation into Gromov's proof of 12.16, van den Dries and Wilkie [239] constructed a computable nondecreasing unbounded $g : \mathbf{N} \to \mathbf{N}$ such that $n^{g(n)}$ is exponentially bounded, and for any constant k, no group has a growth function lying in the gap between all polynomials and $kn^{g(n)}$.

Just as with actions of free groups, the mere presence of a free subsemigroup in a group acting on a set does not guarantee that the set has a paradoxical subset. Proposition 1.9 gives an extra condition on the action that is sufficient; in general, some condition is necessary, because of the trivial actions where $g(x)$ always equals x. Nevertheless, in the special case of groups of Euclidean isometries acting on $\mathbf{R}^n$, no extra condition is necessary. The following result, which is analogous to Theorem 11.19, gives a complete characterization of the subgroups of G_n with respect to which $\mathbf{R}^n$ has a paradoxical subset.

Theorem 12.19 (AC). *Suppose G is a subgroup of G_n. Then the following statements are equivalent to each other and to statements (1)–(4) following Theorem 12.18.*

(1) *G has no free subsemigroup of rank 2.*
(2) *G is supramenable.*
(3) *For any $E \subseteq \mathbf{R}^n$, $E \neq \varnothing$, there is a finitely additive, G-invariant measure on $\mathscr{P}(\mathbf{R})^n$ that normalizes E.*
(4) *No nonempty subset of $\mathbf{R}^n$ is G-paradoxical.*
(5) *No nonempty subset E of $\mathbf{R}^n$ contains two disjoint subsets, each of which is congruent, via G, to E.*

Proof. As pointed out before, Tits's Theorem and 12.18 yield $(1) \Rightarrow (2)$. That $(2) \Rightarrow (3) \Rightarrow (4)$ is just Theorem 12.2, and $(4) \Rightarrow (5)$ is obvious. To prove $(5) \Rightarrow (1)$, suppose that G does contain a free subsemigroup S, freely generated by σ and τ. Let H be the subgroup of G generated by σ and τ. Since H is countable, and since the set of fixed points of a single $w \in H \setminus \{1\}$ has Lebesgue measure zero, there must be some $P \in \mathbf{R}^n$ that is not fixed by any element of $H \setminus \{1\}$ (see the proof of 11.19). Now, let $E = \{s(P) : s \in S\}$, and let $A = \sigma(E)$, $B = \tau(E)$. Then A and B are subsets of E and are each congruent to E. Moreover, $A \cap B = \varnothing$, for suppose $\sigma s_1(P) = \tau s_2(P)$ for some $s_1, s_2 \in S$. By freeness of S, σs_1 and τs_2 must be distinct elements of G, since they have distinct leftmost terms. But then $(\tau s_2)^{-1} \sigma s_1$ is a nonidentity element of H fixing P, in contradiction to the choice of P. Since the conditions (1)–(4) on page 198 are equivalent for groups of isometries, the theorem is proved. $\square$

This result clarifies the Sierpiński-Mazurkiewicz Paradox since it shows that any free subsemigroup of G_2 of rank 2 yields a paradoxical subset of the plane. The proof of 1.7 given in Chapter 1, however, gives an especially succinct description of the paradoxical set. Moreover, the set, E, of 1.7 splits into two sets, each congruent to E. Theorem 12.19, except for the exceptional case where P is fixed by some nonidentity element of S, yields two subsets whose union is $E \setminus \{P\}$, a proper subset of E.

Recall from 4.9 that if a group is paradoxical using four pieces, as it is if it has a free subgroup of rank 2, then, indeed, the group contains such a subgroup. For free semigroups an analogous result is valid, which we derive as a corollary to the more general theorem that follows.

Theorem 12.20. *Suppose a group G acts on X and $\sigma, \tau \in G$. Then the following are equivalent*:

(1) *Some nonempty $E \subseteq X$ is such that $\sigma(E)$ and $\tau(E)$ are disjoint subsets of E.*

(2) *There is some $x \in X$ such that whenever w_1 and w_2 are nonidentity (semigroup) words in σ and τ beginning with σ, τ, respectively, then $w_1(x) \neq w_2(x)$. (It follows that σ, τ are free generators of a free subsemigroup of G.)*

Proof. The fact that (2) implies (1) is just a restatement of Proposition 1.9 and its proof; E is simply the S-orbit of x. For the converse, let x be any point in E. Because of the hypothesis on w_1 and w_2, $w_1(E) \subseteq \sigma(E)$ and $w_2(E) \subseteq \tau(E)$, or vice versa. In any event, $w_1(x)$ and $w_2(x)$ lie in disjoint subsets of E, and are therefore unequal. The fact that the condition on σ, τ expressed in (2) is sufficient to guarantee that they are free generators of a subsemigroup of G was, in essence, proved at the beginning of Theorem 1.8's proof. ☐

Corollary 12.21. *A group G has a nonempty subset E and two elements, σ, τ, such that σE and τE are disjoint subsets of E if and only if σ and τ are free generators of a free subsemigroup of rank 2.*

In fact, a slightly stronger result is true. If $E \subseteq G, E \neq \varnothing, \sigma E \cup \tau E \subseteq E$ and σ, τ fail to be free generators of a subsemigroup, then for some $w \in G$, $wE \subseteq \sigma E \cap \tau E$. To prove this, choose w_1 and w_2, two distinct words in σ, τ, so that $w_1 = w_2$ in G. Repeated left cancellation allows us to assume that either (1) $w_1 = \sigma \cdots$ and $w_2 = \tau \cdots$, (2) $w_1 = \sigma \cdots$ and $w_2 = 1$, or (3) $w_1 = \tau \cdots$ and $w_2 = 1$. In case (1) let g be the common value of w_1 and w_2 in G; in case (2) let $g = \tau$; and in case (3) let $g = \sigma$. It is easy to see, using $\sigma E \cup \tau E \subseteq E$, that these choices work. This result strengthens the corollary by showing that $\sigma E \cap \tau E$ is not only nonempty, but contains a set congruent to E.

We have seen that the major difference between an action of a free group on a set X and an action of a group with a free subsemigroup is that the former (under appropriate additional hypotheses) causes a paradoxical decomposition of X while the latter causes such a decomposition of a subset of X. But this analogy cannot be pushed too far. For instance, it is not the case that whenever a group with a free subsemigroup of rank n ($n \geqslant 2$) acts on X

without nontrivial fixed points, then given a proper system of n congruences, some nonempty subset of X can be partitioned to satisfy the congruences. To see this, consider the system $\{A_2 \cong A_2 \cup A_3 \cup A_4, A_4 \cong A_1 \cup A_2 \cup A_4\}$. If some nonempty subset E of X splits into A_1, A_2, A_3, A_4 such that $\sigma(A_2) = A_2 \cup A_3 \cup A_4$ and $\tau(A_4) = A_1 \cup A_2 \cup A_4$ where σ and τ comes from a group acting on X, then it can be shown that σ, τ are free generators of a free group. The two equations just mentioned yield: $\sigma(A_2) \subseteq X \setminus A_1$, $\sigma^{-1}(A_1) \subseteq X \setminus A_2$, $\tau(A_4) \subseteq X \setminus A_3$, and $\tau^{-1}(A_3) \subseteq X \setminus A_4$; these four relations were proved in 4.8 to guarantee the independence of σ, τ as group elements. So, for example, in the case of the action of the solvable group G_2 on itself, no subset of G_2 can be partitioned to solve the system despite the existence in G_2 of a free subsemigroup of rank 2 and the lack of fixed points in the action.

Nevertheless, these systems can be solved provided we allow the congruences to be witnessed by a piecewise G-transformation; more precisely, we interpret $\cong$ as $\sim_2$.

Theorem 12.22. *Suppose a group G, acting on a set X, contains a free subsemigroup S of rank κ generated by $\{\rho_\alpha : \alpha < \kappa\}$, where κ is an infinite cardinal. Suppose further that there is some $x \in X$ satisfying: whenever w_1 and w_2 are semigroup words in $\{\rho_\alpha\}$ beginning on the left with ρ_α, ρ_β, respectively, where $\alpha \neq \beta$, then $w_1(x) \neq w_2(x)$. Then E, the S-orbit of x, may be partitioned into κ sets, $\{A_\beta : \beta < \kappa\}$, such that for any two nonempty subsets of κ, L and R, $\bigcup\{A_\beta : \beta \in L\}$ and $\bigcup\{A_\beta : \beta \in R\}$ are G-equidecomposable using two pieces.*

Proof. By Proposition 6.10, the hypotheses imply that the sets $\rho_\beta(E)$ are pairwise disjoint subsets of E. Let $A_\beta = \rho_\beta(E)$ except when $\beta = 0$; in order that the sets A_β partition E, A_0 is defined to be $E \setminus \bigcup\{A_\beta : 0 < \beta < \kappa\}$. To show that this partition of E works, let L and R be two nonempty subsets of κ and choose γ, δ to lie in L, R, respectively. Since each $\rho_\beta(E)$ is a subset of A_β, it must be that $\rho_\gamma(\bigcup\{A_\beta : \beta \in R\}) \subseteq A_\gamma \subseteq \bigcup\{A_\beta : \beta \in L\}$ and $\rho_\delta(\bigcup\{A_\beta : \beta \in L\}) \subseteq A_\delta \subseteq \bigcup\{A_\beta : \beta \in R\}$. Hence by the Banach-Schröder-Bernstein Theorem, $\bigcup\{A_\beta : \beta \in L\} \sim_G \bigcup\{A_\beta : \beta \in R\}$ using two pieces. $\square$

Since G_2 satisfies the hypothesis of Theorem 12.22 (see 6.11) it follows that systems of congruences can be solved using subsets of the plane and $\sim_2$. For example, there exists a nonempty $E \subseteq R^2$ such that E has a subset A satisfying: for any κ between 2 and $2^{\aleph_0}$ inclusive, E can be split into κ sets, each of which is $\sim_2$ to A.

COGROWTH AND AMENABILITY

While the growth of a group has connections with supramenability, recent work by Grigorchuk [81] and Cohen [35] has led to a refinement of this concept, called cogrowth, that has proven to be very important because it yields a characterization of amenability. Indeed, this characterization is the key step in the proof of Theorem 1.12, that is, in the verification that a certain group lies in $NF \setminus AG$.

In order to discuss the growth of an arbitrary group, define $\gamma(G)$ to be the supremum of $\lim_{n \to \infty} \gamma_S(n)^{1/n}$ over all finite subsets S of G. By 12.10 $\gamma(G)$ is well defined and either $\gamma(G) = 1$ (G is exponentially bounded) or $\gamma(G) = \infty$ (G has exponential growth). This is a rather coarse classification, and the point of the cogrowth function is that it resolves the class of groups having exponential growth into more levels, perhaps even a continuum of levels.

Suppose a group G is given by generators and relations: $G = \langle a_1, \ldots, a_r : b_1, b_2, \ldots \rangle$ where $r < \infty$. This means that G is the group F_r/N where F_r is the free group with free generators $a_1, \ldots, a_r$ and N is the smallest normal subgroup of F_r containing $\{b_1, b_2, \ldots\}$. Let E_n be the set of all (reduced) words in F_r of length at most n. It is easy to see that if $r \geqslant 2$, E_n has $[r(2r - 1)^n - 1]/(r - 1)$ elements; but the important point here is that $|E_n|^{1/n} \to 2r - 1$ (even if $r = 1$). Let π be the canonical homomorphism, $\pi : F_r \to F_r/N = G$. The growth function $\gamma_S(n)$, where $S = \{a_1, \ldots, a_r\}$, is simply $|\pi(E_n)|$, the size of the image of π restricted to E_n. For finite groups the size of the kernel of a homomorphism is inversely proportional to the size of the image. But E_n is not a group, and so the size of the kernel might behave quite differently than the size of the image. This leads to the following definition.

Definition 12.23. *If G is presented as just discussed, then $\tilde{\gamma}(n)$, the cogrowth function of the presentation, is defined to be $|N \cap E_n|$; that is, $\tilde{\gamma}$ counts the number of words of length at most n that vanish when interpreted in G.*

As with the growth function, it will be the sequence $\tilde{\gamma}(n)^{1/n}$ and its limit that interests us. If there are no relations at all in the presentation, that is, $G = F_r$ and the presentation is the standard presentation of F_r, then $N = \{1\}$ and $\tilde{\gamma}(n) = 1$ for all n. On the other hand, suppose G is exponentially bounded. It is true for all presentations that some N-coset in F_r/N must contain at least $|E_n|/|\pi(E_n)|$ words of length at most n. Hence there are at least this many words in $N \cap E_n$ and so $|E_n| \geqslant \tilde{\gamma}(n) \geqslant |E_n|/\gamma(n)$. Since $\gamma(n)^{1/n} \to 1$, it follows that $\lim \tilde{\gamma}(n)^{1/n} = \lim |E_n|^{1/n} = 2r - 1$.

These two examples illustrate behavior at opposite ends of the cogrowth spectrum. It turns out that, for any presentation (with finitely many generators) of a group, the sequence $\tilde{\gamma}(n)^{1/n}$ tends to a limit that lies between 1 and $2r - 1$. The proof of this is elementary, but more complicated than for the ordinary growth function, and we refer the reader to [35] for details. The proof yields that the limit, when not equal to 1, is at least $\sqrt{2r - 1}$; that this last inequality is in fact strict is a more difficult result. The following theorem summarizes the basic facts about cogrowth.

Theorem 12.24. *Let a group G with a fixed presentation using r generators, where $2 \leqslant r \leqslant \infty$, be given. Then*

(a) $\lim \tilde{\gamma}(n)^{1/n}$ *exists;*
(b) $1 \leqslant \lim \tilde{\gamma}(n)^{1/n} \leqslant 2r - 1$;
(c) $\lim \tilde{\gamma}(n)^{1/n} = 1$ *if and only if there are no relations in the presentation (in which case $G = F_r$);*
(d) *If $\lim \tilde{\gamma}(n)^{1/n} > 1$, then $\sqrt{2r - 1} < \lim \tilde{\gamma}(n)^{1/n} \leqslant 2r - 1$.*

It must be emphasized that $\lim \tilde{\gamma}(n)^{1/n}$ is generally dependent on the choice of a representation. For example, let $G = \langle a_1, a_2, a_3 : a_2 a_3^{-1} \rangle$; G is isomorphic to F_2. Since $\tilde{\gamma}(n) \geqslant |E_n|/\gamma(n)$ and since $\gamma(n)$ is bounded by the number of words of length at most n in the standard presentation of F_2, $\lim \tilde{\gamma}(n)^{1/n} \geqslant 5/3$. This shows that case (d) of Theorem 12.24 can include free groups.

Now we can define η, the *cogrowth* of a presentation, by $\eta = \log(\lim \tilde{\gamma}(n)^{1/n})/\log(2r - 1)$. It then follows that $\eta \in \{0\} \cup (1/2, 1]$; $\eta = 0$ if and only if there are no relations in the presentation (assuming $r \geqslant 2$); and $\eta = 1$ if G is exponentially bounded. Thus the standard presentation of a free group has zero cogrowth (no words get killed), while an exponentially bounded group has full cogrowth (the number of words in E_n that get killed is, in an asymptotic sense, the same as the total number of words in E_n). Now, here is the remarkable theorem.

Theorem 12.25 (AC). *Let G be a finitely generated group. Then G is amenable if and only if $\eta = 1$ for all presentations of G with finitely many generators.*

Thus, for a group G presented as in the theorem, the amenability of G is characterized by the asymptotic behavior of the number of words of length at most n that collapse to the identity. This number, as n approaches infinity, must be very large. Theorem 12.25 is due to Grigorchuk [81] and, independently, Cohen [35]. The proof of both directions appears in [35] and is moderately difficult, using techniques from C^*-algebras and Kesten's characterization of amenability (discussed after the proof of Theorem 10.11). It would be nice if the method of paradoxical decompositions, which yields such a succinct proof that nonamenable groups have exponential growth, could be used to obtain a proof that a nonamenable group has cogrowth less than one.

One consequence of 12.25 is that for a finitely generated amenable group, the cogrowth is independent of the presentation, provided the latter uses finitely many generators. Another consequence concerns a notion of dimension introduced by Cohen [35]. Given a presentation using r generators, we may metrize F_r by setting $d(u, v)$, in the case $u \neq v$, equal to $(2r - 1)^{-k}$, where

k is the smaller of the lengths of u and v. Then η is the entropic dimension of N as a subspace of F_r (see [35] for a definition and proof). This leads to a natural definition of the dimension of a presentation as $1 - \eta$. Thus a finitely generated group is amenable if and only if all of its presentations are zero-dimensional.

In order to apply these ideas to all groups, simply let $\eta(G)$ be defined to be the greatest lower bound of η taken over all presentations with r generators, $2 \leqslant r < \infty$, of a subgroup of G.

Corollary 12.26 (AC).   *A group G is amenable if and only if $\eta(G) = 1$; G has no free subgroup of rank 2 if and only if $\eta(G) \geqslant \frac{1}{2}$; G has a free subgroup of rank 2 if and only if $\eta(G) = 0$.*

Although it is not known whether $\eta(G)$ can take on all values in $[\frac{1}{2}, 1]$, Ol'shanskii [178] has shown that for a certain periodic group G, $\eta(G) = \frac{1}{2}$. Therefore G is nonamenable, but has no free subgroup of rank 2. However, Cohen [35] has observed that the results connecting entropic dimension with amenability make the following conjecture plausible. Recall that a finitely presented group is one that is isomorphic to F_r/N where $r < \infty$ and N is the least normal subgroup of F_r containing some finite subset of F_r.

Conjecture 12.27 (Cohen's Conjecture).   *If G is a finitely presented group and G is nonamenable, then G has a free subgroup of rank 2; that is, $AG = NF$ when restricted to finitely presented groups.*

NOTES

The definition of supramenable groups was introduced by Rosenblatt [193], but the idea goes back to Tarski and Lindenbaum. The idea of using a free subsemigroup to construct a paradox was first used in the Sierpiński-Mazurkiewicz Paradox (see 1.7). However, Proposition 12.3 was not explicitly stated until the paper of Rosenblatt [193], who, in fact, proved the stronger form given in 12.21. This latter result was discovered independently by Sherman [207, Thm. 1.7].

The fact that Abelian groups are supramenable (12.4(b)) was first proved by Lindenbaum and Tarski in the 1920's. A proof appears in [236, pp. 224–227]. Their proof is essentially the same as the one presented here (12.8), and the only property of the Abelian group that is used is the fact that it is exponentially bounded. Moreover, Tarski was aware that the result is valid for more than just Abelian groups; in [234, p. 223] he points out that groups satisfying some "schwachere, aber kompliziertere Voraussetzungen" are supramenable. So even though growth conditions in groups were not introduced until much later, the result that exponentially bounded groups are supramenable must be credited to Tarski and Lindenbaum. Explicit use of the notion of exponential boundedness first appeared in a paper by Adelson-Velsky and Shreider [A1], who proved that exponentially bounded groups are amenable. The result also appears in [147] (where Kesten's characterization of amenability is used) and in [30] and [193] (where linear functionals are used).

Question 12.5 is due to Rosenblatt [193] as is Question 12.9(b). Unpublished work of Frey [72] shows that any group in $AG \cap NS$ has the property that all of its subsemigroups are amenable. Thus Question 12.9(b) is asking whether an amenable group all of whose subsemigroups are amenable is necessarily supramenable.

The second part of Corollary 12.12, asserting that no nonempty subset of the real line is paradoxical, is due to Sierpiński [219, Thm. 19], and his proof is essentially the same as that presented here; that is, it uses the fact that G_1 is exponentially bounded. Theorem 12.14 was first proved by Lindenbaum [124, p. 218, N. 1]; the proof presented here was discovered by Hadwiger and Debrunner [87, p. 80].

The concept of the growth of a group first appeared in the paper of Adelson-Velsky and Shreider [A1], but Milnor and Wolf provided the first detailed analysis of growth conditions in groups. Milnor [147] proved Proposition 12.10 and introduced the term *exponential growth* for finitely generated groups such that $\lim \gamma_S(n)^{1/n} > 1$. Wolf [258] initiated the study of groups with polynomial growth, and generalizing an example of Milnor [147], he proved the fundamental result that almost nilpotent groups have polynomial growth. The proof of 12.11(c) is essentially due to Wolf [258]. Milnor [146] conjectured that the degree of growth of a group that has polynomial growth is a well-defined integer; that is, that for constants c_1 and c_2 there is an integer d such that $c_1 n^d < \gamma_S(n) < c_2 n^d$. He also raised the possibility that all finitely generated groups of polynomial growth are almost nilpotent. Bass [13] improved the results of Wolf to settle Milnor's first conjecture for almost nilpotent groups, and Milnor's second conjecture was proved recently by Gromov [82]. A useful appendix to [82] by Tits contains proofs of enough of the work of Milnor and Wolf to make Gromov's paper independent of theirs. The appendix also contains a proof of Bass's result. Gromov's characterization had been known in some special cases: its validity for solvable groups was proved by Milnor and Wolf [148, 258], and Chou [30] proved it for elementary groups. Indeed, Chou [30], extending results of Rosenblatt [193], proved the even stronger Theorem 12.18. An elementary proof of a different sort of special case of Gromov's Theorem—where it is assumed that the growth function is linear—is given by Wilkie and van den Dries [257].

Milnor had conjectured that all groups of polynomial growth are almost nilpotent, and Wolf [258] conjectured even more, that all finitely generated exponentially bounded groups are almost nilpotent. By 12.16, this is equivalent to the Milnor-Wolf Conjecture as stated here. Theorem 12.17, which provides a counterexample to this conjecture, was announced by Grigorchuk at the 1983 International Congress of Mathematicians in Warsaw; a sketch of his proof appears in [A8].

Theorem 12.20 and Corollary 12.21 are due to Rosenblatt [193]. Theorem 12.22 is due to Mycielski [156].

The notion of cogrowth, which is more important for the theory of amenable groups than is growth, was introduced independently by Grigorchuk [81] and Cohen [35]. Cohen conjectured that Theorem 12.25 could be used to construct a nonamenable group that has no free subgroup of rank 2, and as stated by Ol'shanskii [178], this has turned out to be the case.

See [24] for further work on growth conditions and amenability.

Chapter 13

The Role of the Axiom of Choice

In this final chapter, we give a more detailed account of the role played by the Axiom of Choice (AC) in the theory of paradoxical decompositions. Ever since its discovery, the Banach-Tarski Paradox has caused many mathematicians to look critically at the Axiom of Choice. Indeed, as soon as the Hausdorff Paradox was discovered it was challenged because of its use of AC; E. Borel [21, p. 256] objected because the choice set was not explicitly defined. We shall discuss these criticisms in more detail later in this chapter, but first we deal with several technical points that are essential to understanding the connection between AC and the Banach-Tarski Paradox.

Results of modern set theory can be used to show that AC is indeed necessary to obtain the Banach-Tarski Paradox, in the sense that the paradox is not a theorem of ZF alone. Before we can explain why this is so we need to introduce some notation and discuss some technical points of set theory. If T is a collection of sentences in the language of set theory, for example, $T = $ ZF or $T = $ ZF $+$ AC, then Con(T) is the assertion, also a statement of set theory in fact, that T is consistent, that is, that a contradiction cannot be derived from T using the usual methods of proof. We take Con(ZF) as an underlying assumption in all that follows. Gödel proved in 1938 that Con(ZF) implies (and so is equivalent to) Con(ZF $+$ AC); thus AC does not contradict ZF (see [98, 99]). Let LM denote the assertion that all subsets of $\mathbf{R}$ are Lebesgue measurable; of course, LM contradicts AC (see 1.6). Since LM implies that all subsets of each $\mathbf{R}^n$ are Lebesgue measurable (see the proof of 7.9), ZF + LM implies that there is no Banach-Tarski Paradox.

Ever since nonmeasurable sets were first constructed, mathematicians wondered whether it was possible to construct a nonmeasurable set without using the Axiom of Choice. All known constructions used AC in apparently unavoidable ways. The feeling that AC was necessary was confirmed in 1964 when Solovay, using the just-discovered method of forcing due to Cohen, proved, with a complication to be discussed, Con(ZF + LM) (see [99]). It follows that ZF + "There is no Banach-Tarski Paradox" is consistent, and hence the Banach-Tarski Paradox is not a theorem of ZF.

To understand the extra complication in Solovay's result, we must introduce the notion of an inaccessible cardinal. A cardinal κ is *inaccessible* if it is not a successor cardinal, nor a sum of fewer than κ cardinals, each of which is smaller than κ. For example, $\aleph_1$ is the successor of $\aleph_0$ and so is not inaccessible, and $\aleph_\omega$ is the sum of the $\aleph_n$, $n = 0, 1, 2, \ldots$, and so also is not inaccessible. One example of an inaccessible cardinal is $\aleph_0$; any others must be very large. However, the existence of an uncountable inaccessible cardinal cannot be proved in ZF or ZF + AC. This is because, letting IC denote the assertion that an uncountable inaccessible cardinal exists, it can be shown that ZF + IC implies Con(ZF) (see [61, 99]). Since by a famous theorem of Gödel, a consistent theory such as ZF cannot prove its own consistency, ZF is not strong enough to prove IC. Even more is true: Con(ZF + IC) cannot be proved from Con(ZF). This contrasts IC with AC: ZF can neither prove nor disprove AC, and ZF cannot prove IC, but it is not known (and cannot be proved) that ZF cannot disprove IC. Thus ZF + IC is a substantial strengthening of ZF; while it is presumably a consistent extension, this fact cannot be proved. Much study has been devoted to inaccessible cardinals, and it would be a great shock to most set theorists if in ZF it could be proved that uncountable inaccessible cardinals do not exist. Indeed, set theorists have studied much larger cardinals (e.g., measurable cardinals) whose existence would imply the existence of many inaccessible cardinals, and it is generally felt that these larger large cardinals yield consistent extensions of ZF. But, as pointed out, the consistency of ZF + IC cannot be proved, and one cannot completely ignore the possibility that IC contradicts ZF.

One technical point arises when discussing Lebesgue measure in the absence of AC. Without any form of Choice at all Lebesgue measure might behave in a way that makes it an object much less worthy of study in the first place. For instance, the property that a countable union of countable sets is countable uses the Axiom of Choice in a subtle but necessary way: it is consistent with ZF that $\mathbf{R}$ is a countable union of countable sets! This latter assertion implies that Lebesgue measure is not countably additive. Thus it is customary to allow a modest amount of Choice, enough so that measure and category behave as expected, but not so much that the reals can be well-ordered or a nonmeasurable set constructed. One possibility is to add the

Axiom of Countable Choice, which allows selections to be made from an arbitrary countable collection of sets and yields the countable additivity of both Lebesgue measure and the ideal of meager sets. But a slightly stronger axiom, the Axiom of Dependent Choice (DC), which guarantees the existence of a countable choice set $\{a_n\}$ where a_{n+1} depends on a_n, turns out to be more useful. It allows more of the classical arguments of analysis and descriptive set theory to be retained, but does not yield a well-ordering of the reals. Thus the assertion LM is usually considered in the context of ZF + DC, rather than just ZF. See [99, 150] for more details on these weaker forms of AC and further references to a variety of consistency results.

With the preceding set-theoretical points in hand, we can state the fundamental result that connects IC and LM.

Theorem 13.1. *Con(ZF + IC) is equivalent to Con(ZF + DC + LM).*

Some remarks on the history of this remarkable result are in order. Soon after Cohen invented the method of forcing to prove the independence of both AC and the Continuum Hypothesis from the ZF axioms, Solovay [224] saw how to obtain Con(ZF + DC + LM). The complication of Solovay's result alluded to before is that he needed to assume that the existence of an uncountable inaccessible cardinal was noncontradictory, that is, Con(ZF + IC). With this extra assumption, it follows from Solovay's result that ZF + DC is not strong enough to produce a nonmeasurable set.

For many years it was felt that the hypothesis regarding IC would eventually be eliminated from this result. But in 1980, S. Shelah (see [188]) proved the reverse direction of 13.1, thus establishing the necessity of Solovay's assumption. Although it is most likely that both ZF + IC and ZF + DC + LM are consistent, it is remarkable that these two questions should be so inextricably linked. Reinterpreting Shelah's theorem slightly, it states that if someone does succeed in proving from ZF that uncountable inaccessible cardinals do not exist, then it will follow that there is also a proof from ZF + DC that a nonmeasurable set exists.

Theorem 13.1 shows that, assuming Con(ZF + IC), the Banach-Tarski Paradox is not a theorem of ZF alone, nor of ZF + DC. In fact, this conclusion can be derived without any additional metamathematical assumptions, thus separating it from the status of inaccessible cardinals. This follows from a different, less well known, measure-theoretic consistency result of Solovay, one that deals with measures more general than Lebesgue measure.

Let GM denote the assertion that there is a countably additive, isometry-invariant measure on all subsets of $\mathbf{R}^n$ that normalizes the unit cube. Recall that the classic example of a nonmeasurable set actually negates GM, not just LM (see 1.6). Thus ZF + AC yields that GM is false. Note also that by 9.10, a general measure as in GM must agree with Lebesgue measure on the

Lebesgue measurable subsets of $\mathbf{R}^n$. If, however, the Axiom of Choice is not assumed, then LM and GM differ dramatically; the latter is a much weaker statement.

Theorem 13.2. *Con(ZF) is equivalent to Con(ZF + DC + GM).*

Theorem 13.2 shows that while the theory ZF + DC + LM is equiconsistent with ZF + IC, the theory ZF + DC + GM is equiconsistent with ZF, a theory that is strictly weaker than ZF + IC in consistency strength. Many consequences of ZF + DC + LM can be derived from ZF + DC + GM, since the latter yields measures that are defined on all subsets of $\mathbf{R}^n$. In particular, GM is sufficient to guarantee the nonexistence of the Banach-Tarski Paradox.

Corollary 13.3. *The Banach-Tarski Paradox is not a theorem of ZF, nor of ZF + DC.*

Proof. Suppose ZF + DC is sufficient to derive the Banach-Tarski Paradox (of a ball in $\mathbf{R}^3$). Then ZF + DC would imply that there is no finitely additive, isometry-invariant measure defined on all subsets of $\mathbf{R}^3$ and normalizing the unit cube. This contradicts Theorem 13.2, which guarantees that even GM, which provides a countably additive invariant measure, is not inconsistent with ZF + DC. $\square$

The proofs of 13.1 and 13.2 are similar in that they both use the technique of forcing with a measure algebra to construct a model witnessing the measure-theoretic assertions. The proof of 13.1 can be found in [99]. The proof of 13.2 proceeds by forming $L[G]$, the generic extension of the constructible universe that adds $\aleph_1$ random reals. One then forms the submodel N consisting of all sets in $L[G]$ that are hereditarily ordinal definable over the subclass of $L[G]$ consisting of all countable sequences of ordinals. Then N is a model of ZF + DC (see [99, p. 546]), and Solovay proved that GM is true in N. Note that by 13.2, LM is necessarily false in N; indeed, the set of constructible reals is a nonmeasurable set (see [99, p. 568]). For more details of the proof of Theorem 13.2, see [184, 199].

The preceding discussion shows that some form of Choice is necessary to obtain the Banach-Tarski Paradox. The same is true for many of the results in the other direction, for example, the existence of a finitely additive, translation-invariant measure on $\mathscr{P}(\mathbf{R})$ that normalizes $[0, 1]$. We shall give some indication of how such unprovability results are obtained by showing why ZF + DC is not strong enough to yield the Measure Extension Theorem or the amenability of Abelian groups.

At the same time as he proved 13.2, Solovay showed that Con(ZF + IC) implies Con(ZF + DC + PB), where PB denotes the assertion that all sets of reals have the Property of Baire. (Recall that AC yields a set without the Property of Baire; see the discussion preceding Question 3.12.) As with LM, it was generally expected that the seemingly extraneous hypothesis about inaccessible cardinals would be eliminated; unlike the case with LM, this expectation turned out to be correct. The following recent theorem of Shelah (see [188]), together with Theorem 13.2, shows that there is a deep and unexpected metamathematical distinction between measure and category. A proof of Solovay's version, assuming Con(ZF + IC), can be found in [99]; a proof of the stronger version has not yet been published.

Theorem 13.4. *Con(ZF) implies Con(ZF + DC + PB).*

This theorem asserts that ZF + DC is not strong enough to yield a set of reals that fails to have the Property of Baire. We now show that certain statements about measures—for example, the Measure Extension Theorem or the amenability of Abelian groups—do yield a set without the Property of Baire. It follows that the measure theory statements, which are theorems of ZF + AC, are not theorems of ZF + DC. The next result shows why certain measures yield topologically bad sets.

Theorem 13.5. *If there is a σ-algebra $\mathscr{A}$ that bears a finitely additive measure of total measure one that fails to be countably additive, then there is a set of reals that does not have the Property of Baire.*

Proof. Given μ, a measure on $\mathscr{A}$ as hypothesized, let $\{A_n : n \in \mathbf{N}\}$ witness the failure of countable additivity. Because μ is finitely additive, it follows that $a = \mu(\bigcup A_n) - \sum \mu(A_n) > 0$. Now, define v on $\mathscr{P}(\mathbf{N})$ by $v(X) = (1/a)(\mu(\bigcup\{A_n : n \in X\}) - \sum\{\mu(A_n) : n \in X\})$; v is a finitely additive measure on $\mathscr{P}(\mathbf{N})$ having total measure one and vanishing on singletons. The proof will be complete once it is shown that the collection I of subsets of $\mathbf{N}$ with v-measure zero fails to have the Property of Baire when viewed as a subset of the space $2^{\mathbf{N}}$, topologized as a product. (In this proof we identify 2 with $\{0, 1\}$ and m with $\{0, 1, \ldots, m - 1\}$.) For it then follows that the subset of the Cantor set in $[0, 1]$, consisting of reals that correspond to sequences in I, fails to have the Property of Baire.

Suppose, by way of contradiction, that I does have the Property of Baire. Because v vanishes on finite subsets of $\mathbf{N}$, I is a tail set in $2^{\mathbf{N}}$ and hence (see [181, pp. 84–85]) either I is meager or I^c, the complement of I in $2^{\mathbf{N}}$, is meager. The latter possibility cannot occur, for if I^c is meager, then so is $\{\mathbf{N}\backslash A : A \notin I\}$, since this set is obtained by simply switching 0's and 1's in the sequences in I^c, and this operation preserves nowhere dense sets. But then $I^c \cup \{\mathbf{N}\backslash A : A \notin I\}$, which equals all of $2^{\mathbf{N}}$, is meager, which contradicts the Baire Category Theorem.

Thus we may assume that I is meager, say, $I = \bigcup\{X_n : n \in \mathbf{N}\}$, where each X_n is a nowhere dense subset of $2^{\mathbf{N}}$. We shall obtain a contradiction by constructing uncountably many sets in $\mathscr{P}(\mathbf{N})\backslash I$ with finite pairwise intersections. For such a family there must be a positive integer n such that uncountably many sets in the family have v-measure at least $1/n$, and this contradicts $v(\mathbf{N}) = 1$.

Consider $2^{\mathbf{N}}$ as the set of all branches (i.e., maximal totally ordered subsets) through the full binary tree. Recall that a basis for the product topology on $2^{\mathbf{N}}$ consists of all sets of the form $[s] = \{b \in 2^{\mathbf{N}} : b \text{ extends } s\}$, where $s : m \to 2$ for some finite m. Each X_n, being nowhere dense, has the property that any $s \in 2^m$ may be extended to a longer sequence t such that $[t] \cap X_n = \varnothing$. Now, build a subtree of the full binary tree as follows. Extend the sequence $\langle 0 \rangle$ to some t with $[t]$ disjoint from X_0. Then extend $\langle 1 \rangle$ to $\langle 100 \cdots 0 \rangle$, with the same length as t, and extend this new sequence further to t', where $[t']$ is disjoint from X_0. Then replace t by the sequence obtained by adding 0's to t to bring its length up to that of t'. Note that no extension of t or of t' lies in X_0. Now, consider the two sequences $t \frown 0$ and $t \frown 1$ and extend them in the same way that $\langle 0 \rangle$ and $\langle 1 \rangle$ were extended, but this time avoid X_1 rather than X_0. Then add enough 0's to t' to bring its length up to that of the two sequences just constructed, split the resultant sequence by adding a 0 and a 1, and extend these two sequences in the same way as $t \frown 0$ and $t \frown 1$ were extended, that is, to avoid X_1. Then bring the two extension of $t \frown 0$ and $t \frown 1$ up to the same length by adding zeros. We now have four sequences of the same length, and no extension of any of them lies in $X_0 \cup X_1$. Moreover, the insertion of all the zeros guarantees that any two branches through the subtree under construction will yield subsets of $\mathbf{N}$ that are disjoint past the level at which the branches diverge.

Continue the construction, dodging each X_n in turn, and always adding enough zeros to preserve the property that each level of the tree has at most one 1. This yields a subtree of $2^{\mathbf{N}}$ with the property that every node has an extension that splits. It follows that there are $2^{\aleph_0}$ branches through the subtree, and by the construction the subsets of $\mathbf{N}$ corresponding to these branches have finite pairwise intersections and fail to lie in $\bigcup X_n = I$, as desired.   □

Corollary 13.6. *The following assertions, which are theorems of* $ZF + AC$, *are not theorems of* $ZF + DC$.

(a) *The Measure Extension Theorem* (10.7).
(b) *Abelian groups are amenable* (10.4(b)).
(c) *A direct union of a directed system of amenable groups is amenable* (10.4(f)).

Proof.   Because of Theorems 13.4 and 13.5 it is sufficient to show that each of (a)–(c) yields the existence of a measure as in the hypothesis of 13.5. For (a), let $\mathscr{A} = \mathscr{P}(\mathbf{N})$ and let $\mathscr{A}_0$ be the subalgebra of $\mathscr{A}$ consisting of all finite sets and their complements. If μ is the $\{0, 1\}$-valued measure on $\mathscr{A}_0$ that vanishes on the finite sets, then the Measure Extension Theorem yields an extension of

μ to all of $\mathscr{A}$. This finitely additive extension vanishes on singletons, and so fails to be countably additive, as required. For (b), simply consider $\mathbf{Z}$. Any measure witnessing the amenability of $\mathbf{Z}$ must, by invariance, vanish on singletons, and so is not countably additive. For (c), let G be the direct union of $\aleph_0$ copies of $\mathbf{Z}_2$. Then G is the direct union of groups isomorphic to $\mathbf{Z}_2$, $\mathbf{Z}_2 \times \mathbf{Z}_2, \mathbf{Z}_2 \times \mathbf{Z}_2 \times \mathbf{Z}_2, \ldots$, each of which is finite and hence amenable. But a measure witnessing G's amenability cannot be countably additive. $\square$

The proof of Corollary 13.6 shows that ZF + DC does not yield the amenability of $\mathbf{Z}$. A much stronger result, due to Pincus and Solovay [184], is that ZF + DC + NM is consistent, where NM states that no infinite set bears a finitely additive measure of total measure one that is defined on all subsets and vanishes on singletons. This yields the unprovability in ZF + DC of the existence of a finitely additive, translation-invariant measure on $\mathscr{P}(\mathbf{R}^1)$ or $\mathscr{P}(\mathbf{R}^2)$ that normalizes an interval (Corollary 10.9): the restriction of such a measure to subsets of the interval would violate NM. Their result also yields the unprovability in ZF + DC of the existence of an infinite amenable group: NM yields that AG consists of precisely the finite groups.

The proofs of three of the six closure properties of the class of amenable groups (Theorem 10.4) use AC. Corollary 13.6 shows that two of the three are not theorems of ZF + DC. The remaining assertion—that a subgroup of an amenable group is not amenable—must be handled differently. Solovay's original work on Theorems 13.1 and 13.4 showed that the assertions were jointly consistent; that is, he proved that Con(ZF + IC) implies Con(ZF + DC + LM + PB). Now, it is a consequence of ZF + DC + LM that S^1 is amenable: Lebesgue measure witnesses amenability since all subsets of the circle are measurable. But, as shown before, ZF + DC + PB implies that $\mathbf{Z}$ is not amenable. Since S^1 has a subgroup isomorphic to $\mathbf{Z}$, this means that it is a theorem of ZF + DC + LM + PB that some amenable group has a nonamenable subgroup. Hence assuming the consistency of an uncountable inaccessible cardinal, Theorem 10.4(c) is not a theorem of ZF + DC. The presumably unnecessary hypothesis about inaccessible cardinals would be eliminated if Theorems 13.2 and 13.4 could be combined; that is, if it could be proved that Con(ZF) implies Con(ZF + DC + GM + PB). This is because GM, like LM, yields the amenability of S^1.

These ideas also yield the unprovability of Tarski's Theorem (9.1–9.3) in ZF. For let G be the group of all permutations of $\mathbf{Z}$ that equal the identity on all but a finite set. It is easy to see that for each $n = 1, 2, \ldots, (n + 1)\mathbf{Z} \not\leq n\mathbf{Z}$ in $\mathscr{S}(\mathbf{Z})$, the type semigroup for G-equidecomposability. But, assuming ZF + DC + PB, there is no finitely additive, G-invariant measure on $\mathscr{P}(\mathbf{Z})$ with total measure one, since such a measure would vanish on singletons.

In analyzing the necessity of the Axiom of Choice for certain results, two approaches are possible. The first, which we have followed in the previous discussion, is to show that the result is not provable in ZF alone. The second approach is to show that the result is provably equivalent to AC (in ZF). It is not reasonable to expect that the Banach-Tarski Paradox is fully equivalent to AC since it requires only a well-ordering of **R**, while AC is equivalent to the assertion that *all* sets can be well-ordered. But the question arises whether some of the more general theorems, such as the Measure Extension Theorem, are equivalent to AC. Such questions have been well studied, and the situation is summarized in Figure 13.1. The statements in each group are provably equivalent, and no statement implies a statement of a group above it; by 13.6 none of the statements are theorems of ZF + DC. For proofs of these and many other related results see [98].

The Axiom of Choice

Zorn's Lemma

Every set can be well ordered

$|X|^2 = |X|$ for all infinite sets X

Tychonoff's Theorem (the product of compact topological spaces is compact)

$\downarrow\!\not\uparrow$

Boolean Prime Ideal Theorem (all Boolean algebras bear a two-valued measure)

The product of compact Hausdorff spaces is compact

$\{0,1\}^I$ is compact for all sets I

Stone Representation Theorem (all Boolean algebras are isomorphic to a field of sets)

Compactness Theorem for first-order logic

$\downarrow\!\not\uparrow$

Hahn-Banach Theorem

Measure Extension Theorem

All Boolean algebras bear a [0,1]-valued measure

Figure 13.1.

ELIMINATING THE AXIOM OF CHOICE

The previous discussion concerns the noneliminability of AC from various results concerning measures. But AC can be eliminated in some special cases of a geometric nature. The most important consequence of Corollary 10.9 for paradoxical decompositions is that neither an interval in $\mathbf{R}^1$ nor a square in $\mathbf{R}^2$ is paradoxical. We have already seen (12.12) that a proof of the linear case can be given in ZF, using the fact that G_1 is exponentially bounded. In fact, as we shall show, AC can also be eliminated from the proof that a square is not paradoxical. The main idea is the observation that one does not require a measure on all of $\mathscr{P}(\mathbf{R}^2)$, but only on a countable subalgebra, and such a measure can be constructed without having to use the Axiom of Choice. The proof requires the following result that shows that, despite the inability of ZF to yield the amenability of Abelian groups, ZF is strong enough to get a weakened form of amenability for solvable groups.

Theorem 13.7. *If G is a solvable group and $\mathscr{A}$ is a countable left-invariant subalgebra of $\mathscr{P}(G)$, then there is a left-invariant, finitely additive measure $\mu: \mathscr{A} \to [0, 1]$ with $\mu(G) = 1$.*

Proof. Let $\{1\} = H_0 \lhd H_1 \lhd \cdots \lhd H_n = G$, where each H_i/H_{i-1} is Abelian, witness the solvability of G. We use induction on n. Since $\mathscr{A}_0 = \{A \cap H_{n-1} : A \in \mathscr{A}\}$ is a left-invariant subalgebra of $\mathscr{P}(H_{n-1})$, we may assume that there is an H_{n-1}-invariant measure $v: \mathscr{A}_0 \to [0, 1]$. Define $\bar{v}$ on $\mathscr{A}$ by $\bar{v}(A) = v(A \cap H_{n-1})$; $\bar{v}$ is an H_{n-1}-invariant measure on $\mathscr{A}$.

Now, it is sufficient to show that for each finite $W \subseteq G$ and each $\varepsilon > 0$, there is a measure $\mu_{W,\varepsilon}$ on $\mathscr{A}$ that is invariant within ε for left multiplication by members of W. For then the usual compactness technique can be applied to the family of subsets $\mathscr{M}_{W,\varepsilon}$ of $[0, 1]^{\mathscr{A}}$ (see the proof of 10.4(b)) to obtain the desired measure. Note that because $\mathscr{A}$ is countable, the compactness of $[0, 1]^{\mathscr{A}}$ does not require any Choice (Proposition 9.4).

The key point in the construction of $\mu_{W,\varepsilon}$ is the observation that for each $A \in \mathscr{A}$ and $g_1, g_2 \in G$, $\bar{v}(g_1 g_2 A) = \bar{v}(g_2 g_1 A)$. This holds because the commutativity of G/H_{n-1} implies that $H_{n-1} g_1 g_2 = H_{n-1} g_2 g_1$, whence $g_1 g_2 g_1^{-1} g_2^{-1} \in H_{n-1}$. Now,

$$\bar{v}(g_1 g_2 A) = v(H_{n-1} \cap g_1 g_2 A) = v(H_{n-1} \cap g_1 g_2 g_1^{-1} g_2^{-1} g_2 g_1 A)$$
$$= v(g_1 g_2 g_1^{-1} g_2^{-1}(H_{n-1} \cap g_2 g_1 A)) = v(H_{n-1} \cap g_2 g_1 A) = \bar{v}(g_2 g_1 A).$$

We may now construct $\mu_{W,\varepsilon}$ in a manner similar to the proof of 10.4(b). Let $W = \{g_1, \ldots, g_m\}$ and choose N so that $2/N \leqslant \varepsilon$. Then define $\mu_{W,\varepsilon}(A)$ to be $(1/N^m) \sum \{\bar{v}(g_1^{i_1} \cdots g_m^{i_m} A) : 1 \leqslant i_1, \ldots, i_m \leqslant N\}$. This yields a finitely additive measure, and because $\bar{v}(g_i g_j A) = \bar{v}(g_j g_i A)$, $\mu_{W,\varepsilon}(g_k A)$ differs from $\mu_{W,\varepsilon}(A)$ by no more than

$$\frac{1}{N^m} \sum_{1 \leqslant i \leqslant m} \mu(g_1^{i_1} \cdots g_{k-1}^{i_{k-1}} g_{k+1}^{i_{k+1}} \cdots g_m^{i_m} A) + \mu(g_1^{i_1} \cdots g_k^{N+1} \cdots g_m^{i_m} A),$$

which is at most $2N^{m-1}/N^m = 2/N \leqslant \varepsilon$. This completes the proof. $\square$

The previous result can be used to obtain a Choiceless version of the Invariant Extension Theorem for solvable groups.

Theorem 13.8. *Suppose G is a solvable group of automorphisms of a countable Boolean algebra $\mathscr{A}$, and suppose $\mu_0 : \mathscr{A}_0 \to [0, \infty]$ is a G-invariant measure on $\mathscr{A}_0$, a G-invariant subalgebra of $\mathscr{A}$. Then μ_0 has a G-invariant extension to $\mathscr{A}$.*

Proof. Because $\mathscr{A}$ is countable, we may use the Measure Extension Theorem (see the remarks following its proof) to extend μ_0 to a measure $\bar{\mu} : \mathscr{A} \to [0, \infty]$. Now, for $a \in \mathscr{A}$ and rationals r, s, let $S_{a,r,s} = \{g \in G : r \leqslant \bar{\mu}(g^{-1}a) < s\}$, and let $\mathscr{C}$ be the subalgebra of $\mathscr{P}(G)$ generated by these countably many sets $S_{a,r,s}$. Then $\mathscr{C}$ is countable, and because $hS_{a,r,s} = S_{ha,r,s}$, $\mathscr{C}$ is G-invariant. By Theorem 13.7, then, there is a left-invariant measure $\nu : \mathscr{C} \to [0, 1]$.

Now, for any $a \in \mathscr{A}$ let $f_a : G \to [0, \infty]$ be defined by $f_a(g) = \bar{\mu}(g^{-1}a)$. By the way $\mathscr{C}$ was defined, f_a is a $\mathscr{C}$-measurable function and so we may define $\mu(a)$ to be $\int f_a \, d\nu$ if f_a is bounded and let $\mu(a) = \infty$ otherwise. It is easy to check that μ is a measure on $\mathscr{A}$, μ extends μ_0, and μ is G-invariant. $\square$

By applying the preceding theorem to the (countable) algebra generated by the pieces of a purported paradoxical decomposition, one obtains a proof in ZF that a square is not paradoxical.

Corollary 13.9. *If G is a solvable group of isometries of $\mathbf{R}^n$ and A, B are two measurable sets that are G-equidecomposable, then $\lambda(A) = \lambda(B)$. In particular, a square in the plane is not G_2-paradoxical.*

Proof. Suppose the equidecomposability of A and B uses m pieces and is witnessed by $A = \bigcup A_i$ and $B = \bigcup g_i(A_i)$. Let H be the subgroup of G generated by $\{g_i : 1 \leqslant i \leqslant m\}$ (H is countable and solvable), and let $\mathscr{A}$ be the H-invariant subalgebra of $\mathscr{P}(G)$ generated by the $2m$ sets $A_i, g_i(A_i)$. Then $\mathscr{A}$ contains A, B, and all the pieces of the decomposition. By Theorem 13.8, Lebesgue measure on the H-invariant subalgebra $\mathscr{A} \cap \mathscr{L}$ may be extended to a measure on all of $\mathscr{A}$. Since A and B are H-equidecomposable in $\mathscr{A}$, this extension must assign them the same measure, and therefore $\lambda(A) = \lambda(B)$. $\square$

The preceding proof also yields that $\mathbf{R}^n$ is not G-paradoxical if G is any solvable group acting on $\mathbf{R}^n$. If a paradox exists, let $\mathscr{A}$ be the H-invariant algebra generated by the pieces and extend the two-valued measure on $\{\varnothing, \mathbf{R}^n\}$ to an H-invariant measure on $\mathscr{A}$.

FOUNDATIONAL IMPLICATIONS OF THE BANACH-TARSKI PARADOX

What are the implications of the Banach-Tarski Paradox for the axiomatic basis of set theory? Does the paradox, so clearly false in physical reality, mean that the Axiom of Choice yields unreliable results and should be discarded? Many critics of AC have buttressed their arguments by citing the Banach-Tarski Paradox. For example, E. Borel [22, p. 210], using probability theory to make his point, argued as follows. "Hence we arrive at the conclusion that the use of the axiom of choice and a standard application of the calculus of probabilities to the sets* A, B, C, which this axiom allows to be defined, leads to a contradiction: therefore the axiom of choice must be rejected." Before turning to a defense of AC, we point out how the technical results of this chapter bear upon the issue.

To those who feel that the Banach-Tarski Paradox is absurd, it seems evident that the Axiom of Choice is the culprit. The whole situation would have to be considered in an entirely different light if somehow AC could be eliminated from the proof and the paradox turned out to be a theorem of ZF. That this cannot happen is precisely the content of Corollary 13.3. Thus it is fair to say that AC is indeed to blame. The rejection of AC does yield a system in which no proof of a result that contradicts naive expectations about the measure of sets—such as the Banach-Tarski Paradox—is possible.

In their original paper [11], Banach and Tarski anticipated the controversy that their counterintuitive result would spawn, and they analyzed their use of the Axiom of Choice as follows. "It seems to us that the role played by the axiom of choice in our reasoning deserves attention. Indeed, consider the following two theorems, which are consequences of our research:

I. Any two polyhedra are equivalent by finite decomposition.
II. Two different polygons, with one contained in the other, are never equivalent by finite decomposition.

Now, it is not known how to prove either of these theorems without appealing to the Axiom of Choice: neither the first, which seems perhaps paradoxical, nor the second, which agrees fully with intuition. Moreover, upon

---

* The sets A, B, and C here denote sets that are simultaneously a half and a third of a sphere, that is, the pieces of a Hausdorff decomposition of S^2.

analysing their proofs, one could state that the Axiom of Choice occurs in the proof of the first theorem in a more limited way than in the proof of the second."

Thus Banach and Tarski pointed out that if AC is discarded, then not only would their paradox be lost, but also the result that such paradoxes do not exist in the plane. The last sentence of the excerpt refers to the fact that statement II uses choices from a larger family of sets than does statement I. But Corollary 13.9, which was proved by A. P. Morse [152] in 1949, shows that statement II is provable in ZF, and is therefore not relevant to a discussion of AC. The exact role of AC can therefore be loosely summarized as follows. It is necessary to disprove the existence of various invariant measures on $\mathscr{P}(\mathbf{R}^n)$ and to construct such measures, but it is not necessary to disprove the existence of paradoxes.

Despite the Banach-Tarski Paradox and other objections to the Axiom of Choice, often focusing on its nonconstructive nature, the great majority of contemporary mathematicians fully accept the use of AC. It is generally understood that nonmeasurable sets (either of the Vitali type or of the sort that arise in the Banach-Tarski Paradox) lead to curious situations that contradict physical reality, but the mathematics that is brought to bear on physical problems is almost always mathematics that takes place entirely in the domain of measurable sets. And even in this restricted domain of sets, the Axiom of Choice is useful, and ZF + AC provides a more coherent foundation than does ZF alone.

Perhaps the most important use of AC in the domain of measurable sets has to do with Lebesgue measure. As pointed out earlier in this chapter, it is consistent with ZF that Lebesgue measure fails to be countably additive; some form of Choice is needed to make Lebesgue measure the useful concept that it is. While either DC or the Axiom of Countable Choice suffices, these are somewhat unnatural weakenings of AC. If some nonconstructive choice is allowed, why not permit all nonconstructive choice? One fear of this liberal attitude, that unlimited choice will lead to a contradiction, is unfounded because of Gödel's famous theorem that ZF + AC is noncontradictory. Of course, choosing AC over DC means that duplications of the sphere are possible, but this is simply irrelevant to the mathematics of the Lebesgue measurable sets.

It is worth noting that the fact that Choice creeps into the basic results on Lebesgue measure is a subtle one, and it went unnoticed by the early practitioners in the field, some of whom (Lebesgue and Borel) went on to become strong critics of the use of nonconstructive choice. Moore [151] points out that Lebesgue's "work reveals how a mathematician of the first rank may subtly fail to see that he is fundamentally violating his philosophical scruples in his own work."

The preceding remarks should not be construed as implying that the countable additivity of Lebesgue measure is more self-evident than other alternatives. The point is that Lebesgue measure, in its usual form, has become a valuable tool of modern mathematics, and the use of spaces of Lebesgue integrable functions provides a simpler and clearer approach to topics of classical analysis, for example, Fourier series. An analogous problem arises with the question of the existence of infinite sets. Suppose all mathematicians and scientists become firmly convinced that all aspects of our universe are finite, that a true continuum of points simply does not exist. This does not mean that infinite sets or real numbers must be discarded as mathematical tools. It may well be that calculus is best understood and taught in the context of the real numbers, whether or not the set **R** has physical existence. The case for AC is similar: ZF + AC is a natural and simple foundation for mathematics that is rich enough to provide mathematicians and scientists with the tools that have proved most useful in practice.

Despite the general acceptance of AC, it is recognized that this axiom has a different character than the others. Because of the nonconstructivity it introduces, AC is avoided when possible, and proofs in ZF are considered more basic (and are often more informative) than proofs using AC. A purist might argue that once accepted as an axiom, AC should be accorded the same status as the other axioms: a statement that is proved in ZF + AC is just as mathematically true as one proved in ZF. But there are sometimes beneficial side effects of a discriminating attitude toward the use of AC. For instance, consider the Schröder-Bernstein Theorem for cardinality, which is much easier to prove in ZFC than in ZF (see remarks following Corollary 8.10). But the ZF-proof, essentially presented here in Theorem 3.5, is much more valuable because it generalizes in a way that the ZFC-proof does not. In particular, one obtains the Banach-Schröder-Bernstein Theorem for the equidecomposability relation in an arbitrary countably complete Boolean algebra. On the other hand, the Cancellation Law for equidecomposability (using arbitrary pieces) uses the Axiom of Choice (despite the fact that the corresponding result for cardinality avoids AC), and therefore it is not clear that it can be generalized to more general equidecomposability, such as the case where the pieces are Borel sets.

The interplay between ZF and ZF + AC works the other way as well. For instance, the ZF-proof that a square is not paradoxical (Corollary 13.9) was found by closely analyzing the original ZFC-proof. If a more restrictive attitude toward AC had prevailed, the original ZFC-proof might not have been discovered so soon, and the proof (in ZF) that squares are not paradoxical might have been delayed.

Despite how often it is cited in discussions of the Axiom of Choice, the Banach-Tarski Paradox is more important to pure mathematics than it is to foundational questions. The role of the Banach-Tarski Paradox in discussions about AC is, for the most part, the same as the role of nonmeasurable sets; indeed, the paradox of the sphere simply reinforced the views of those whose ideas were shaped by the question of nonmeasurable subsets of the real line.

But, as we have emphasized throughout this book, the Banach-Tarski Paradox has spawned some valuable mathematical ideas (principally, amenability in groups). Thus the Banach-Tarski Paradox has a historical importance for mathematics that is completely independent of foundational questions about the axioms of set theory.

NOTES

See [99] for an exposition of several fundamental consistency results such as Con(ZF + AC), Con(ZF + $\neg$AC), Con(ZFC + $\neg$IC), Con(ZF + **R** is a countable union of countable sets), and a proof of Solovay's theorem: Con(ZF + IC) implies Con(ZF + DC + LM + PB). More technical consistency results concerning the interplay between AC, its variants, and certain mathematical statements (such as those in Figure 13.1) may be found in [98]. For a discussion of the role of DC and the Axiom of Countable Choice in the theory of measure and category see [151]. This topic is also discussed in Moore's book [150], which provides a valuable historical account of many of the controversies concerning AC.

Solovay's theorem (the forward direction of 13.1 and the result of 13.4 using Con(ZF + IC)) was proved in 1964; complete details were published in [224]. The problem of using forcing to prove the consistency of LM was suggested by Paul Cohen, the discoverer of forcing. It is noteworthy that Solovay first proved, assuming only Con(ZF), Con(ZF + LM) (see [150, p. 304]), but DC failed in his model of ZF + LM. Thus he persisted to get Con(ZF + DC + LM), which, however, required the consistency of Con(ZF + IC). Shelah's refinements of Solovay's results (the reverse direction of 13.1 and Theorem 13.4 without IC) were proved in 1980 [206]. His proof that Con(ZF + DC + LM) implies Con(ZF + IC) was improved by Raisonnier [188] (see also [187]). It is worth noting that the full strength of LM is not required. Shelah proved that Con(ZF + DC + "Every Σ_3^1 set of reals is Lebesgue measurable") yields that $\aleph_1$ is inaccessible in L (which yields Con(ZFC + IC)). Moreover, Raisonnier showed that Con(ZF + DC + "Every Σ_2^1 set is Lebesgue measurable and every Σ_3^1 set has the Property of Baire") also yields the inaccessibility of $\aleph_1$ in L. The latter result is noteworthy because Con(ZF + DC + PB) can be proved assuming only Con(ZF). This is Theorem 13.4, the details of which have not yet been published.

Theorem 13.2, which is important for us because it yields 13.3, was proved by Solovay in 1964. A proof appeared in [199], while a proof of a similar result, which contains arguments very similar to those in the sketch of Theorem 13.2's proof given here, appears in [184]. Originally Solovay proved only Con(ZF + DC + GM$_1$), where GM$_n$ denotes the assertion of GM for **R**^{n}

only. But, as he pointed out to the author, only a slight modification is needed to get that each GM_n holds in the model N discussed after Corollary 13.3.

Corollary 13.6 was known to Solovay [224, p. 3]. A proof of Theorem 13.5 along a somewhat different line than the proof presented here (which is due to A. Taylor) is presented by Pincus [183].

Corollary 13.9 is due to A. P. Morse [152]. The proof presented here was inspired by an argument of Mycielski [168, Thm. 3.6], who showed that the conclusion of 13.7 is valid for countable groups satisfying Følner's Condition.

Appendix A

Euclidean Transformation Groups

The types of transformations that are used to produce paradoxes in Euclidean spaces and on spheres are usually the Euclidean isometries, but occasionally more general affine maps arise. Since the affine group is useful in studying and classifying isometries, we summarize the relevant facts about affine transformations. The book by Hausner [92] is a good reference for a more detailed presentation.

Definition A.1. *A bijection* $f : \mathbf{R}^n \to \mathbf{R}^n$ *is called* affine *if for all* P, $Q \in \mathbf{R}^n$ *and reals* α, β *with* $\alpha + \beta = 1$, $f(\alpha P + \beta Q) = \alpha f(P) + \beta f(Q)$. *The affine transformations of* $\mathbf{R}^n$ *form a group, which is denoted by* A_n.

Geometrically, a bijection is affine if and only if it carries lines to lines and preserves the ratio of distances along a line. Any nonsingular linear transformation is affine, since a linear transformation satisfies Definition A.1 for all α, β, not just pairs summing to one. The group of nonsingular linear transformations of $\mathbf{R}^n$ is denoted by GL_n (general linear group). Linear maps leave the origin fixed, but affine maps need not do so; all translations of $\mathbf{R}^n$ are affine. Let T_n denote the group of translations of $\mathbf{R}^n$. T_n is isomorphic to the additive group of $\mathbf{R}^n$ because composition of translations corresponds to addition of the translation vectors. It is an extremely useful fact that every affine map has a canonical representation in terms of linear maps and translations.

Theorem A.2. *If* $f \in A_n$, *then there are uniquely determined maps* $\tau \in T_n$ *and* $\ell \in GL_n$ *such that* $f = \tau\ell$. *Moreover, the map* $\pi : A_n \rightarrow GL_n$ *defined by* $\pi(f) = \ell$ *is a group homomorphism with kernel* T_n. *Hence* T_n *is a normal subgroup of* A_n *and* $A_n/T_n \cong GL_n$.

To prove the first part of this theorem one shows τf is linear, where τ is the translation by $-f(0)$; the rest follows in a straightforward manner. (For details, see [92, Ch. 8].) A useful consequence of this theorem and the fact that a member of GL_n is determined by its values at n linearly independent vectors, is that an affine map is completely determined by its values at $n + 1$ points, no three of which are collinear.

Theorem A.2 allows us to define the determinant of an affine map to be the determinant of the corresponding linear map, that is, $\det f = \det \pi(f)$. Let SL_n (special linear group) denote the group of linear transformations of determinant 1; SA_n the group of affine maps of determinant 1. Note that $T_n \subseteq SA_n$.

Definition A.3. *An* isometry *of* $\mathbf{R}^n$ *(or of any metric space) is a distance-preserving bijection of* $\mathbf{R}^n$ *to itself. Let* G_n *denote the n-dimensional isometry group.*

We are using the word isometry in a global sense. Occasionally we refer to *partial isometries*: bijections from a subset A to a subset B that preserve distance. In Euclidean space, and on spheres, any partial isometry may be extended to an isometry.

Theorem A.4. (See [92, §9.4].) *Every isometry of* $\mathbf{R}^n$ *is affine, that is,* $G_n \subseteq A_n$.

The group of isometries of $\mathbf{R}^n$ that are also linear transformations is quite important, because it is the group of isometries of the unit sphere, S^{n-1}. Such linear transformations are characterized by their representation by *orthogonal matrices*, matrices A for which $A^T = A^{-1}$ (A^T denotes the transpose of A). Hence O_n is used to denote this group; that is, $O_n = G_n \cap GL_n$, and O_n is called the *orthogonal group*. The transformations in O_n are exactly those linear transformations of $\mathbf{R}^n$ that preserve the dot product, that is, $\ell(P) \cdot \ell(Q) = P \cdot Q$. Since $\det A = \det A^T$, and if A is orthogonal, $AA^T = I$, it follows that if $\ell \in O_n$, then $(\det \ell)^2 = 1$. This yields the following theorem.

Theorem A.5. *If* $\ell \in O_n$, *then* $\det \ell = \pm 1$.

The homomorphism π takes the isometry group G_n to O_n, and since $T_n \subseteq G_n$, it follows that T_n is a normal subgroup of G_n and $G_n/T_n \cong O_n$. Thus every isometry is an orthogonal transformation followed by a translation. We use SO_n (special orthogonal) to denote the orthogonal transformations having determinant $+1$; these are the orientation-preserving orthogonal transformations. By analogy with $\mathbf{R}^2$, transformations in SO_n are sometimes called rotations, and SO_n may be called the rotation group of the sphere S^{n-1}. It is sometimes useful to have a notation for the orientation-preserving isometries of $\mathbf{R}^n$; thus let $SG_n = \{\sigma \in G_n : \det \sigma = +1\}$.

Since for any $f \in A_n$, $|\det f|$ is the factor by which f changes area (or Lebesgue measure; see [92, §9.2] or [256, §6.3]), Theorem A.5 implies that distance-preserving maps preserve area as well. Thus G_n and SA_n contain only measure-preserving transformations. Because det is a homomorphism, O_n/SO_n and G_n/SG_n are each isomorphic to $\mathbf{Z}_2$. This allows us to extend results proved for certain maps of determinant $+1$ to maps of the same sort, but with determinant ± 1.

A useful consequence of the preceding representation of G_n is that G_n is isomorphic to a subgroup of SL_{n+1}. If $\sigma \in G_n$ has the form $\tau \ell$ where τ is the translation by $(v_1, \ldots, v_n)$ and ℓ is represented by the orthogonal matrix (a_{ij}), then let $M(\sigma)$ be the $(n+1) \times (n+1)$ matrix:

$$\begin{bmatrix} & & & v_1 \\ & & & v_2 \\ & a_{ij} & & \vdots \\ & & & v_n \\ 0 & 0 \cdots 0 & & 1 \end{bmatrix}.$$

The mapping M is an isomorphism of G_n with a subgroup of SL_{n+1}.

In low dimensions we can be more explicit about these groups; proofs may be found in [92, Ch. 9].

In $\mathbf{R}^1$ the only orthogonal maps are the identity, I, and the flip taking x to $-x$. Hence $SO_1 = \{I\}$ and $O_1 = \{I, x \mapsto -x\}$. By A.2 and A.4 it follows that $G_1 = \{x \mapsto a + x : a \in \mathbf{R}\}$. Since $G_1/T_1 \cong O_1 \cong \mathbf{Z}_2$ and $T_1 \cong \mathbf{R}$, the normal series $\{I\} \lhd T_1 \lhd G_1$ shows that G_1 is solvable. Since A_1 consists of transformations of the form $x \mapsto ax + b$, it follows that the length-preserving affine maps (those with $|a| = 1$) are just the isometries.

Let ρ_θ denote the counterclockwise rotation about the origin in $\mathbf{R}^2$ through θ radians; ρ_θ is represented by $\begin{bmatrix} \cos \theta & -\sin \theta \\ \sin \theta & \cos \theta \end{bmatrix}$. Let ϕ_θ denote the reflection in the line through the origin making an angle θ with the x-axis. Then $SO_2 = \{\rho_\theta : 0 \leqslant \theta < 2\pi\}$, which is Abelian and isomorphic to the circle group (the set of complex numbers of unit modulus under multiplication), and $O_2 = SO_2 \cup \{\phi_\theta : 0 \leqslant \theta < \pi\}$. To see that G_2 is solvable, consider the sequence $\{I\} \lhd T_2 \lhd SG_2 \lhd G_2$. Because T_2 is the kernel of $\pi : SG_2 \to SO_2$, SG_2/T_2 is isomorphic to the Abelian group SO_2. Since $G_2/SG_2 \cong \mathbf{Z}_2$, this sequence witnesses the solvability of G_2.

A planar isometry, ψ, may be written uniquely as $\tau \ell$ with $\tau \in T_2$ and $\ell \in O_2$, and ψ can be characterized according to the choice of τ and ℓ:

(i) If $\ell = 1$, then ψ is a translation.

(ii) If $\ell = \rho_\theta$, then ψ is a counterclockwise rotation through θ radians about some point in $\mathbf{R}^2$.

(iii) If $\ell = \phi_\theta$, then ψ is a *glide reflection*, that is, a reflection in some line followed by a translation in the direction parallel to the line of reflection. If the component of τ in that direction is the identity (i.e., τ is a translation perpendicular to the line of reflection), then ψ is simply a reflection.

If we let $SL_2(\mathbf{Z})$ denote the subgroup of SL_2 that carries $\mathbf{Z} \times \mathbf{Z}$ to itself (matrices with coefficients in $\mathbf{Z}$), then there are elements of $SL_2(\mathbf{Z})$ that are not orthogonal. For example, $\begin{bmatrix} 1 & 1 \\ 0 & 1 \end{bmatrix} \in SL_2(\mathbf{Z})\backslash O_2$. Thus, unlike $\mathbf{R}^1$, the area-preserving affine transformations consist of more than just the isometries. In fact, G_2 is solvable, while $SL_2(\mathbf{Z})$ has a free non-Abelian subgroup (see Notes to Chapter 2 and Chapter 7).

A solvable group cannot contain a free group of rank 2, since it follows from the characterization of solvability in terms of the finiteness of the series of commutator subgroups that a solvable group universally satisfies a nontrivial relation $w = 1$, where w is a nontrivial reduced word in the variables x, x^{-1}, y, y^{-1}. But we can be more explicit for G_1 and G_2. Since $\sigma^2 \in T_1$ for any $\sigma \in G_1$, the equation $x^2 y^2 x^{-2} y^{-2} = 1$ is satisfied by all pairs x, y in G_1. In G_2, the normal series $\{I\} \lhd T_2 \lhd SG_2 \lhd G_2$ yields that $\sigma^2 \rho^2 \sigma^{-2} \rho^{-2}$ lies in T_2 for any $\sigma, \rho \in G_2$. Hence $\sigma^2 \rho^2 \sigma^{-2} \rho^{-2}$ commutes with $\sigma^{-2} \rho^{-2} \sigma^2 \rho^2$, yielding a nontrivial relation in G_2. This shows directly why neither G_1 nor G_2 contains a pair of independent elements.

A complete classification of isometries gets more complicated as the dimension increases. In $\mathbf{R}^3$, we do not need such a classification, but we do require more information about SO_3. Using the fact that any element of SO_3 has $+1$ as an eigenvalue, it can be shown (see [92]) that SO_3 consists of rotations about a line (axis) through the origin. Hence the composition of two rotations of $\mathbf{R}^3$ that fix the origin is another such rotation. Moreover, using the fact that orthogonal transformations preserve angles, it follows that if $\rho, \sigma \in SO_3$, then $\rho\sigma\rho^{-1}$ is a rotation whose rotation angle is the same as that of σ. Unlike SO_1 and SO_2, SO_3 is not solvable, because it contains a free subgroup of rank 2 (see Theorem 2.1).

The matrix representation of an element of SO_3 with respect to an orthonormal basis whose first element is the axis is simply

$$\begin{bmatrix} 1 & 0 & 0 \\ 0 & \cos\theta & -\sin\theta \\ 0 & \sin\theta & \cos\theta \end{bmatrix}$$

where θ is the angle of rotation. In fact, one can compute the axis and the angle of a rotation from its matrix representation with respect to any orthonormal basis, as the following theorem shows.

Theorem A.6.   *Let (a_{ij}) be the matrix of a rotation ρ of $\mathbf{R}^3$ through an axis containing $\mathbf{0}$. Let θ be ρ's angle of rotation $(0 \leqslant \theta < 2\pi)$ and let $\vec{A}$ be a unit vector along ρ's axis so oriented that the rotation obeys the right-hand rule. Then $2\vec{A}\sin\theta = (a_{32} - a_{23}, a_{13} - a_{31}, a_{21} - a_{12})$.*

Proof.   Let $\vec{A} = (b_1, b_2, b_3)$ and ρ_1 be the rotation represented by

$$\begin{bmatrix} b_1 & 0 & b \\ b_2 & -b_3/b & -b_1 b_2/b \\ b_3 & b_2/b & -b_1 b_3/b \end{bmatrix}$$

where $b = \sqrt{b_2^2 + b_3^2}$; ρ_1 takes $(1, 0, 0)$ to (b_1, b_2, b_3). Then

$$\rho = \rho_1 \begin{bmatrix} 1 & 0 & 0 \\ 0 & \cos\theta & -\sin\theta \\ 0 & \sin\theta & \cos\theta \end{bmatrix} \rho_1^{-1},$$

and a computation, using the fact that ρ_1 is orthogonal and hence $\rho_1^{-1} = \rho_1^T$, shows that the right-hand side of the desired equation is

$$\left(2(b_1 b_2^2 + b_1 b_3^2)\sin\theta/b^2, \; 2b_2\sin\theta, \; 2b_3\sin\theta\right),$$

which equals $2\vec{A}\sin\theta$ as required.   □

One important fact about orientation-preserving isometries of $\mathbf{R}^3$ is that it is easy to recognize when there is a fixed point. For if $\sigma \in SG_3$ has the representation $\tau\ell$ where τ is the translation by a vector $\vec{V}$ and ℓ is a rotation about the axis $\vec{A}$, then σ has a fixed point in $\mathbf{R}^3$ if and only if $\vec{V}$ is perpendicular to $\vec{A}$.

Finally, we note that the groups considered here contain the corresponding groups in lower dimensions. If $m < n$, then simply decompose $\mathbf{R}^n$ into $\mathbf{R}^m \times \mathbf{R}^{n-m}$ and extend the transformations of $\mathbf{R}^m$ by letting them be the identity in the $n - m$ new coordinates.

Appendix B

Jordan Measure

This appendix contains some of the basic facts about Jordan measure, which is a more elementary notion than Lebesgue measure. Lebesgue measure uses countable collections of intervals (or cubes) to cover a set, but Jordan measure uses only finite collections of intervals.

Definition B.1. *Let A be a bounded subset of $\mathbf{R}^n$. Define $v^*(A)$ to be $\inf\{\sum_{i=1}^n \text{ volume } (K_i) : \{K_i\}$ is a pairwise interior-disjoint collection of finitely many cubes such that $A \subseteq \bigcup K_i\}$ and $v_*(A)$ to be $\sup\{\sum_{i=1}^n \text{ volume } (K_i) : \{K_i\}$ is a pairwise interior-disjoint collection of finitely many cubes such that $\bigcup K_i \subseteq A\}$. If $v^*(A) = v_*(A)$, then A is called* Jordan measurable *and this common value, the Jordan measure of A, is denoted by $v(A)$. The collection of bounded Jordan measurable sets is denoted by $\mathscr{J}$.*

Some authors use rectangles rather than cubes in the definition of v^* and v_* but since any rectangle may be approximated arbitrarily closely by finitely many cubes, the two definitions are equivalent. Since

$$v_*(A) \leqslant \lambda_*(A) \leqslant \lambda^*(A) \leqslant v^*(A),$$

where λ^*, λ_* denote Lebesgue outer and inner measure, respectively, $\mathscr{J} \subseteq \mathscr{L}$, the collection of Lebesgue measurable sets, and λ agrees with v on the sets in $\mathscr{J}$.

But $\mathcal{J}$ is much smaller than $\mathcal{L}$: if A is the set of rationals in $[0, 1]$, then $v_*(A) = 0$ and $v^*(A) = 1$. Or, choose an open set E containing A with $\lambda(E) < 1$; then $v_*(E) < 1 = v^*(E)$, so $\mathcal{J}$ does not even contain all open sets. The following characterization is central to the study of Jordan measurable sets; for a proof see [177, §1305]. The boundary of a set A $(\text{i.e., } \bar{A} \setminus \text{int}(A))$ is denoted by ∂A.

Theorem B.2. *A bounded subset A of $\mathbf{R}^n$ is in $\mathcal{J}$ if and only if* $v(\partial A) = 0$.

One consequence of this characterization is that if $A, B \in \mathcal{J}$, then $A \cup B$, $A \setminus B \in \mathcal{J}$; that is, $\mathcal{J}$ is a subring of $\mathcal{P}(\mathbf{R}^n)$. Moreover, if $A \in \mathcal{J}$, then $\bar{A} \setminus A$, which is contained in ∂A, is nowhere dense; therefore A differs from $\bar{A}$ by a meager set, which implies that $A \in \mathcal{B}$, the collection of sets with the Property of Baire. So $\mathcal{J} \subseteq \mathcal{L} \cap \mathcal{b}$. It was pointed out before that $\mathcal{J}$ does not contain the Borel sets; but neither is $\mathcal{J}$ contained in the Borel sets. If C is the Cantor subset of $[0, 1]$, then the usual proof that $\lambda(C) = 0$ shows that, in fact, $v(C) = 0$. It follows that all subsets of C lie in $\mathcal{J}$; but $|C| = 2^{\aleph_0}$ so the number of subsets of C is greater than $2^{\aleph_0}$, the number of Borel sets.

From the definition of v_* it follows that a set $A \in \mathcal{J}$ has Jordan measure zero if and only if A has no interior. Since for $A \in \mathcal{J}$, $v(A) = v(\bar{A})$, this yields that if $A \in \mathcal{J}$, then $v(A) = 0$ if and only if $\lambda(A) = 0$ if and only if A is nowhere dense if and only if A is meager. Thus v is unbiased as regards Lebesgue measure and category. A set in $\mathcal{J}$ has Jordan measure zero if either the set has Lebesgue measure zero or the set is meager.

See Proposition 9.6 for a result about Jordan measure's uniqueness. The historical importance of Jordan measure stems from its connection with the Riemann integral. A function is Jordan measurable if and only if it is Riemann integrable (if and only if the set of points at which it is discontinuous has Lebesgue measure zero).

Appendix C

Unsolved Problems

The following is a list of unsolved problems in the area of paradoxical decompositions, equidecomposability, and finitely additive measures. The order represents the author's view as to their interest and importance.

1. Marczewski's Problem, Circa 1930 (3.12, 9.9)

Is there a finitely additive, isometry-invariant measure on the Borel sets in S^n, $n \geq 2$ (or $\mathbf{R}^n$, $n \geq 3$), that has total measure one (or, normalizes the unit cube) and vanishes on meager sets? Such a measure cannot be countably additive (9.15), and so by 13.5, it is consistent with ZF + DC that no such finitely additive Borel measure exists. An equivalent problem in the case of $\mathbf{R}^3$ is: Is the unit cube paradoxical using pieces that have the Property of Baire?

2. Tarski's Circle-Squaring Problem, 1924 (7.5)

Is a circle (with interior) in the plane equidecomposable to a square (necessarily of the same area)*?

Variations

(a) (p. 102) Can a negative solution be obtained if the pieces are restricted to the Borel sets? It is known that restriction to pieces

* See note on page 101.

that are parts of Jordan curves or two-cells (interior of a Jordan curve) yields a negative solution.

(b) (3.14) Is a regular tetrahedron in $\mathbf{R}^3$ equidecomposable to a cube using measurable pieces? A restriction to polyhedral pieces yields a negative solution (Hilbert's Third Problem).

(c) Is $\left(\frac{1}{3}, \frac{2}{3}\right) \cup \left(\frac{7}{9}, \frac{8}{9}\right) \cup \left(\frac{25}{27}, \frac{26}{27}\right) \cup \cdots$ equidecomposable to $(0, 1/2)$? (p. 119)

3. Mycielski's Problem on Geometric Bodies, 1977 (8.15)

Is volume a complete invariant for equidecomposability in the Boolean ring of geometric bodies in $\mathbf{R}^n$? That is, if $v(A) = v(B)$ for $A, B \in \mathscr{R}_v$, then are A and B equidecomposable in $\mathscr{R}_v$? This is unsolved even for $\mathbf{R}^1$.

4. Cohen's Conjecture, 1982 (12.27)

A nonamenable finitely presented group has a free subgroup of rank 2; that is, the classes AG and NF coincide when restricted to finitely presented groups. It is known that for groups in general, AG is properly contained in NF.

5. Rosenblatt's Problem, 1981 (11.7)

Suppose $(X, \mathscr{A}, m)$ is a nonatomic measure space with $m(X) = 1$, and G acts on X in a measure-preserving way. If G is amenable, is there an exotic measure on $\mathscr{A}$? Does amenability guarantee the existence of a G-invariant mean on $L^\infty(X)$ that is not equal to the m-integral? This problem has been solved if G is assumed to be countable.

6. Greenleaf's Problem, 1969 (11.26)

Suppose a nonamenable group G acts on a set X. Find conditions on the action that are necessary and sufficient for there to be no finitely additive, G-invariant measure $\mathscr{P}(X)$ having total measure one.

7. Ruziewicz's Problem for Borel Sets (11.13)

Is Lebesgue measure the only finitely additive, isometry-invariant measure on the Borel subsets of S^n, $n \geq 2$ (or $\mathbf{R}^n$, $n \geq 3$), that has total measure one (or, normalizes the unit cube)? Any other measure must fail to be absolutely continuous with respect to Lebesgue measure. A measure as in Problem 1 would provide a negative answer to this problem.

8. Ruziewicz's Problem for Subgroups of O_{n+1} (11.12)

Does the nonamenability of a subgroup G of O_{n+1} guarantee the uniqueness of the Lebesgue integral as a G-invariant mean on $L^\infty(S^n)$? This is known if $G = O_{n+1}$.

9. The Banach-Ulam Conjecture, 1935 (3.13)

A compact metric space is not paradoxical with respect to congruences (i.e., isometries whose domain is a possibly proper subset of the space), using Borel sets as pieces. Slightly stronger: A compact metric space carries a finitely additive, congruence-invariant Borel measure of total measure one. The latter assertion has been proved for countable metric spaces and for locally homogeneous metric spaces.

10. Chuaqui's Problem, 1982 (9.20)

Let G be a locally compact topological group, and let A and B be Borel subsets of G with nonempty interior and finite Haar measure. Are A and B countably equidecomposable using left multiplication and Borel sets as pieces? Equivalently, can a Borel set of Haar measure zero be packed into any nonempty open subset of G using left multiplication and countable many Borel pieces? This is unsolved even for Lie groups, although it is known in the case of $\mathbf{R}^n$, and in the general case provided measurable sets may be used as pieces.

11. Rosenblatt's Supramenability Conjecture, 1974 (12.9(b))

Is an amenable group with no free subsemigroup of rank 2 always supramenable? That is, is the containment $SG \subseteq NS \cap AG$ in fact an equality? A related question is whether the direct product of two supramenable groups is also a supramenable group.

12. Another Supramenability Problem (12.9(a))

Is a supramenable group necessarily exponentially bounded? That is, is the containment $EB \subseteq SG$ in fact an equality?

13. Amenability of Burnside Groups (10.5)

Is $B(2, 665)$ amenable? Does any $B(m, n)$ fail to be amenable or supramenable. A nonamenable $B(m, n)$ would lie in $NF \backslash AG$; an amenable but infinite $B(m, n)$ would lie in $AG \backslash EG$. In fact, an amenable infinite $B(m, n)$ would lie in $NS \cap AG \backslash EB$, and thus it would answer one of Problems 11 or 12, depending on whether or not it is supramenable.

14. Cancellation Law Problems (8.12, 10.12)

Is there an example of a group G acting on a set X and a proper G-invariant subalgebra $\mathscr{A}$ of $\mathscr{P}(X)$ such that the cancellation law fails for equidecomposability in $\mathscr{A}$? Does the cancellation law hold for Borel equidecomposability in a locally compact topological group? This is equivalent to Emerson's

Conjecture: A locally compact topological group is amenable if and only if it is not paradoxical using left multiplication and Borel pieces. Another unsolved special case is: Does the cancellation law hold for the equidecomposability relation in the ring of geometric bodies in $\mathbf{R}^n$?

15. Chuaqui's Conjecture (9.13)

Suppose G acts on X and I consists of all sets A such that infinitely many copies of A can be packed into X using countable G-equidecomposability. Suppose further that there is a countably additive measure on $\mathscr{P}(X)$ of total measure one that vanishes on I. Then there is a countably additive, G-invariant measure on $\mathscr{P}(X)$ of total measure one. This conjecture is an attempt to generalize Tarski's Theorem to countable equidecomposability.

16. De Groot's Problem, 1958 (3.15)

Can the Banach-Tarski duplication of a ball be carried out so that the pieces are moved continuously in $\mathbf{R}^3$ in a way that they never overlap?

17. Foundational Problems

(a) Show that ZF is not strong enough to prove that a subgroup of an amenable group is amenable. This is known (p. 213) under the extra assumption that the existence of an uncountable inaccessible cardinal is consistent with ZF.

(b) (Mycielski, 4.15) Show that it cannot be proved in ZF that the sphere S^2 splits into three congruent pieces. Or, show that S^2 cannot be split into three congruent sets, each of which is Lebesgue measurable.

18. Geometric Problems

(a) (12.13) Is there a bounded nonempty subset of the plane that is paradoxical? It is known that no such set can be paradoxical using just two pieces, although an unbounded set admitting such a two-piece decomposition does exist.

(b) (p. 48) Can a disc in the plane be split into three congruent pieces? It is known that a ball in $\mathbf{R}^n$ cannot be split into n (or fewer) congruent pieces.

(c) (p. 88) Is there a subset of the hyperbolic plane H^2 that is simultaneously a half of H^2 and a $2^{\aleph_0}$th part of H^2? It is known that there is a Borel set that is simulataneously a half and an $\aleph_0$th part of H^2; also, there is a set that is simultaneously a κth part of H^2, for all κ satisfying $3 \leqslant \kappa \leqslant 2^{\aleph_0}$.

19. Problems on the Existence of Free Subgroups

(a) (p. 58) Does SO_n $(n \geqslant 6)$ have a free subgroup of rank 2, some element of which has -1 as an eigenvalue? No such subgroup of SO_n exists if $n \leqslant 5$.

(b) (11.24) Does SO_4 have a subgroup G that has no locally commutative free subgroup of rank 2, but yet has the property that if H is a normal subgroup of G with G/H amenable, then no point on S^3 is fixed by all of H?

(c) (p. 98) Does SA_2 have a free subgroup of rank 2 that is locally commutative on $\mathbf{R}^2$? There is no such free subgroup whose action on $\mathbf{R}^2$ is fixed-point free.

References

1. Adams, J. F., On decompositions of the sphere, *J. London Maths. Soc.* **29** (1954), 96–99.

2. Adian, S. I., *The Burnside Problem and Identities in Groups*, Berlin: Springer-Verlag, 1978.

3. Agnew, R. P., and A. P. Morse, Extensions of linear functionals, with applications to limits, integrals, measures, and densities, *Ann. Math.* **39** (1938), 20–30.

4. Akemann, C. A., Operator algebras associated with Fuchsian groups, *Houston J. Math.* **7** (1981), 295–301.

5. Balcerzyk, S., and J. Mycielski, On the existence of free subgroups in topological groups, *Fund. Math.* **44** (1957), 303–308.

6. ———, On faithful representations of free products of groups, *Fund. Math.* **50** (1961), 63–71.

7. Balfour, M. *Magic Snake Shapes*, New York: Pocket Books, 1981.

8. Banach, S. Sur le problème de la mesure, *Fund. Math.* **4** (1923), 7–33. Reprinted in S. Banach, *Oeuvres*, vol. I, Warsaw: Éditions Scientifiques de Pologne, 1967.

9. ———, Un théorème sur les transformations biunivoques, *Fund. Math.* **6** (1924), 236–239. Reprinted, loc. cit.

10. Banach, S., and C. Kuratowski, Sur une gèneralization du problème de la mesure, *Fund. Math.* **14** (1929), 127–131. Reprinted, loc. cit.

11. Banach, S., and A. Tarski, Sur la decomposition des ensembles de points en parties respectivement congruents, *Fund. Math.* **6** (1924), 244–277. Reprinted, loc. cit.

12. Banach, S., and S. Ulam, Problème 34, *Coll. Math.* **1** (1948), 152–153.

13. Bass, H., The degree of polynomial growth of finitely generated nilpotent groups, *Proc. London Math. Soc.* **25** (1972), 603–614.

14. Beardon, A. F., *The Geometry of Discrete Groups*, New York: Springer-Verlag, 1983.

15. Belley, J.-M., and V. S. Prasad, A measure invariant under group endomorphisms, *Mathematika* **29** (1982), 116–118.

16. Bernstein, F., Untersuchungen aus der Mengenlehre, *Math. Ann.* **61** (1905), 117–155.

17. Boltianskii, V. G., *Hilbert's Third Problem*, trans. by R. Silverman, Washington, D.C.: Winston, 1978.

18. Bolzano, B., *Paradoxes of the Infinite*, trans. by D. Steele, London: Routledge and Kegan Paul, 1950. Originally, *Paradoxien des Unendlichen*, Leipzig: Reclam, 1851.

19. Bondy J. A., and U. S. R. Murty, *Graph Theory with Applications*, New York: American Elsevier, 1976.

20. Borel, A., On free subgroups of semi-simple groups, *Ens. Maths.* **29** (1983), 151–164.

21. Borel, É., *Leçons sur la Théorie des Fonctions*, 2d ed., Paris: Gauthier-Villars, 1914.

22. ———, *Les Paradoxes de l'Infini*, 3d ed., Paris: Gallimard, 1946.

23. Bottema, O., Orthogonal isomorphic representations of free groups, *Nieuw Arch. v. Wiskunde* (3) **5** (1957), 71–74.

24. Bożejko, M., Uniformly amenable discrete groups, *Math. Ann.* **251** (1980), 1–6.

25. Brenner, J. L., Quelques groupes libre de matrices, *C. R. Acad. Sci. Paris* **241** (1955), 1689–1691.

26. Cater, F. S., Problem 284, *Can. Math. Bull.* **24** (1981), 382–383.

27. Chang, B., S. A. Jennings, and R. Ree, On certain pairs of matrices which generate free groups, *Can. J. Math.* **10** (1958), 279–284.

28. Chen, S., On nonamenable groups, *Internat. J. Math. and Math. Sci.* **1** (1978), 529–532.

29. C. Chou, The exact cardinality of the set of means on a group, *Proc. Amer. Math. Soc.* **55** (1976), 103–106.

30. ———, Elementary amenable groups, *Ill. J. Math.* **24** (1980), 396–407.

31. Chuaqui, R. B., Cardinal algebras and measures invariant under equivalence relations, *Trans. Amer. Math. Soc.* **142** (1969), 61–79.

32. ———, The existence of an invariant countably additive measure and paradoxical decompositions, *Not. Amer. Math. Soc.* **20** (1973), A-636–637.

33. ———, Measures invariant under a group of transformations, *Pac. J. Math.* **68** (1977), 313–329.

34. ———, Simple cardinal algebras and their applications to invariant measures, preprint.

35. Cohen, J., Cogrowth and amenability of discrete groups, *J. Func. Anal.* **48** (1982), 301–309.

36. Cohn, L., *Measure Theory*, Boston: Birkhäuser, 1980.

37. Comfort, W., and S. Negrepontis, *The Theory of Ultrafilters*, Berlin: Springer-Verlag, 1974.

38. Connes, A., J. Feldman, and B. Weiss, An amenable equivalence relation is generated by a single transformation, *Ergod. Th. and Dynam. Sys.* **1** (1981), 431–450.

39. Connes, A., and B. Weiss, Property *T* and asymptotically invariant sequences, *Isr. J. Math.* **37** (1980), 209–210.

40. Dani, S. G., On invariant finitely additive measures for automorphism groups acting on tori, *Trans. Amer. Math. Soc.* (forthcoming).

41. Davies, R. O., and A. Ostaszewski, Denumerable compact metric spaces admit isometry-invariant finitely additive measures, *Mathematika* **26** (1979), 184–186.

42. Day, M. M., Amenable groups, *Bull. Amer. Math. Soc.* **55** (1949), 1054.

43. ———, Amenable semigroups, *Ill. J. Math.* **1** (1957), 509–544.

44. ———, Fixed-point theorems for compact convex sets, *Ill. J. Math.* **5** (1961), 585–590. Correction, *Ill. J. Math.* **8** (1964), 713.

45. ———, Convolutions, means and spectra, *Ill. J. Math.* **8** (1964), 100–111.

46. de Groot, J., Orthogonal isomorphic representations of free groups, *Can. J. Math.* **8** (1956), 256–262.

47. de Groot, J., and T. Dekker, Free subgroups of the orthogonal group, *Comp. Math.* **12** (1954), 134–136.

48. Dekker, T. J., Decompositions of sets and spaces, I, *Indag. Math.* **18** (1956), 581–589.

49. ———, Decompositions of sets and spaces, II, *Indag. Math.* **18** (1956), 590–595.

50. ———, Decompositions of sets and spaces, III, *Indag. Math.* **19** (1957), 104–107.

51. ———, *Paradoxical Decompositions of Sets and Spaces*, Academisch Proefschrift, Amsterdam: van Soest, 1958.

52. ———, On free groups of motions without fixed points, *Indag. Math.* **20** (1958), 348–353.

53. ———, On reflections in Euclidean spaces generating free products, *Nieuw Arch. v. Wiskunde* (3) **7** (1959), 57–60.

54. ———, On free products of cyclic rotation groups, *Can. J. Math.* **11** (1959), 67–69.

55. Dekker, T. J., and J. de Groot, Decompositions of a sphere, *Proc. Int. Math. Cong.* **2** (1954), 209.

56. ———, Decompositions of a sphere, *Fund. Math.* **43** (1956), 185–194.

57. Deligne, P., and D. Sullivan, Divison algebras and the Hausdorff-Banach-Tarski Paradox, *End. Math.* **29** (1983), 145–150.

58. del Junco, A., and J. Rosenblatt, Counterexamples in ergodic theory and number theory, *Math. Ann.* **245** (1979), 185–197.

59. Dixmier, J., Les moyennes invariantes dans les semigroupes et leur applications, *Acta Sci. Math. (Szeged)* **12** (1950), 213–227.

60. Dixon, J., Free subgroups of linear groups, *Lecture Notes in Math.* **319** (1973), Conference on group theory, Univ. of Wisconsin, Parkside, 1972, ed. Gatterdam, 45–55.

61. Drake, F. R., *Set Theory, An Introduction to Large Cardinals*, Amsterdam: North-Holland, 1974.

62. Dubins, L., M. Hirsch, and J. Karush, Scissor congruence, *Isr. J. Math.* **1** (1963), 239–247.

63. Dunford, N., and J. Schwartz, *Linear Operators, Part I*, New York: Wiley, Interscience, 1967.

64. Edelstein, M., On isometric complementary subsets of the unit ball, preprint.

65. Emch, A., Endlichgleiche Zerscheidung von Parallelotopen in gewöhnlichen und höhern Euklidischen Räumen, *Comm. Math. Helv.* **18** (1946), 224–231.

66. Emerson, W., Characterizations of amenable groups, *Trans. Amer. Math. Soc.* **241** (1978), 183–194.

67. ———, The Hausdorff Paradox for general group actions, *J. Func. Anal.* **32** (1979), 213–227.

68. Enderton, H. B., *Elements of Set Theory*, New York: Academic Press, 1977.

69. Erdös, P., and R. D. Mauldin, The nonexistence of certain invariant measures, *Proc. Amer. Math. Soc.* **59** (1976), 321–322.

70. Eves, H., *A Survey of Geometry*, vol. 1, Boston: Allyn and Bacon, 1963.

71. Følner, E., On groups with full Banach mean value, *Math. Scand.* **3** (1955), 243–254.

72. Frey, A. H., *Studies in Amenable Semigroups*, Ph.D. diss., Univ. of Washington, Seattle, 1960.

73. Freyhoffer, H., *Medidas en grupos topólogicos*, Notas Mathemáticas, Universidad Catolica de Chile, No. 8, 1978.

74. Galileo, *Dialogues Concerning Two New Sciences*, New York: Macmillan, 1914. Trans. of original text published in 1638.

75. Gerwien, P., Zerschneidung jeder beliebigen Anzahl von gleichen geradlinigen Figuren in dieselben Stücke, *J. für die Reine und Angew. Math. (Crelle's J.)* **10** (1833), 228–234.

76. ———, Zerschneidung jeder beliebigen Menge verschieden gestalteter Figuren von gleichem Inhalt auf der Kugelfläche in dieselben Stücke, *J. für die Reine und Angew. Math. (Crelle's J.)* **10** (1883), 235–240.

77. Goldberg, K., and M. Newman, Pairs of matrices of order two which generate free groups, *Ill. J. Math.* **1** (1957), 446–448.

78. Granirer, E., On amenable semigroups with a finite-dimensional set of invariant means, I, *Ill. J. Math.* **7** (1963), 32–48.

79. Greenberg, M. J., *Euclidean and Non-Euclidean Geometries*, 2d ed., San Francisco: Freeman, 1980.

80. Greenleaf, F. P., *Invariant Means on Topological Groups*, New York: van Nostrand, 1969.

81. Grigorchuk, R. I., Symmetric random walks on discrete groups, *Uspekhi Mat. Nauk* **32** No. 6 (1977), 217–218. (Russian).

82. Gromov, M., Groups of polynomial growth and expanding maps, with an appendix by J. Tits, *Inst. Hautes Etudes Sci. Publ. Math.* No. 53 (1981), 53–78.

83. Grünbaum, B., and G. C. Shephard, *Tilings and Patterns*, San Francisco: Freeman, forthcoming.

84. Gustin, W., Partitioning an interval into finitely many congruent parts, *Ann. Math.* **54** (1951), 250–261.

85. Gysin, W., Aufgabe 51: Lösung, *Elem. der Math.* **4** (1949), 140.

86. Hadwiger, H., *Vorlesungen über Inhalt, Oberfläche und Isoperimetrie*, Berlin: Springer-Verlag, 1957.

87. Hadwiger, H., H. Debrunner, and V. Klee, *Combinatorial Geometry in the Plane*, New York: Holt, Rinehart and Winston, 1964.

88. Hanf, W., On some fundamental problems concerning isomorphism of Boolean algebras, *Math. Scand.* **5** (1957), 205–217.

89. Hausdorff, F., *Grundzüge der Mengenlehre*, Leipzig: Veit, 1914. Reprinted, New York: Chelsea.

90. ——, Bemerkung über den Inhalt von Punktmengen, *Math. Ann.* **75** (1914), 428–433.

91. ——, *Set Theory*, trans. by J. Aumann et al., New York: Chelsea, 1957.

92. Hausner, M., *A Vector Space Approach to Geometry*, Englewood Cliffs, N.J.: Prentice-Hall, 1965.

93. Heath, T. L., *The Thirteen Books of Euclid's Elements*, vol. 3, New York: Dover, 1956.

94. Henle, J., and S. Wagon, Problem 6353, *Amer. Math. Monthly* **90** (1983), 64.

95. Hewitt, E., and K. Ross, *Abstract Harmonic Analysis*, vol. 1, Berlin: Springer-Verlag, 1963.

96. Horn, A., and A. Tarski, Measures in Boolean algebras, *Trans. Amer. Math. Soc.* **64** (1948), 467–497.

97. Jackson, W. H., Wallace's theorem concerning plane polygons of the same area, *Amer. J. Math.* **34** (1912), 383–390.

98. Jech, T. J., *The Axiom of Choice*, Amsterdam: North-Holland, 1973.

99. ——, *Set Theory*, New York: Academic Press, 1978.

100. Jørgensen, T., A note on subgroups of $SL(2,C)$, *Quart. J. Math. Oxford* (2) **28** (1977), 209–211.

101. Jůza, M., Remarque sur la mesure intérieure de Lebesgue, *Časopis pro Pěstování Matematiky* **107** (1982), 422–424.

102. Kazhdan, D. A., Connection of the dual space of a group with the structure of its closed subgroups, *Func. Anal. Appl.* **1** (1967), 63–65. English trans. of Funktsional'nyi Analizi Ego Prilozheniya **1** (1967), 71–74.

103. Kesten, H., Symmetric random walks on groups, *Trans. Amer. Math. Soc.* **92** (1959), 336–354.

104. ——, Full Banach mean values on countable groups, *Math. Scand.* **7** (1959), 146–156.

105. Kinoshita, S., A solution of a problem of R. Sikorski, *Fund. Math.* **40** (1953), 39–41.

106. Klee, V., Invariant extensions of linear functionals, *Pac. J. Math.* **4** (1954), 37–46.

107. ——, Some unsolved problems in geometry, *Math. Mag.* **52** (1979), 131–145.

108. Klein, F., and R. Fricke, *Vorlesungen über die Theorie der Elliptischen Modulfunctionen*, Erster Band, Leipzig: Teubner, 1890.

109. König, D., Über graphen und ihre Anwendung auf Determinantentheorie und Mengenlehre, *Math. Ann.* **77** (1916), 454–465.

110. ———, Sur les correspondances multivoque des ensembles, *Fund. Math.* **8** (1926), 114–134.

111. ———, Über eine Sclussweise aus dem Endlichen uns Unendliche, *Acta Sci. Math. (Szeged)* **3** (1927), 121–130.

112. König, D., and S. Valkó, Über mehrdeutige Abbildungen von Mengen, *Math. Ann.* **95** (1925), 135–138.

113. Krivine, J.-L., *Introduction to Axiomatic Set Theory*, trans. by D. Miller, Dordrecht: Reidel, 1971.

114. Kummer, H., Translative Zerlegungsgleichheit k-dimensionaler Parallelotopen, *Arch. der Math.* **7** (1956), 219–220.

115. Kunen, K., *Set Theory, An Introduction to Independence Proofs*, Amsterdam: North-Holland, 1980.

116. Kuratowski, K., Une propriété des correspondances biunivoques, *Fund. Math.* **6** (1924), 240–243.

117. ———, *Topology*, vol. 1, trans. by J. Jaworowski, New York: Academic Press. 1966.

118. Kuratowski, K., and A. Mostowski, *Set Theory*, trans. by M. Mączyński, Amsterdam: North-Holland, 1968.

119. Kurosh, A. G., *The Theory of Groups*, vol. 2, trans. by K. Hirsch, New York: Chelsea, 1956.

120. Lang, R., *Kardinalalgebren in der Masstheorie*, Diplomarbeit, Universität Heidelberg, 1970.

121. Lebesgue, H., *Leçons sur l'Intégration et la Recherche des Fonctions Primitives*, Paris: Gauthier-Villars, 1904.

122. Lehner, J., *Discontinuous Groups and Automorphic Functions*, Providence, R.I.: American Mathematical Society, 1964.

123. ———, *A Short Course in Automorphic Functions*, New York: Holt, Rinehart and Winston, 1966.

124. Lindenbaum, A., Contributions à l'étude de l'espace métrique I, *Fund. Math.* **8** (1926), 209–222.

125. Lindenbaum, A., and A. Tarski, Communication sur les recherches de la théorie des ensembles, *C. R. Séances Soc. Sci. Lettres Varsovie*, **19** (1926), 299–330.

126. Łoś, J., and C. Ryll-Nardzewski, On the application of Tychonoff's theorem in mathematical proofs, *Fund. Math.* **38** (1951), 233–237.

127. Losert, V., and H. Rindler, Almost invariant sets, *Bull. Lon. Math. Soc.* **13** (1981), 145–148.

128. Lovasz, L., *Combinatorial Exercises and Problems*, Amsterdam: North-Holland, 1979.

129. Luxembourg, W. A. J., Reduced direct products of the real number system and equivalents of the Hahn-Banach extension theorem, in *Applications of Model*

Theory to Algebra, Analysis, and Probability, ed. by W. A. J. Luxembourg, New York: Holt, Rinehart and Winston, 1969.

130. Lyndon, R. C., and J. L. Ullman, Pairs of real 2-by-2 matrices that generate free products, *Mich. Math. J.* **15** (1968), 161–166.

131. ——, Groups generated by two parabolic linear fractional transformations, *Can. J. Math.* **21** (1969), 1388–1403.

132. Magnus, W., Rational representations of Fuchsian groups and non-parabolic subgroups of the modular group, *Nachr. Akad. Wiss. Göttingen Math.-Phys. Kl. II* (1973), 179–189.

133. ——, *Noneuclidean Tesselations and Their Groups*, New York: Academic Press, 1974.

134. ——, Two generator subgroups of *PSL*(2,**C**), *Nachr. Akad. Wiss. Göttingen, Math.-Phys. Kl. II* (1975), 81–94.

135. Magnus, W., A. Karass, and D. Solitar, *Combinatorial Group Theory*, New York: Interscience, 1966.

136. Maharam, D., On measure in abstract sets, *Trans. Amer. Math. Soc.* **51** (1942), 413–433.

137. Margulis, G. A., Some remarks on invariant means, *Mh. Math.* **90** (1980), 233–235.

138. ——, Finitely additive invariant measures on Euclidean spaces, *Ergod. Th. and Dynam. Sys.* **2** (1982), 383–396.

139. Mason, J. H., Can regular tetrahedra be glued together face to face to form a ring?, *Math. Gaz.* **56** (1972), 194–197.

140. Mauldin, R. D., *The Scottish Book*, Boston: Birkhäuser, 1981.

141. Mazurkiewicz, S., Problème 8, *Fund. Math.* **1** (1920), 224.

142. ——, Sur la décomposition d'un segment en une infinité d'ensembles non mesurables superposables deux a deux, *Fund. Math.* **2** (1921), 8–14.

143. Mazurkiewicz, S., and W. Sierpiński, Sur un ensemble superposables avec chacune de ses deux parties, *C. R. Acad. Sci. Paris* **158** (1914), 618–619.

144. Meschkowski, H., *Unsolved and Unsolvable Problems in Geometry*, trans. by J. Burlak, New York: Ungar, 1966.

145. Millman, R. S., and G. D. Parker, *Geometry: A Metric Approach with Models*, New York: Springer-Verlag, 1981.

146. Milnor, J., Problem 5603, *Amer. Math. Monthly* **75** (1968), 685–686.

147. ——, A note on curvature and fundamental group, *J. Diff. Geom.* **2** (1968), 1–7.

148. ——, Growth of finitely generated solvable groups, *J. Diff. Geom.* **2** (1968), 447–449.

149. Moise, E. E., *Elementary Geometry from an Advanced Standpoint*, 2d ed., Reading, Mass.: Addison-Wesley, 1974.

150. Moore, G. H., *Zermelo's Axiom of Choice*, New York: Springer-Verlag, 1982.

151. ——, Lebesgue's measure problem and Zermelo's axiom of choice: The mathematical effects of a philosphical dispute, *Ann. N. Y. Acad. Sci.* **412** (1983), 129–154.

152. Morse, A. P., Squares are normal, *Fund. Math.* **36** (1949), 35–39.

153. Mycielski, J., On a problem of Sierpiński concerning congruent sets of points, *Bull. Acad. Pol. Sc. Cl. III* **2** (1954), 125–126.

154. ———, About sets with strange isometrical properties (I), *Fund. Math.* **42** (1955), 1–10.

155. ———, On the paradox of the sphere, *Fund. Math.* **42** (1955), 348–355.

156. ———, On the decompositions of Euclidean spaces, *Bull. Acad. Pol. Sc. Cl. III* **4** (1956), 417–418.

157. ———, On the congruence of sets, *Bull. Acad. Pol. Sc. Cl. III* **4** (1956), 419–421.

158. ———, On the decomposition of a segment into congruent sets and related problems, *Coll. Math.* **5** (1957), 24–27.

159. ———, Problème 166, *Coll. Math.* **4** (1957), 240.

160. ———, About sets with strange isometrical properties (II), *Fund. Math.* **45** (1958), 292–295.

161. ———, About sets invariant with respect to denumerable changes, *Fund. Math.* **45** (1958), 296–305.

162. ———, Independent sets in topological algebras, *Fund. Math.* **55** (1964), 139–147.

163. ———, Commentary on the paper of Banach, "Sur le problème de la mesure," in S. Banach, *Oeuvres*, vol. I Warsaw: Éditions Scientifiques de Pologne.

164. ———, Almost every function is independent, *Fund. Math.* **81** (1973), 43–48.

165. ———, Remarks on invariant measures in metric spaces, *Coll. Math.* **32** (1974), 105–112.

166. ———, Two problems on geometric bodies, *Amer. Math. Monthly* **84** (1977), 116–118.

167. ———, Can one solve equations in groups?, *Amer. Math. Monthly* **84** (1977), 723–726.

168. ———, Finitely additive measures, I, *Coll. Math.* **42** (1979), 309–318.

169. ———, Problems on finitely additive invariant measures, in *General Topology and Modern Analysis*, ed. by McAuley and Rao, New York: Academic Press, 1980, 431–436.

170. Mycielski, J., and S. Świerczkowski, On free groups of motions and decompositions of the Euclidean space, *Fund. Math.* **45** (1958), 283–291.

171. Mycielski, J., and S. Wagon, Large free groups of isometries and their geometrical uses, *Ens. Math.* (forthcoming).

172. Namioka, I., Følner's condition for amenable semi-groups, *Math. Scand.* **15** (1964), 18–28.

173. Nash-Williams, C. St. J. A., Infinite graphs—a survey, *J. Comb. Th.* **3** (1967), 286–301.

174. Neumann, B., Über ein gruppentheoretisch-arithmetisch Problem, *Sitzungber. Preuss. Akad. Wiss. Phys.-Math. Klasse* No. X (1933), 18 pp.

175. Nisnewitsch, V. L., Über gruppen, die durch matrizen über einem kommutativen Feld isomorph darstellbar sind, *Mat. Sbornik* **50** (N. S. **8**) (1940), 395–403. Russian, with German summary.

176. Niven, I., *Irrational Numbers*, Carus Mathematical Monographs, No. 11, The Mathematical Association of America. Distributed by Wiley, New York, 1967.

177. Olmsted, J. M. H., *Real Variables*, New York: Appleton-Century-Crofts, 1959.

178. Ol'shanskii, A. Y., On the problem of the existence of an invariant mean on a group, *Russian Math. Surveys (Uspekhi)* **35** No. 4 (1980), 180–181.

179. Osofsky, B., Problem 6102, *Amer. Math. Monthly* **83** (1976), 572.

180. Osofsky, B., and S. Adams, Problem 6102 and solution, *Amer. Math. Monthly* **85** (1978), 504.

181. Oxtoby, J. C., *Measure and Category*, New York: Springer-Verlag, 1971.

182. Pincus, D., Independence of the prime ideal theorem from the Hahn-Banach theorem, *Bull. Amer. Math. Soc.* **78** (1972), 766–770.

183. ———, The strength of the Hahn-Banach theorem, *Lecture Notes in Math.* **369** (1974), Victoria Symposium on Nonstandard Analysis, 203–248.

184. Pincus, D., and R. Solovay, Definability of measures and ultrafilters, *J. Sym. Logic* **42** (1977), 179–190.

185. Playfair, J., *Elements of Geometry*, 8th ed., with additions by William Wallace, Edinburgh: 1831.

186. Promislow, D., Nonexistence of invariant measures, *Proc. Amer. Math. Soc.* **88** (1983), 89–92.

187. Raisonnier, J., Ensembles non mesurables et filtres rapides, *C. R. Acad. Sci. Paris* **294** (1982), 285–287.

188. ———, A mathematical proof of S. Shelah's theorem on the measure problem and related results, *Isr. J. Math.* (forthcoming).

189. Rajagopalan, M., and K. G. Witz, On invariant means which are not inverse invariant, *Can. J. Math.* **20** (1968), 222–224.

190. Ree, R., On certain pairs of matrices which do not generate a free group, *Can. Math. Bull.* **4** (1961), 49–52.

191. Rickert, N. W., Amenable groups and groups with the fixed point property, *Trans. Amer. Math. Soc.* **127** (1967), 221–232.

192. Robinson, R. M., On the decomposition of spheres, *Fund. Math.* **34** (1947), 246–260.

193. Rosenblatt, J., Invariant measures and growth conditions, *Trans. Amer. Math. Soc.* **193** (1974), 33–53.

194. ———, Finitely additive invariant measures, II, *Coll Math.* **42** (1979), 361–363.

195. ———, Uniqueness of invariant means for measure-preserving transformations, *Trans. Amer. Math. Soc.* **265** (1981), 623–636.

196. Royden, H. L., *Real Analysis*, 2d ed., New York: Macmillan, 1968.

197. Ruziewicz, S., Sur un ensemble non dénombrable de points, superposables avec les moitiés de sa parties aliquote, *Fund. Math.* **2** (1921), 4–7.

198. ———, Une application de l'équation functionelle $f(x + y) = f(x) + f(y)$ à la décomposition de la droite en ensembles superposables non mesurables, *Fund. Math.* **5** (1924), 92–95.

199. Sacks, G., Measure theoretic uniformity in recursion theory and axiomatic set theory, *Trans. Amer. Math. Soc.* **142** (1969), 381–420.

200. Sageev, G., An independence result concerning the Axiom of Choice, *Ann. Math. Logic* **8** (1975), 1–184.

201. Sah, C.-H., *Hilbert's Third Problem: Scissors Congruence*, Research Notes in Mathematics 33, San Francisco: Pitman, 1979.

202. Sallee, G. T., Are equidecomposable plane convex sets convex equidecomposable?, *Amer. Math. Monthly* **76** (1969), 926–927.

203. Sanov, I. N., A property of a representation of a free group, *Doklady Akad. Nauk SSSR (N.S.)* **57** (1947), 657–659. (Russian)

204. Schmidt, K., Asymptotically invariant sequences and an action of $SL(2,\mathbf{Z})$ on the 2-sphere, *Isr. J. Math.* **37** (1980), 193–208.

205. ———, Amenability, Kazhdan's property T, strong ergodicity and invariant means for ergodic group-actions, *Ergod. Th. and Dynam. Sys.* **1** (1981), 223–236.

206. Shelah, S., Going to Canossa, *Abstracts Amer. Math. Soc.* **1** (1980), 630.

207. Sherman, J., *Paradoxical sets and amenability in groups*. Ph.D. diss., UCLA, 1975.

208. ———, A new characterization of amenable groups, *Trans. Amer. Math. Soc.* **254** (1979), 365–389.

209. Siegel, C. L., Bemerkung zu einem Satze von Jakob Nielsen, *Mat. Tidsskrift, Ser. B* (1950), 66–70.

210. Sierpiński, W., Sur l'égalité $2m = 2n$ pour les nombres cardinaux, *Fund. Math.* **3** (1922), 1–6.

211. ———, Sur le paradoxe de MM. Banach et Tarski, *Fund. Math.* **33** (1945), 229–234.

212. ———, Sur le paradoxe de la sphère, *Fund. Math.* **33** (1945), 235–244.

213. ———, Sur la congruence des ensembles de points et ses généralisations, *Comm. Math. Helv.* **19** (1946–7), 215–226.

214. ———, Sur un ensemble plan qui se décompose en $2^{\aleph_0}$ ensembles disjoints superposables avec lui, *Fund. Math.* **34** (1947), 9–13.

215. ———, Sur l'équivalence des ensembles par décomposition en deux parties, *Fund. Math.* **35** (1948), 151–158.

216. ———, Sur un paradoxe de M. J. von Neumann, *Fund. Math.* **35** (1948), 203–207.

217. ———, Sur quelques problèmes concernant la congruence des ensembles de points, *Elem. der Math.* **5** (1950), 1–4.

218. ———, Sur un ensemble plan singulier, *Fund. Math.* **37** (1950), 1–4.

219. ———, *On the Congruence of Sets and Their Equivalence by Finite Decomposition*, Lucknow, 1954. Reprinted, New York: Chelsea, 1967.

220. ———, Sur une relation entre deux substitutions linéaires, *Fund. Math.* **41** (1955), 1–5.

221. Sikorski, R., *Boolean Algebras*, 3d ed., New York: Springer-Verlag, 1969.

222. Silverman, R. J., Invariant linear functions, *Trans. Amer. Math. Soc.* **81** (1956), 411–424.

223. ———, Means on semigroups and the Hahn-Banach Extension Property, *Trans. Amer. Math. Soc.* **83** (1956), 222–237.

224. Solovay, R. M., A model of set-theory in which every set of reals is Lebesgue measurable, *Ann. Math.* **92** (1970), 1–56.

225. Straus, E. G., On a problem of W. Sierpiński on the congruence of sets, *Fund. Math.* **44** (1957), 75–81.

226. ———, On Sierpiński sets in groups, *Fund. Math.* **46** (1959), 332–333.

227. Stromberg, K., The Banach-Tarski Paradox, *Amer. Math. Monthly* **86** (1979), 151–161.

228. Sullivan, D., For $n > 3$, there is only one finitely-additive measure on the n-sphere defined on all Lebesgue measurable sets, *Bull. Amer. Math. Soc.* (*N. S.*) **1** (1981), 121–123.

229. Świerczkowski, S., On a free group of rotations of the Euclidean space, *Indag. Math.* **20** (1958), 376–378.

230. Tarski, A., O równoważności wielokatów, *Przegląd Matematyczno-Fizyczny* No. 1–2 (1924), 54 (Polish, with French summary).

231. ———, Sur quelques théorèmes qui équivalent à l'axiome du choix, *Fund. Math.* **5** (1924), 147–154.

232. ———, Problème 38, *Fund. Math.* **7** (1924), 381.

233. ———, Sur les fonctions additives dans les classes abstraites et leur applications au problème de la mesure, *C. R. Séances Soc. Sci. Lettres Varsovie, Cl. III* **22** (1929), 114–117.

234. ———, Über das absolute Mass linearen Punktmengen, *Fund. Math.* **30** (1938), 218–234.

235. ———, Algebraische Fassung des Massproblems, *Fund. Math.* **31** (1938), 47–66.

236. ———, *Cardinal Algebras*, New York: Oxford, 1949.

237. Tits, J., Free Subgroups in linear groups, *J. Alg.* **20** (1972), 250–270.

238. Ulam, S., *A Collection of Mathematical Problems*, New York: Interscience, 1960.

239. van den Dries, L., and A. J. Wilkie, Gromov's theorem on groups of polynomial growth and elementary logic, *J. Alg.* **89** (1984), 349–374.

240. van der Waerden, B. L., Aufgabe 51, *Elem. der Math.* **4** (1949), 18.

241. van Douwen, E., Measures invariant under actions of F_2 (forthcoming).

242. Viola, T., Su un problema riguardante le congruenze degli insiemi di punta, I, II, *Rendic. Accad. Naz. Lincei, Ser. VIII*, **20** (1956), 290–293, 431–438.

243. Vitali, G., *Sul problema della mesura dei gruppi di punti di una retta*, Bologna, 1905.

244. von Neumann, J., Ein System algebraisch unabhängiger Zahlen, *Math. Ann.* **99** (1928), 134–141. Reprinted in J. von Neumann, *Collected Works*, vol. I, Oxford: Pergamon, 1961.

245. ———, Die Zerlegung eines Intervalles in abzählbar viele kongruente Teilmengen, *Fund. Math.* **11** (1928), 230–238. Reprinted, loc. cit.

246. ———, Zur allgemeinen Theorie des Masses, *Fund. Math.* **13** (1929), 73–116. Reprinted, loc. cit.

247. Wagon, S., Invariance properties of finitely additive measures in $\mathbf{R}^n$, *Ill. J. Math.* **25** (1981), 74–86.

248. ———, Circle-squaring in the twentieth century, *Math. Intelligencer* **3** (1981), 176–181.

249. ———, The use of shears to construct paradoxes in $\mathbf{R}^2$, *Proc. Amer. Math. Soc.* **85** (1982), 353–359.

250. ———, Partitioning intervals, spheres and balls into congruent pieces, *Can. Math. Bull.* **26** (1983), 337–340.

251. Wang, S. P., The dual space of semi-simple Lie groups, *Amer. J. Math.* **91** (1969), 921–937.

252. ———, On the first cohomology group of discrete groups with Property (*T*), *Proc. Amer. Math. Soc.* **42** (1974), 621–624.

253. ———, On isolated points in the dual spaces of locally compact groups, *Math. Ann.* **218** (1975), 19–34.

254. ———, A note on free subgroups in linear groups, *J. Alg.* **71** (1981), 232–234.

255. Wehrfritz, B. A. F., *Infinite Linear Groups*, Ergeb. der Math. vol. 76, Berlin: Springer-Verlag, 1973.

256. Weir, A. J., *Lebesgue Integration and Measure*, Cambridge: Cambridge University Press, 1973.

257. Wilkie, A. J., and L. van den Dries, An effective bound for groups of linear growth, *Arch. Math.* (forthcoming).

258. Wolf, J., Growth of finitely-generated solvable groups and curvature of Riemannian manifolds, *J. Diff. Geom.* **2** (1968), 421–446.

A1. Adelson-Velsky, G. M., and Yu. A. Shreider, The Banach mean on groups (Russian), *Uspehi Mat. Nauk* **12** (78) (1957), 131–136. (*Math. Rev.* **20** (1959) #1238.)

A2. Armstrong, T., and K. Prikry, κ-Finiteness and κ-additivity of measures on sets and left invariant measures on discrete groups, *Proc. Amer. Math. Soc.* **80** (1980), 105–112.

A3. Bandt, C., Metric invariance of Haar measure, *Proc. Amer. Math. Soc.* **87** (1983), 65–69.

A4. ———, Sets of equal Haar measure admit congruent partitions (forthcoming).

A5. Bandt, C., and G. Baraki, Metrically invariant measures on locally homogeneous spaces and hyperspaces (forthcoming).

A6. De La Harpe, P., Free groups in linear groups, *Ens. Math.* **29** (1983), 129–144.

A7. Gardner, R. J., A problem of Sallee on equidecomposable convex bodies, Proc. Amer. Math. Soc. (forthcoming).

A8. Grigorchuk, R. I., On Milnor's problem of group growth, *Soviet Math. Dokl.* **28** (1983), 23–26.

A9. Kuranishi, M., On everywhere dense imbedding of free groups in Lie groups, *Nag. Math. J.* **2** (1951), 63–71.

A10. Vershik, A., Amenability and approximation of infinite groups, *Sel. Math. Sov.* **2** (1982), 311–330. Revision of an appendix to the Russian translation of [80].

List of Symbols

λ	Lebesgue measure
λ_*	Lebesgue inner measure
λ^*	Lebesgue outer measure
v	Jordan measure
$\mathbf{R}^n$	Euclidean n-space
S^n	The unit sphere in $\mathbf{R}^{n+1}$
H^n	Hyperbolic n-space
L^n	Elliptic n-space
J	The unit cube in $\mathbf{R}^n$
$\mathbf{C}$	The field of complex numbers
$\mathbf{N}$	The natural numbers (nonnegative integers)
$\mathbf{Z}$	The set of all integers
$\mathscr{L}$	The set of Lebesgue measurable subsets of $\mathbf{R}^n$ or of S^n
$\mathscr{J}$	The set of Jordan measurable subsets of $\mathbf{R}^n$
$\mathscr{B}$	The set of subsets of $\mathbf{R}^n$ or of S^n having the Property of Baire
$\mathscr{R}$	The Boolean ring of bounded regular-open subsets of $\mathbf{R}^n$ or S^n
$\mathscr{R}_v$	The Boolean ring of bounded, regular-open, Jordan measurable subsets of $\mathbf{R}^n$ or S^n
$\aleph_0$	The cardinality of $\mathbf{N}$
$2^{\aleph_0}$	The cardinality of the continuum
$\sim$	The equidecomposability relation

246

$\sim_n$	Equidecomposability using n pieces
$\lesssim$	A is equidecomposable to a subset of B
$\mathscr{S}$	The semigroup of equidecomposability types
1	The identity in a group
T_n	The group of all translations of $\mathbf{R}^n$
O_n	The group of $n \times n$ real orthogonal matrices
SO_n	The group of matrices in O_n with determinant $+1$.
G_n	The group of all isometries of $\mathbf{R}^n$
SG_n	The group of all orientation-preserving isometries of $\mathbf{R}^n$
A_n	The group of all affine transformations of $\mathbf{R}^n$
SA_n	The group of all affine transformations in $\mathbf{R}^n$ with determinant $+1$
GL_n	The group of all real matrices with nonzero determinant
SL_n	The group of all real matrices with determinant $+1$
π	The canonical homomorphism from A_n to GL_n
$SL_n(\mathbf{Z})$	The group of matrices in SL_n with integer entries
$SL_n(\mathbf{C})$	The group of complex matrices with determinant $+1$
PSL_n	$SL_n/\pm 1$
$PSL_n(\mathbf{Z})$	$SL_n(\mathbf{Z})/\pm 1$
$PSL_n(\mathbf{C})$	$SL_n(\mathbf{C})/\pm 1$
AG	The class of amenable groups
EG	The class of elementary groups
NF	The class of groups without a free subgroup of rank 2
SG	The class of supramenable groups
NS	The class of groups without a free subsemigroup of rank 2
EB	The class of exponentially bounded groups
$B(m, n)$	The Burnside group with m generators and having exponent n
ZF	The Zermelo-Fraenkel axioms of set theory
AC	The Axiom of Choice
ZFC	ZF + AC
LM	The assertion "all sets of reals are Lebesgue measurable"
PB	The assertion "all sets of reals have the Property of Baire"
DC	The Axiom of Dependent Choice
IC	The assertion "an uncountable inaccessible cardinal exists"

Index

GIAN-CARLO ROTA, *Editor*
ENCYCLOPEDIA OF MATHEMATICS AND ITS APPLICATIONS

Other volumes in preparation